Invasive Plant Ecology and Management: Linking Processes to Practice

---

CABI INVASIVE SPECIES SERIES

Invasive species are plants, animals or microorganisms not native to an ecosystem, whose introduction has threatened biodiversity, food security, health or economic development. Many ecosystems are affected by invasive species and they pose one of the biggest threats to biodiversity worldwide. Globalization through increased trade, transport, travel and tourism will inevitably increase the intentional or accidental introduction of organisms to new environments, and it is widely predicted that climate change will further increase the threat posed by invasive species. To help control and mitigate the effects of invasive species, scientists need access to information that not only provides an overview of and background to the field, but also keeps them up to date with the latest research findings.

This series addresses all topics relating to invasive species, including biosecurity surveillance, mapping and modeling, economics of invasive species and species interactions in plant invasions. Aimed at researchers, upper-level students and policy makers, titles in the series provide international coverage of topics related to invasive species, including both a synthesis of facts and discussions of future research perspectives and possible solutions.

Titles Available

1. Invasive Alien Plants: An Ecological Appraisal for the Indian Subcontinent
   Edited by J.R. Bhatt, J.S. Singh, R.S. Tripathi, S.P. Singh and R.K. Kohli
2. Invasive Plant Ecology and Management: Linking Processes to Practice
   Edited by T.A. Monaco and R.L. Sheley

Invasive Plant Ecology and Management: Linking Processes to Practice

Edited by

THOMAS A. MONACO

US Department of Agriculture, Agricultural Research Service, Forage and Range Research Laboratory, Utah State University, Logan, Utah, USA

ROGER L. SHELEY

US Department of Agriculture, Agricultural Research Service, Eastern Oregon Agricultural Research Center, Burns, Oregon, USA

www.cabi.org

CABI is a trading name of CAB International

CABI
Nosworthy Way
Wallingford
Oxfordshire OX10 8DE
UK

Tel: +44 (0)1491 832111
Fax: +44 (0)1491 833508
E-mail: cabi@cabi.org
Website: www.cabi.org

CABI
875 Massachusetts Avenue
7th Floor
Cambridge, MA 02139
USA

Tel: +1 617 395 4056
Fax: +1 617 354 6875
E-mail: cabi-nao@cabi.org

A catalogue record for this book is available from the British Library, London, UK.

Library of Congress Cataloging-in-Publication Data
Invasive plant ecology and management : linking processes to practice / edited by Thomas A. Monaco, Roger L. Sheley.
    p. cm. -- (CABI invasive species series ; 2)
  Includes bibliographical references and index.
  ISBN 978-1-84593-811-6 (hbk. : alk. paper) 1. Invasive plants--Ecology. 2. Invasive plants--Control. I. Monaco, Thomas A. II. Sheley, Roger L.

  SB613.5.I552 2012
  581.62--dc23

                    2011037079

ISBN-13: 978 1 84593 811 6

Commissioning Editor: David Hemming
Editorial Assistant: Gwenan Spearing
Production Editor: Simon Hill

Typeset by Columns Design XML Limited, Reading, UK.
Printed and bound in the UK by MPG Books Group.

Contents

Contributors

_____

Brady W. Allred, Natural Resource Ecology and Management, Oklahoma State University, 008C Agricultural Hall, Stillwater, Oklahoma 74077, USA; Email: brady.allred@okstate.edu

Jacob N. Barney, Department of Plant Pathology, Physiology and Weed Science, Virginia Tech, 435 Old Glade Road (0330), Glade Road Research Center, Blacksburg, Virginia 24061, USA; Email: jnbarney@vt.edu

Brandon T. Bestelmeyer, US Department of Agriculture, Agricultural Research Service, Jornada Experimental Range, New Mexico State University, Box 30003 MSC 3JER, Las Cruces, New Mexico 88003, USA; Email: bbestelm@nmsu.edu

Joel R. Brown, US Department of Agriculture, Natural Resources Conservation Service and Jornada, Experimental Range, New Mexico State University, Box 30003 MSC 3JER, Las Cruces, New Mexico 88003, USA; Email: joelbrow@nmsu.edu

Jaepil Cho, US Department of Agriculture, Agricultural Research Service, Northwest Watershed Research Center, 800 Park Blvd, Suite 105, Boise, Idaho 83712, USA; Email: jaepil.cho@ars.usda.gov

Joseph M. DiTomaso, Department of Plant Sciences, Mail Stop 4, University of California, Davis, California 95616, USA; Email: jmditomaso@ucdavis.edu

R. Dwayne Elmore, Natural Resource Ecology and Management, Oklahoma State University, 008C Agricultural Hall, Stillwater, Oklahoma 74077, USA; Email: dwayne.elmore@okstate.edu

David M. Engle, Natural Resource Ecology and Management, Oklahoma State University, 008C Agricultural Hall, Stillwater, Oklahoma 74077, USA; Email: david.engle@okstate.edu

Valerie T. Eviner, Department of Plant Sciences, University of California 1210 PES, Mail Stop 1, One Shields Ave, Davis, California 95616, USA; Email: veviner@ucdavis.edu

Samuel D. Fuhlendorf, Natural Resource Ecology and Management, Oklahoma State University, 008C Agricultural Hall, Stillwater, Oklahoma 74077, USA; Email: sam.fuhlendorf@okstate.edu

Thomas A. Grant III, Department of Forest, Rangeland, and Watershed Stewardship, Colorado State University, 230 Forestry Building, Campus Delivery 1472, Fort Collins, Colorado 80523-1472, USA; Email: metag3@gmail.com

Stuart P. Hardegree, US Department of Agriculture, Agricultural Research Service, Northwest Watershed Research Center, 800 Park Blvd, Suite 105, Boise, Idaho 83712, USA; Email: stuart.hardegree@ars.usda.gov

Christine V. Hawkes, Section of Integrative Biology, The University of Texas, 1 University Station C0930, Austin, Texas 78712, USA; Email: chawkes@mail.utexas.edu

Jeremy J. James, US Department of Agriculture, Agricultural Research Service, Eastern Oregon Agricultural Research Center, 67826-A Hwy 205, Burns, Oregon 97720, USA; Email: jeremy.james@oregonstate.edu

A. Joshua Leffler, US Department of Agriculture, Agricultural Research Service, Utah State University Forage and Range Research Laboratory, 696 N 1100 E, Logan, Utah 84322-6300, USA; Email: josh.leffler@usu.edu

Jane M. Mangold, Department of Land Resources and Environmental Sciences, Montana State University PO Box 173120, Bozeman, Montana 59717-3120, USA; Email: jane.mangold@montana.edu

Lesley R. Morris, US Department of Agriculture, Agricultural Research Service, Utah State University, Forage and Range Research Laboratory, 696 N 1100 E, Logan, Utah 84322-6300, USA; Email: lesleyrmorris@gmail.com

Mark W. Paschke, Department of Forest, Rangeland, and Watershed Stewardship, Colorado State University, 230 Forestry Building, Campus Delivery 1472, Fort Collins, Colorado 80523-1472, USA; Email: Mark.Paschke@colostate.edu

Ronald J. Ryel, Department of Wildland Resources and the Ecology Center, Utah State University, Logan, Utah 84322-5230, USA; Email: ron.ryel@usu.edu

Jeanne M. Schneider, US Department of Agriculture, Agricultural Research Service, Great Plains Agroclimate and Natural Resources Research Unit, 7207 W Cheyenne Street, El Reno, Oklahoma 73036, USA; Email: jeanne.schneider@ars.usda.gov

Steven G. Whisenant, Department of Ecosystem Science and Management, Texas A&M University 2138 TAMU, College Station, Texas 77843, USA; Email: s-whisenant@tamu.edu

Foreword

———————————

Invasive plant species can substantially alter ecosystems. Direct economic losses to agriculture, livestock production, forestry, and recreation are well known consequences of certain invasive species. Less understood, yet probably more important, are changes to community and ecosystem processes that are caused by invasive plant species. These altered ecosystems, sometimes created by large-scale plant invasions, are not simply structural reorganizations with new species. Many are novel ecosystems that look, function, and react quite differently than their predecessors. Invasive plant species that diminish biological diversity, alter nutrient and hydrologic processes, or dramatically change fire regimes have serious consequences. Ecosystem management strategies based on this knowledge should have significant advantages over strategies simply focused on removing invasive species.

The ultimate test of ecological restoration is our ability to understand ecosystems and apply relevant science to solve real-world environmental problems such as the detrimental impacts of invasive plant species. My experiences in ecological restoration lead me to believe that invasive species are among our most difficult and perplexing challenges. We may be very good at killing and removing individual plants. Yet ultimately we lose the battle by failing to understand the underlying processes or even the scales at which they operate. This occurs whether we apply traditional chemical or mechanical plant removal tools to relatively small areas or use less direct strategies over large spatial scales. Too often we treat the symptoms of invasive plant problems rather than identifying and managing the underlying cause of those problems. Not addressing and removing the underlying causes of invasions usually reduces treatment longevity. Clearly, recognizing and removing these underlying causes is more easily stated than done. This difficulty is why I view this book as a seminal contribution to the science and practice of invasive plant management.

In Invasive Plant Ecology and Management: Linking Processes to Practice, editors Thomas Monaco and Roger Sheley have assembled an impressive group of authors for the purpose of applying contemporary ecological knowledge to the practice of invasive plant management. Collectively, at all major levels of ecological organization, they address the attributes that make plants invasive as well as those characteristics that make ecosystems receptive to invasions. Significantly, the authors then describe how this knowledge can be used to both enhance the competitive ability of native and other desirable species while limiting the success of invasive species. These are substantial contributions to the science and practice of invasive plant management.

Steven G. Whisenant
College Station, Texas
July 2011

Preface

———————————

The primary objective of this book is to illustrate how understanding ecological processes will foster scientifically based approaches to invasive plant management in semi-arid ecosystems. Ecological processes serve as the underpinning and common ground within the scientific literature that bridges the gap between researchers and land managers. Our focus on ecological processes is also justified based on the overwhelming realization that invasive plant management must move beyond treating symptoms of damaged lands to repairing and influencing the processes responsible for plant community change. While our emphasis is clearly slanted towards semi-arid wildlands, we believe the ecological principles outlined by the contributing authors can easily be applied to many other systems impacted by invasive plant species.

Assessing ecological processes and how they are impacted by invasive plant species is a critical aspect of land management, which can easily be overlooked when the impetus to "do something" overshadows sound decision-making. Part 1 of this book, comprising Chapters 1 through 5, provides a compelling justification to assess ecosystem and landscape heterogeneity and historical land-use legacy effects before a process-based understanding of how ecosystems operate can be realized. In this same vein, Part 1 also showcases how the emerging concept of resource pool dynamics provides a much more adequate mechanism to assess ecological processes associated with plant resource use than traditional emphasis on competition. Concluding Part 1 with a comprehensive assessment of how invasive species impact soils, nutrient cycling, and microbial communities emphasizes that effective invasive plant management will also require designing management practices that influence processes that operate within soils.

The idea of embracing successional-ecological processes to initiate and direct plant community change is not new, but the principles and practices to achieve greater success when managing invasive plant species continue to emerge. If the underlying origin for the persistence of invasive species lies with damaged ecological processes, intervening with restorative actions to repair these ecological processes is our responsibility and should be emphasized in management. Part 2 of this book illustrates a mechanistic approach at presenting principles and practices to optimize soil microclimate conditions, eradicate adverse plant–soil feedbacks, and systematically alter plant species performance. Ultimately, it is our hope that adopting these ideas will assist you in achieving desired outcomes and greater predictability when applying land management practices.

Thomas A. Monaco and Roger L. Sheley

Acknowledgements

The conception and preparation of this book was only possible through financial support from the US Department of Agriculture, Agricultural Research Service and funding they provided to the Areawide Pest Management Program for Annual Grasses in the Great Basin Ecosystem. The unique collaboration of this program inspired much of the conceptual and scientific organization of the chapters. We also recognize the organizations that supported the authors of this book. Finally, we are grateful to all the authors and publishers who have granted permission for us to reproduce material in this book.

Thomas A. Monaco and Roger L. Sheley

Part I

Assessing Ecosystem Processes and Invasive Plant Impacts

1

Managing Invasive Species in Heterogeneous Ecosystems

Joel R. Brown and Brandon T. Bestelmeyer

US Department of Agriculture, New Mexico State University, USA

Introduction

Ecologically based invasive plant management (Sheley *et al.*, 2010) provides a mechanistic framework for diagnosing causes of plant invasions and selecting management responses. This approach requires the organization of multiple sources of information, much of which is highly dependent upon spatial and temporal context. Although there have been substantial efforts to identify the characteristics of successful invaders as a means to predict which plants are likely to be successful when introduced into new environments, the use of species attributes alone is a poor predictor of which plants will invade a particular landscape (Mack *et al.*, 2000). In fact, many of the plants currently defined as 'invaders' in ecological terms and 'noxious weeds' in legal terms are native to the regions and landscapes, if not the plant communities, they invade. This combination of species attributes (invasiveness) and plant community or landscape susceptibility (invasibility) complicates the development of a universal set of principles for prediction, and even *post hoc* analysis, of the interactions of invasive plants and landscapes. Because the information necessary for successful implementation of management responses is so highly variable in both time and space, as well as by invasive species, a systematic approach to organization, analysis, and decision-making is essential.

In this chapter, we first discuss the challenge of managing invasive species in complex landscapes. In particular, how managing invasive species in wildland environments differs from more intensively managed systems, such as croplands, where the spatiotemporal relationships are much more clearly defined. Then we present and demonstrate the use of ecological sites (soil-based landscape subunits) and state-and-transition models (graphic and text descriptions of soil:vegetation dynamics) as a means of organizing and synthesizing information about plant community and landscape susceptibility to invasion. These tools, in combination with a basic understanding of invasive species biology, can be used to predict the pattern of invasion, develop management responses with an increased probability of success, and devise monitoring systems to assess the effectiveness of management actions. Finally, we provide examples of the application of ecological sites and state and transition models to the diagnosis of the ecological process drivers of invasions and the development of management responses. We believe that an Ecological Site-based approach offers an improvement in the management of invasive species on wildlands by increasing the spatial and temporal precision of monitoring, analysis, and responses, thereby increasing the chances of success of ecologically based invasive plant management.

Managing Invasive Plants in Complex Landscapes

The spatial and temporal patterns of invasive species movement are critical to understanding the processes of invasion and to developing effective prediction models and decision-support systems for management responses (Masters and Sheley, 2001). Such patterns, if properly described and quantified, can provide insight into the causes and drivers of invasions and are the foundation for predictive models essential in developing effective responses, both for reducing the rates and impacts of invasion and for restoring invaded landscapes.

Accurate predictive models for the movement of invasive species are especially important in wildlands (rangeland, forests, and deserts). Although many important agronomic and economic principles have emerged from research in crop and timber production systems (weed science), they may be of limited value in the management of invasive species in wildlands. Agro-economic models used for decision making in production systems, where the focus is largely on recognition of economic impact thresholds, are generally based on the assumption that control technologies are effective against target species (Cousens and Mortimer, 1995). While the relationships between weed density and crop yield are relatively straightforward and can be used as a solid basis for decisions about the deployment of weed control technologies, these decisions are typically based on an annual timeframe and considerations are usually limited to an individual field or management unit.

Invasive species management in extensively managed lands is much more complex, requiring considerations that extend to greater spatial and temporal scales (Brown et al., 1999). There is often a substantial amount of time, in a managerial context, between the presence of invasive species propagules or juveniles and the ability of land managers to detect nascent populations, and it is relatively common for years to elapse between the appearance of detectable populations and the impacts on economic yields. This time lag between the initial, largely irreversible stages of invasion (in the form of persistent banks of seeds or juveniles) and an economic threshold when yield is reduced may further complicate management responses. When the com-

plexities of decision-making and priority setting for management actions are considered, it is easy to see how invasive species populations can be well established long before there is a clear motivation to respond (Brown et al., 1999).

While exotic species have been the focus of much research and analysis, many of the 'invasive species' with measured negative economic impacts are actually native with long histories as a part of the landscape, but changes in land use, land management or climate have resulted in increased populations. Radosevich et al. (2007) enumerated the broad range of behaviors, characteristics, and impacts ascribed to 'weeds' or invasive plants, which in many cases, could just as easily apply to native species. Thus, in the context of natural resource management on wildlands, the definition of invasive species must go beyond 'exotic' to include 'undesirable' species. Regardless of the geographic source of the invasive plant species, the management of undesirable species requires: (i) detection methodologies and response protocols; (ii) selection of the appropriate control technology to limit their impact; and (iii) knowledge of the treatment cost:benefit ratio (Radosevich et al., 2007). Quantitative information about these relationships allows land managers to make informed decisions and evaluate progress toward objectives. However, because wildland natural resource management typically encompasses large (>100 km^2) spatial scales and long (decades) timeframes, changes in the composition and arrangement of plant communities on a landscape can greatly influence vulnerability to invasive species and can also alter the success of management responses. Management tools must be able to systematically organize these spatio-temporal changes if they are to be of value.

Invasive species detection in wildland environments commonly involves detecting very low densities of plant populations that may reside as seeds or juveniles within communities for years before reaching a detection threshold. While remote sensing applications have been developed, they currently lack the precision needed to detect populations of target plants early in the invasion process (Mitchell and Glenn, 2009). Even well-trained managers struggle to detect invasive species by direct observation and initiate a timely response. For the immediate future, wildland invasive species management will not benefit from remote sensing applications intended to detect low level populations (Blumenthal *et al.*, 2007). However, remote sensing will remain a core technology for inventory and for the development of restoration strategies and tactics when populations are more easily observed (Washington-Allen *et al.*, 2006).

Without the benefit of a reliable, cost-effective early detection system for invasive species, wildland managers must rely more heavily on predictive models to identify which invasive species are threats and which plant communities, landscapes, and regions are most likely to be invaded. Such a prediction approach would allow land managers and technical advisors to target locations for detection and treatment. This approach must address both the spatial and temporal dynamics of invasions and requires a well-developed understanding of: (i) dispersal vectors; (ii) species life histories and performance attributes; and (iii) a range of vulnerabilities in individual landscape components (e.g. soil units) that may change the pattern of invasion across landscapes (Sheley *et al.*, 2010). Managing low-level populations of invasive species in hetero-geneous landscapes requires either precise application of intensive technologies or widespread application of extensive, low-cost technology. Selecting and implementing the tactics that follow from either of these approaches not only requires a thorough knowledge of both invasive species life history and plant community characteristics, but also an understanding of how the targeted invasive species and the plant community will respond to the treatment and the removal of the species. Managing invasive species in wildlands clearly requires knowledge about the landscape matrix, and individual plant communities that comprise it. Landscapes have differing vulnerabilities and differing responses to invasive species that, in turn, have different vulnerabilities and impacts on landscapes.

Differential vulnerabilities to invasive species can result from the complex interactions of climate, geomorphology, landscape position, soils, and vegetation. These factors govern the movement of invasive species propagules across the landscape via a variety of vectors; directly by wind and water or indirectly by endo- and ecto-zoochory. The availability of resources at different points in the landscape and in different time periods can affect the recruitment and survival of invasive species (Davis *et al.*, 2000). These shifting resources can vary on time scales of years, months, and even days. In addition, the changes in vegetation occupying the site can change the vulnerabilities of landscapes to invasive species over time. Varying soil:vegetation interactions result in functionally different ecological configurations that can change the availability of resources and the vulnerability of sites to invasion. The varying availability of soil resources (water and nutrients) governs the rate and success of ecological processes such as seed germination, seedling survival, juvenile growth rates, interspecies competition, and the effects of disturbance at different stages of the life cycle.

Changes in the patterns of vulnerability at the community scale and the arrangement of those communities can also affect the way a species spreads across the landscape and how the impacts are manifested. Within this complex opportunity matrix, the vulnerability to invasion, the impacts of the invasion, and the range of successful responses and restoration tactics will change with time. In a management context, property owners must select and prioritize specific landscape components to target with management actions. As the vulnerability of landscape components changes over time (either seasonally or across seasons), the vulnerability to invasion changes and the deployment of techniques to detect invasions would also be affected. Likewise, the selection and use of management treatments to address the invasion would change.

Developing responses to invasive species also more often than not involves multiple landowners. When landscape-scale respon-ses include landowners with different objectives, resources, and skills, the role of policy becomes increasingly important. In this case, policy can be considered as encompassing the full range of government responses: regulations, rules, programs, and financial and technical assistance. Depending upon the spatiotemporal patterns of invasion and restoration needs, success in responding to invasive species may require multi-scale information to implement successful responses. The ability of policy-developers to incorporate relevant spatio-temporal information is thus critical to responding successfully to the challenges of managing invasive species.

Understanding the effects of spatio-temporal heterogeneity on invasion processes is a necessary first step in developing a guide for estimating impacts and selecting and implementing responses to invasive species. While individual species performance under a range of environmental conditions is relatively easy to define and investigate, the range of possibilities associated with the spatiotemporal distribution of favorable establishment conditions (site availability), propagule presence (dispersal) and species performance is staggeringly large. An ability to organize knowledge of this 'spatial heterogeneity' as a source of conditions necessary for invasive plant establishment, growth, and reproduction is a key component in designing assessment, prediction, and treatment approaches.

Ecological Sites and State and Transition Models

Wildlands and their management are generally defined more by their limitations than their uses. These limitations (aridity, shallow soils, steep slopes) generally preclude the use of intensive management technologies because of the high cost of implementation and the relatively low probability of achieving a favorable economic return. Thus, the successful management of wildlands, regardless of the objectives, depends on land managers being able to

reach their goals within those highly variable conditions (spatial and temporal heterogeneity) by directing ecological processes with extensive management (Hammitt and Cole, 1998). Understanding this heterogeneity in a management framework requires subdividing the landscape into components that behave similarly in response to management. These groupings of similarly responding landscape units are especially important in the detection, prediction, and treatment of invasive species. These landscape subdivisions are referred to as 'sites.' The concept of site and its application has been one of the central tenets of modern natural resource management on rangelands and forestlands, incorporating the evolving ideas of ecosystem behavior and management responses as they developed (Brown, 2010). Grouping portions of a landscape based on their climatic, geomorphic, and edaphic similarities to predict the behavior of soil, vegetation, and related resources has been the basis for organizing information and decision-making. The site concept has been valuable to managers, advisors, and policy makers in assessing current conditions, setting objectives for change (or avoiding change), selecting and implementing management practices to achieve objectives, allocating scarce resources, and evaluating the effects of practices, programs, and policies (see Briske, 2011). The site, as a means of organizing information, was also an effective means of communicating professionally with land managers and with the public.

As the science of land management has advanced, the concepts used to define and describe the dynamics of sites have become more sophisticated and complex (e.g. Rumpff *et al.*, 2011). The availability of more precise observational technologies (e.g. remotely sensed landscape imagery, monitoring of ecosystem gas exchange) and nearly two centuries of direct observation has led to several refinements and improvements. The initial concept of site, developed in the early 20th century, was focused on the 'climax' vegetation, defined as the endpoint of ecological succession processes, and assumed to be the most stable and productive. In this approach, invasive species, and undesirable plants in general, were classified as 'invaders' (Dykesterhuis, 1949) that resulted from inappropriate management, but were not regarded as particularly influential in altering ecosystem-driving processes. The resulting weed management practices and policies had two serious shortcomings: (i) the plant community representing the 'climax' or successional endpoint was supposed to be the most competitive, and competition was the dominant ecological process controlling invasive plant populations (climax communities were less invasible); and (ii) once an invasive species was established, management responses emphasized the elimination of the 'problem' species rather than an evaluation of the altered disturbance regime (fire, grazing, drought, flood). This agronomic approach was relatively effective in communities and ecosystems where the invasive species had not affected ecological processes, but failed to have an effect when the processes had been altered (i.e. shrub invasion controlled water and nutrient cycles), and a new disturbance regime was in place (Brown et al., 1999). An enhanced understanding of the role of disturbance and climate variability in structuring plant communities has led to the acceptance of the existence of multiple stable 'states' (soil:vegetation combinations) for each site. Even subtle shifts in the seasonal distribution of precipitation (Archer and Predick, 2008) and atmospheric chemistry (Morgan et al., 2008) can result in changes in dominant species and create opportunities for invasive species.

This change in perspective, which recognizes the overriding influence of controlling variables (climate regime, landscape position, soil properties) on the dynamic behavior of plant communities, has resulted in an increased emphasis on static properties as a means to classify and group subunits of the landscape (Duniway et al., 2010). These landscape subunits are termed 'ecological sites' (USDA NRCS, 2007) and are unique groupings of soil properties and landscape position within climate-based regions called land resource units (LRUs) or major land resource areas (MLRAs). MLRAs are similar in extent to ecoregion sections used by the US Department of Agriculture, Forest Service (Cleland et al., 1997). For each ecological site, the physical setting, ecological dynamics, and management interpretations are contained in an ecological site description (ESD; Bestelmeyer et al., 2010). This document serves as the repository for information supporting the conceptual delineation of the landscape subunit.

Ecological sites are typically not mapped directly, but rather they are spatially defined by their correlation to mapped soils as a part of the National Cooperative Soil Survey (http://soils.usda.gov/partnerships/ncss/). Because of the inherent heterogeneity of wildlands and the more extensive nature of soil mapping, soil map units often contain much more variability than soils mapped in croplands. These 'complexes' often contain sufficient variability that an assumption of homogeneity of ecological processes, plant community dynamics, and response to management can lead to flawed decision-making. Thus, within the boundaries of an ecological site, displayed as part of a hardcopy or electronic soil property-based map, there may be substantial heterogeneity. For the purposes of planning invasive species assessment and management, users can assume that a particular soil map unit will respond as defined by the spatially-dominant, ecological site it contains. However, management-practice implementation and assessment should not be based on the assumption of within-map unit homogeneity. Assessment and management-practice implementation necessarily require closer inspection to ensure success. Field inspections of soil properties (depth, surface texture) are necessary to insure that assumptions about applying the appropriate site description are valid.

In addition to the defining spatial factors, every ecological site has unique temporal soil:vegetation dynamics that result in different plant communities, different relationships among the plant communities, and different responses to management over time (Bestelmeyer et al., 2010). An important part of the ESD is the graphical and textual description of soil:vegetation dynamics called a 'state and transition model.' The primary purpose of a state and transition model (STM) is to display and describe the possible soil:vegetation configurations (states) on a particular site, and the relationships (transitions) among them. STMs describe the range of dynamics of plant communities and ecological processes possible on the site (Briske et al., 2005). Figure 1.1 presents an example STM to illustrate the components and relationships. The large boxes represent ecological states. States are defined as different plant communities that may exist on the same site. Each state has different ecological process rates and functional relationships and possesses sufficient resistance such that significant management effort or disturbance is required to change (Stringham et al., 2003). Within each state, embedded boxes represent plant community phases that are similar in their ecological functional attributes and among which change is relatively simple and can occur in response to minor management changes or climatic fluctuations (Beisner et al., 2003). An individual state may contain multiple community phases (a range of variability), or they may have only one phase.

A recent innovation in the information in STMs is the inclusion of dynamic soil properties (DSPs). DSPs characterize near soil surface attributes that are tightly coupled through feedbacks with plant community structure and function. In most wildland plant communities, surface soils and vegetation are so closely linked that they must be considered a single unit when described (Duniway et al., 2010). For example, the amount and distribution of soil resources described by DSPs are particularly important in determining conditions and site availability for potential invasive species establishment. In particular, soil moisture and nutrient availability are important controls on seedling recruitment. In more mesic environments, bare soil patch size and distribution are often important

predictors of sites prone to invasive species establishment.

Each of the alternative states has a characteristic resistance to change and should persist under the defined disturbance regime. For each state, there are a set of characteristic diagnostic attributes and indicators. These are quantifiable parameters that can be used to distinguish states and should reflect differences in ecological processes. Plant attributes (species composition, cover, basal area, canopy height/ cover) and soil surface attributes (erosion, litter distribution, and surface stability) are typically used (Herrick et al., 2006). Resistance to invasion and resilience after disturbance are key state attributes, which may be helpful in conveying an overall sense of community dynamics. However, it could be misleading to assume that resistance or resilience somehow imparts low invasibility. The likelihood of a particular species invading a particular state or community can only be evaluated within the context of both species attributes and community characteristics (Heger and Trepl, 2003). The community phases within a state represent a characteristic range of variability in the processes and structure for that state. In particular for invasive species management, each state has an inherent resistance to invasion based on the manner in which plant species use the available resources and how disturbance is likely to alter those utilization patterns and create new opportunities. If a particular disturbance (e.g. fire, grazing, drought, or insect attack) creates new opportunities within a plant community, which allow an invasive species to establish, persist, and, more importantly, alter ecological functions, then a new, novel state is reached. If the disturbance does not result in invasive plant establishment and altered ecological function, the changes can be described merely as another community phase within the original state. The distinction between a new state and a new phase within an existing state is critical because of the implications for future management and restoration.

Transitions are the pathways of change among states. Unlike community pathways,

(a)

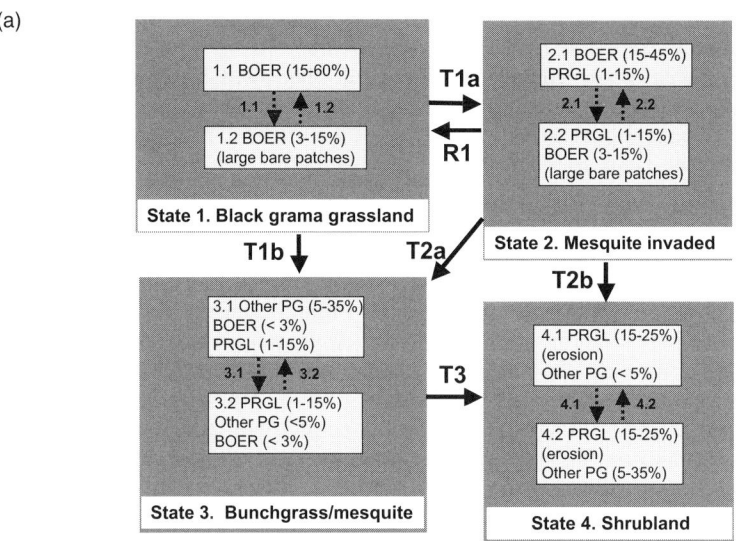

T1a. Mesquite establishment facilitated by seed transport by cattle, bare patches > 50 cm, and relatively wet springs
R1. Shrub removal via herbicide or fire followed by black grama recovery to > 15%
T1b, T2a. Black grama is reduced below ca. 3% cover by heavy grazing in drought
T2b, T3. At perennial grass cover < 5%, wind and storm events, trigger deep, spreading soil erosion

(b)

Fig. 1.1. (a) State-and-transition model for the Sandy ecological site (reference # R042XB012NM) of the southern desertic basins, plains, and mountains major land resource area (MLRA 42) in southern New Mexico, USA. BOER=*Bouteloua eriopoda* (black grama); PRGL=*Prosopis glandulosa* (honey mesquite); PG=perennial grasses. Arrows labeled T and R are transitions. Dotted lines are community pathways. Redrawn from Bestelmeyer *et al.* (2011). (b) Plant community scale photos of different states in Chihuahuan desert grassland.

transitions represent changes in ecological processes, not merely changes in structure that require substantial management input or changes in disturbance regime to result in a new state. Transitions generally represent changes in plant life-form (e.g. grass to shrub dominance) and sometimes near-surface soil properties. Transitions can be caused both by gradual change associated with slow variables and rapid change associated with distinct triggers (Bestelmeyer et al., 2010). Transitions caused by gradual change may stem from increasing invasive species recruitment rates or loss of herbaceous species due to chronic over-grazing. In contrast, transitions caused by triggers are relatively rapid, discrete events, such as an exceedingly hot fire followed by a drought period, or a flooding event that causes massive soil erosion.

A key point in the process of transition between two states is the threshold. Thresholds are specific points in time in which negative feedback mechanisms switch from reinforcing the inertia inherent within a state (resistance) to positive feedbacks accelerating that change (Brown et al., 1999). When describing thresholds in STMs, it is important to identify conditions that immediately precede a change. However, precise descriptions of thresholds are unlikely to be accurate given limited pre-dictive capability for most ecosystems, which could impart a false sense of confidence to managers (Brown, 2010). Perhaps the most important application of transition and threshold concepts for invasive species management is the appli-cation to ecological restoration (Bestelmeyer, 2006). Restoration technologies have typically been developed based on the assumption that important ecosystem pro-cesses remained intact, or in the terminology of a state and transition model, there has not been a state change (Young et al., 2001). For example, the Conservation Reserve Program that has been implemented on more than 15 million ha in the USA establishes perennial vegetation (mostly grasses) on marginal cropland. The program has been successful in reestablishing exotic and native perennial grasses and trees on

relatively fertile soils (usually mollisols) in comparatively mesic (>25 cm annual precipitation) regions. Although the result-ing perennial vegetation communities have value for a variety of ecosystem services (erosion control, carbon sequestration, wildlife habitat), it would indeed be a mistake to assume that the primary limiting factors in the restoration of all plant communities is the availability of seed of adapted plant species and appropriate cultural practices to suppress competition. A far greater challenge is the restoration of partially intact wildland communities that have experienced significant changes in life-form dominance and the distribution of soil properties. Bestelmeyer et al. (2003) demonstrated the challenges of restoration in a desert grassland ecosystem (Fig. 1.1). The loss of seed-producing, native tussock grasses from the system and the difficulty of reestablishing grass populations via seeding, the loss and redistribution of fine textured soil resources via wind erosion, and the increased competition from invasive shrubs present a substantial barrier to the restoration of important native species and ecological processes. There are many other ecosystems with similar challenges caused by redistribution of soil nutrients, changes in climatic patterns, and alteration of groundwater availability in addition to the competitive effects of invasive species (Society for Range Management, 2010).

Because of the information they contain and the relationships they describe, STMs can be used to assist the development of management strategies, choose appropriate assessment and monitoring approaches, and focus research and modeling efforts on key ecosystem properties and processes (Herrick et al., 2006). These applications are especially appropriate when applied to invasive species management. STMs organize, and put into context, information that allows managers to: (i) assess vulnerability to invasive species with dif-fering life history characteristics and alter management to increase the resistance to invasion; (ii) devise management practices and systems to limit the impacts of invasive species; and (iii) develop strategies to restore invaded plant communities, landscapes, and regions. Because they are conceptual, draft versions can be developed quickly and updated as new information is available. In addition to the management applications, STMs can be used to improve the link between research and management by focusing hypothesis development, experimental design and data collection to refine the models. This link to field research in a hypothesis-testing framework allows them to guide the development of quantitative models, which can then be used to extend the understanding gained through experimentation to other similar sites. Thus, community dynamics in response to changes in resource availability (e.g. precipitation), disturbance, and post-disturbance recovery can be described (Westoby *et al.*, 1989; Briske *et al.*, 2003). STMs constructed using these principles are especially valuable in estimating the effects of changes in life-history traits associated with invasive species.

The application of STMs to invasive species management is limited primarily by the fact that our understanding of the dynamic interactions among soil, vegetation, climate, and animals is incomplete. Unfortunately, existing datasets are frequently inadequate to provide the information necessary to define states and community phases, transitions, and thresholds. In many cases information is adequate to construct initial STMs, but details and subtleties necessary for complex management decisions are lacking. Perhaps the greatest limitation of STMs is that the dynamics represented within them are not applicable beyond the boundaries of the mapped soil properties that serve as the spatial link to specific land units (Briske *et al.*, 2005). Few invasions or invasive species distributions are limited to one ecological site, and the landscape scale patterns of ecological sites/states may be just as important in determining the rates and patterns of invasions as the within-community attributes. However, the information in ESDs and STMs can be used to calibrate models of invasive plant dynamics and behaviors at larger scales that quantify the

community scale interactions that are necessary to develop credible predictions, hypotheses, and experiments. As STMs become more frequently used, and implementation extends to more ecosystems, the quality of the logic, and data that goes into them, will become more critical.

ESDs and STMs are valuable tools in applying ecologically based invasive species management. Ecological processes that affect invasive species dynamics and influence the success of management responses are highly variable in both space and time. The information in ESDs and STMs can aid managers in subdividing the landscape according to soil:vegetation behaviors, and develop and test hypotheses about the dynamics of invasive plant populations. They also can integrate information gained from experimentation, monitoring, and treatment case studies. Together, these technologies can be combined into a powerful system to generate, interpret, and distribute relevant ecological information for managing invasive species and natural resources.

Applying Ecological Site Information to Invasive Species Management

Ecologically based invasive plant management relies on the application of 'adaptive management' to guide the process of identifying problems, selecting appropriate technologies, implementing responses, and assessing impacts (Fig. 1.2). At the core of adaptive management is the relevant and timely use of scientific information. In particular, the approach requires a format for organizing information to 'understand the linkages among ecological processes, vegetation dynamics, management practices and assessment' (see Sheley *et al.*, 2010). In this section, we provide examples of the application of ecological site/state and transition models to the process of defining problems, selecting and implementing management responses, and assessing impacts of a common invasive species problem: the invasion of arid and semi-arid grasslands by shrubs.

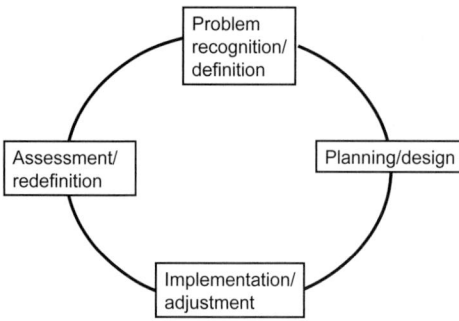

Fig. 1.2. The adaptive management cycle. Some steps have been combined to more accurately reflect the application of adaptive management to invasive species.

While invasive species management often focuses on the conditions necessary for plant establishment and reproduction, the spatial distribution of these conditions, at the scale of seed dispersal and seedling establishment, has been poorly described (Brown *et al.*, 2002). For example, while the climate and soils of a region may be adequate, on average, for plant recruitment and growth, the availability in both space and time of those conditions is what ultimately determines the ability of a species to establish and become abundant in the plant community (Davis *et al.*, 2000; Masters and Sheley, 2001). In essence, spatial heterogeneity describes the patterning of environmental conditions in which an organism resides or through which it moves. Thus, spatial heterogeneity is the most important factor in determining how a particular species invades a community, landscape, or region (Theoharides and Dukes, 2007). These patterns are critical in recognizing and defining the problem of plant invasions (Fig. 1.2). In the planning phase, accounting for spatial heterogeneity is a necessary component in setting management priorities and allocating resources to achieve results. The use of geographic information systems (GIS) has aided tremendously in the development of invasive species management plans. As more information about the spatiotemporal distribution of ecological processes that

influence plant invasions is incorporated into GIS tools, their utility in decision-making will greatly improve.

In the implementation phase (Fig. 1.2), identifying spatial heterogeneity is a critical requirement in achieving the desired outcomes of a management strategy (Rumpff *et al.*, 2011). Soils vary in their inherent water and nutrient holding capacity, in their susceptibility to wind and water erosion, and in their ability to tolerate and recover from disturbance; all of which help define the relative susceptibility of an ecosystem to invasion. Likewise, changes in vegetation over time can create a variety of different conditions. Periods of drought or heavy livestock grazing can create a patchwork of bare ground or open spaces among vegetation patches that can vary both within and across seasons.

The conversion of grasslands to shrublands in a variety of locations around the world over the past century is an excellent model for illustrating the importance of multi-scale heterogeneity that can lead to ecosystem-scale change. The importance of spatiotemporal heterogeneity in determining the rates and patterns of invasive species spread and impact is illustrated by desert grassland invasion by the shrub species honey mesquite (*Prosopis glandulosa*). In a Chihuahuan desert grassland, increasing distance between patches of relatively uniform perennial grass ground cover (black grama grass; *Bouteloua eriopoda*) in response to extended drought (a rapid, discrete trigger) or grazing-initiated (a slow, gradual change), grass loss and soil redistribution favors shrub seedling establishment, recruitment, and growth (Okin *et al.*, 2006). The transition from a black grama grassland state to a mesquite invaded state represents the establishment of processes that favor continued shrub recruitment, limit grass competition and fire frequency/intensity (see Bestelmeyer *et al.*, 2006). As individual grass plants succumb to drought stress, excessive defoliation, or burial by locally transported sediment, the loss of cover and increase in bare ground leads to black grama loss, dominance of shorter-lived bunch-grasses, and further wind and water redistribution of soil resources to shrub patches in the bunchgrass/mesquite state. In the bare interspaces, invasive shrubs are more likely to establish and persist than perennial grass plants, garnering below-ground resources via their extensive root system, and capturing aboveground resources via their shrub structure. As shrubs capture more resources and enhance their performance (growth and seed pro-duction), the availability of their propagules increases within the community and positively reinforces the process of change, enabling the transition to the shrubland state. Ultimately, these processes increase in scale to encompass the entire plant community and cascade across larger scales, resulting in landscape-scale conversions (Peters and Havstad, 2006). In this case, mesquite invasion restructures not only the plant community, but also the soil resources (Schlesinger et al., 1990).

The redistribution of vegetation pro-duction and soil resources creates multi-scale heterogeneity, enhancing runoff–run-on relationships and creating increased opportunities for shrub seedling recruit-ment (Rango et al., 2006). The redistribution of resources at multiple scales acts as a positive feedback on the processes imparting resilience to the mesquite shrubland state by concentrating resources around shrubs and limiting resource availability in the bare interspaces for grass recruitment. At any point in the change from one state to another, different disturbances (e.g. grazing or fire) can interact with climatic drivers (e.g. drought, high rainfall) in a variety of ways. Assessing the effects of management intervention (Fig. 1.2, step 4) can only be done within the context of the particular site × season interactions. In particular, adapting management responses to chang-ing conditions requires explicit knowledge of these unique site × season interactions and how they impact invasion patterns.

Each of these states (Fig. 1.1) features a set of attributes that dramatically alter how management of invasive species is accomplished. In the black grama grassland state (Fig. 1.1), water and nutrient cycles are controlled largely by the herbaceous peren-nial grass component (Fredrickson et al., 2006). Relatively infrequent fire and competition from a continuous warm season grass layer are thought to be sufficient to limit the recruitment and growth of shrubs. In the mesquite-invaded or bunchgrass/mesquite states (Fig. 1.1), a management strategy would have to include a disturbance regime that suppresses existing shrubs, allowing the existing grasses to recolonize bare spaces (Havstad and James, 2010). This approach would necessarily include both intensive measures to target shrubs via more frequent and intense fires, herbicide application, and more extensive grazing management, e.g. low stocking rate to ensure adequate herbaceous fuel accu-mulation and exclusion from grazing during critical periods for grass recruitment and expansion. This management regime represents a substantial redirection of resources and management to a level that precludes decision-making based primarily on economic considerations related to live-stock production. Finally, if the shrub:grass relationship has shifted entirely in favor of shrubs (Fig. 1.1; shrubland state), there is relatively little opportunity to implement a management-based strategy, and the emphasis of management actions shifts to either intensive mechanical manipulation of the soil surface or alternative land uses (Havstad and James, 2010). The transition away from mesquite dominance, via intensive management actions, is certainly not cost effective, and may not even be possible.

The preceding example illustrates how the various soils × vegetation interactions on a single site can have a tremendous influence on the development of manage-ment strategies and tactics to control an invasive species. However, the model of community dynamics presented in Fig. 1.1 is representative of only a relatively small portion of the entire landscape (Fig. 1.3). Each of the geomorphic surfaces (limestone, gravelly, clayey, sandy, loamy) has different soil properties, vegetation dynamics and different vulnerabilities to invasion and other forms of degradation. Limestone soils are relatively stable and only moderately

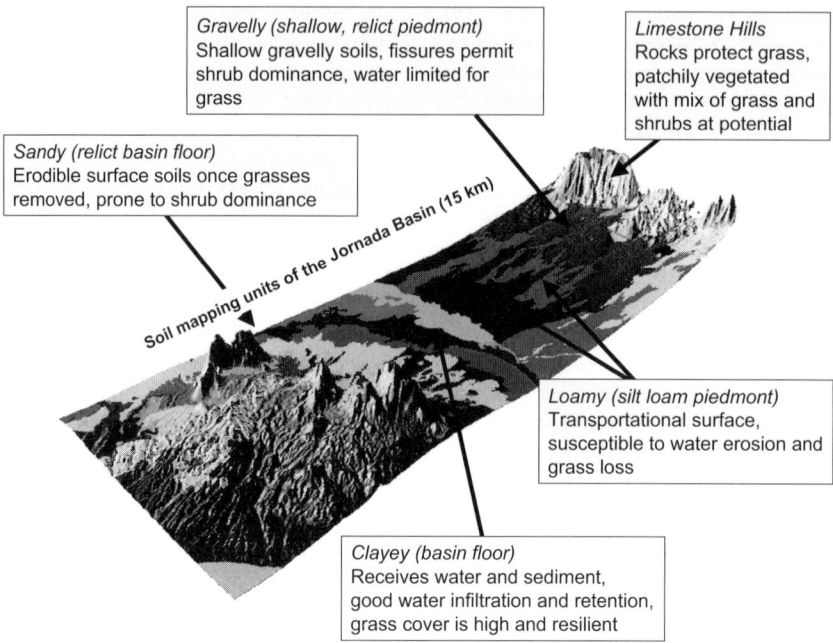

Gravelly (shallow, relict piedmont)
Shallow gravelly soils, fissures permit
shrub dominance, water limited for
grass

Limestone Hills
Rocks protect grass,
patchily vegetated
with mix of grass and
shrubs at potential

Sandy (relict basin floor)
Erodible surface soils once grasses
removed, prone to shrub dominance

Soil mapping units of the Jornada Basin (15 km)

Loamy (silt loam piedmont)
Transportational surface,
susceptible to water erosion and
grass loss

Clayey (basin floor)
Receives water and sediment,
good water infiltration and retention,
grass cover is high and resilient

Fig. 1.3. The soil-geomorphic template of a southern New Mexico, USA landscape. Each of the soil-geomorphic units possesses unique characteristics that impart differing resistance (tolerance to disturbance) to invasive species and differing resilience (ability to recover after disturbance). Reproduced from Bestelmeyer *et al.* (2006).

susceptible to degradation. Gravelly soils are shallow and susceptible to erosion and the invasion of the shrub *Larrea tridentata* (creosote bush). Sandy and loamy soils have similar levels of vulnerability to shrub invasion by mesquite via the processes described previously. Thus, the distribution of plant communities across the landscape can have a profound impact on invasive species spread and impact. In the above example, the processes driving the increase in both density and cover of honey mesquite on the sandy and loamy soils are different than the processes driving the increase of creosote bush on the gravelly soils. Given the differing susceptibility to shrub invasion and soil patterns within a landscape, substantially different management responses are required depending on which state the plant community is in (Bestelmeyer *et al.*, 2011).

Managing ecosystems for invasion resistance among individual landscape components and for landscape patterns that

confer the greatest resistance to invasion is certainly desirable. When management favors plant communities that use available resources as efficiently as possible, the availability of those resources to potential invaders will be limited. However, it is possible that a particular invasive species' ability to disperse, establish, and thrive across the landscape may overwhelm passive management designed to maximize invasion resistance. For example, in another semi-arid grassland, density and cover increases in the same invasive species, honey mesquite, result from a much different set of processes and presents a different set of challenges for management. In more mesic south Texas and east Texas ecosystems, USA, the conversion from grasslands to shrublands is essentially a case of invasion driven by dispersal rather than loss of resistance (Archer *et al.*, 1988). The change in plant community dominance from grasses to shrubs is attributed to a change in dispersal vectors (increase in domestic cattle) as the driving cause of the invasion, not the changes within community attributes (competition from grasses) or the spatial patterning of plant communities across the landscape (Brown and Archer, 1999). Competition from surrounding grasses was largely ineffective in limiting honey mesquite ingress and increase because of its ability to germinate and establish a taproot that extended beyond the zone of competition with grass roots within one season (Brown and Archer, 1989). The availability of soil water and nutrients was not a limiting factor because of the ability of the invasive shrub to escape competition with grasses within a time period in which soil moisture was abundant. Although honey mesquite recruitment in this ecosystem seemed to transcend community scale controls, Wu and Archer (2005) identified relatively fine-scale community topoedaphic variability that could be used as a reliable predictor of the spatial distribution of mesquite recruitment success.

The same processes can drive exotic species invasion patterns at the regional and continental scales. For example, an exotic shrub, prickly acacia (*Acacia nilotica*) has invaded subtropical perennial grassland in a relatively short time in Australia (Brown and Carter, 1998). Prickly acacia was introduced to the Mitchell Grass Plains as early as the late 19th century, but was not considered an invasive species until the mid-1970s. Its increase in density within areas where it had been planted, and expansion to new areas was associated with the replacement of sheep by cattle as the dominant domestic livestock species. Both grazers consumed acacia seed pods as seasonally important parts of their diets, but sheep were far more likely to destroy the seed during chewing than cattle. Many of the ingested seeds passed through cattle digestive system intact and were deposited in dung across a much wider area than seed could be transported by wind or water (Radford *et al.*, 2001). The patterns of species movements through the region and landscapes were entirely predictable based on the knowledge of livestock movements. Within-community patterns of shrub recruitment were also

predictable with knowledge of the distribution of features that attracted livestock (water development). Once viable populations were established proximate to water facilities, livestock dispersed the seeds into the uplands. Prior to the introduction of cattle as the dominant grazer in the region, the fire-, drought-, and grazing regimes that were in place throughout the early 20th century were sufficient to limit the expansion of prickly acacia (Radford *et al.*, 2001). When the domestic grazing regime changed (cattle to sheep; chronic over-grazing), the factors dampening or slowing change (fire, competition, and insect control of seed production in sparse populations) were diminished in their importance and another set of factors (dispersal, lack of fire, and seasonal soil moisture abundance) that favored the change to shrub-increase was ascendant.

Conclusion

These above examples illustrate the necessity of considering both the attributes of potential invasive species as well as the attributes of the plant communities, landscapes, and regions vulnerable to invasion. Seemingly minor changes in dispersal vectors, weather, vegetation composition and distribution, soil properties, and landscape position can result in completely different outcomes. While the understanding of plant attributes is critical to predicting invasion processes and designing responses, other authors in this volume address species performance in detail (see Eviner and Hawkes, Chapter 7, this volume; James, Chapter 8, this volume). However, knowledge of how landscape and regional heterogeneity govern patterns of livestock movement and seed dispersal is essential to predicting species movements and devising management strategies.

The patterns of invasive species abundance can be defined by processes that occur within communities (competition) or by processes that link communities (dispersal) and control spread. In either case, spatial structure is a key piece of information that is necessary to make predictions, devise responses, and assess impacts (see Masters and Sheley, 2001). In this chapter, we have presented two tools (ecological sites, state and transition models), which appear critical for the implementation of ecologically based invasive plant management. The ever-changing spatiotemporal dynamics of invasive plants in wildlands requires a systematic way to organize, display, and analyze information that serves as the basis for decision-making. While a particular species' characteristics are an important part of that knowledge base, the ability of an individual to germinate, survive, grow, and reproduce can only be determined by the resources available to it within the context of the surrounding community and landscape. An organized knowledge of these relationships is the basis for reliable prediction of a species' movements and for the development of management responses to slow or stop that movement.

References

Archer, S.R. and Predick, K.I. (2008) Climate change and ecosystems of the southwestern United States. *Rangelands* 30, 23–28.

Archer, S.R., Scifres, C.J., Bassham, C. and Maggio, R. (1988) Autogenic succession in a subtropical savanna, rates, dynamics and processes in the conversion of grassland to thorn woodland. *Ecological Monographs* 58, 111–127.

Beisner, B.E., Haydon, D.T. and Cuddington, K. (2003) Alternative stable states in ecology. *Frontiers in Ecology and the Environment* 1, 376–382.

Bestelmeyer, B.T. (2006) Threshold concepts and their use in rangeland management and restoration: the good, the bad and the insidious. *Restoration Ecology* 14, 325–329.

Bestelmeyer, B.T., Brown, J.R., Havstad, K.M., Alexander, R., Chavez, G. and Herrick, J.E. (2003) Development and use of state-and-transition models for rangelands. *Journal of Range Management* 56, 114–126.

Bestelmeyer, B.T., Ward, J.P. and Havstad, K.M. (2006) Soil geomorphic heterogeneity governs patchy vegetation dynamics at an arid ecotone. *Ecology* 87, 963–973.

Bestelmeyer, B.T., Moseley, K., Shaver, P.L.,

Sanchez, H., Briske, D.D. and Fernandez-Gimenez, M.E. (2010) Practical guidance for developing state-and-transition models. *Rangelands* 32, 23–30.

Bestelmeyer, B.T., Brown, J.R., Fuhlendorf, S., Fults, G. and Wu, X.B. (2011) A landscape approach to rangeland conservation practices. In: Briske, D.D. (ed.) *Conservation Benefits of Rangeland Practices: Assessment, Recommendations, and Knowledge Gaps*. Allen Press, Lawrence, Kansas/USA, (in press).

Blumenthal, D., Booth, D.T., Cox, S.E. and Ferrier, C.E. (2007) Large-scale aerial images capture details of invasive plant populations. *Rangeland Ecology and Management* 60, 523–528.

Briske, D.D. (2011) Introduction. In: Briske, D.D. (ed.) *Conservation Benefits of Rangeland Practices: Assessment, Recommendations, and Knowledge Gaps*. Allen Press, Lawrence, Kansas/USA, pp. 1–8.

Briske, D.D., Fuhlendorf, S.D. and Smeins, F.E. (2003) Vegetation dynamics on rangelands: a critique of the current paradigms. *Journal of Applied Ecology* 40, 601–614.

Briske, D.D., Fuhlendorf, S.D. and Smeins, F.E. (2005) State-and-transition models, thresholds, and rangeland health: a synthesis of ecological concepts and perspectives. *Rangeland Ecology and Management* 58, 1–10.

Brown, J.R. (2010) Ecological sites, their history, status and future. *Rangelands* 32, 5–8.

Brown, J.R. and Archer, S.R. (1989) Woody plant invasion of grasslands, establishment of honey Mesquite (*Prosopis glandulosa* var. *glandulosa*) on sites differing in herbaceous biomass and grazing history. *Oecologia* 80, 19–26.

Brown, J.R. and Archer, S.R. (1999) Shrub invasion of grassland, recruitment is continuous and not regulated by herbaceous biomass or density. *Ecology* 80, 2385–2396.

Brown, J.R. and Carter, J. (1998) Spatial and temporal patterns of exotic shrub invasion in an Australian tropical grassland. *Landscape Ecology* 13, 93–102.

Brown, J.R., Herrick, J. and Price, D. (1999) Managing low-output agroecosystems sustainably, the importance of ecological thresholds. *Canadian Journal of Forest Research* 29, 1112–1119.

Brown, J.R., Svejcar, T., Brunson, M., Dobrowolski, J., Fredrickson, E., Krueter, U., Launchbaugh, K., Southworth, J. and Thurow, T. (2002) Are range sites the appropriate spatial unit for measuring and monitoring rangelands? *Rangelands* 24, 7–12.

Cleland, D.T., Avers, R.E., McNab, W.H., Jensen, M.E., Bailey, R.G., King, T. and Russell, W.E.

(1997) National hierarchical framework of ecological units. In: Boyce, M.S. and Haney, A. (eds) *Ecosystem Management, Applications for Sustainable Forest and Wildlife Resources.* Yale University Press, New Haven, Connecticut, pp. 181–200.

Cousens, R. and Mortimer, M. (1995) *Dynamics of Weed Populations.* Cambridge University Press, New York.

Davis, M.A., Grime, J.P. and Thompson, K. (2000) Fluctuating resources in plant communities, a general theory of invisibility. *Journal of Ecology* 88, 528–534.

Duniway, M.C., Bestelmeyer, B.T. and Tugel, A. (2010) Soil properties that distinguish ecological site and states. *Rangelands* 32, 9–15.

Dykesterhuis, E.J. (1949) Condition and management of rangelands based on quantitative ecology. *Journal of Range Management* 2, 104–115.

Fredrickson, E.L., Estell, R.E., Laliberte, A. and Anderson, D.M. (2006) Mesquite recruitment in the Chihuahuan desert, historic and prehistoric patterns with long-term impacts. *Journal of Arid Environments* 65, 285–295.

Hammitt, W.E. and Cole, D.N. (1998) *Wildland Recreation, Ecology and Management.* John Wiley and Sons, New York.

Havstad, K.M. and James, D. (2010) Prescribed burning to affect a state transition in a shrub encroached desert grassland. *Journal of Arid Environments* 74, 1324–1328.

Heger, T. and Trepl, L. (2003) Predicting biological invasions. *Biological Invasions* 5, 313–321.

Herrick, J.E., Bestelmeyer, B.T., Archer, S., Tugel, A.J. and Brown, J.R. (2006) An integrated framework for science-based arid land management. *Journal of Arid Environments* 65, 319–335.

Mack, R.N., Simberloff, D., Lonsdale, W.M., Evans, H., Clout, M. and Bazzaz, F.A. (2000) Biotic invasions, causes, epidemiology, global consequences and control. *Ecological Applications* 10, 689–710.

Masters, R.A. and Sheley, R.L. (2001) Principles and practices for managing rangeland invasive plants. *Journal of Range Management* 54, 502–517.

Mitchell, J.J. and Glenn, N.F. (2009) Leafy spurge (*Euphorbia esula*) classification performance using hyperspectral and multispectral sensors. *Rangeland Ecology and Management* 62, 16–27.

Morgan, J.A., Derner, J.D., Milchunas, D.G. and Pendall, E. (2008) Management implications of global change for Great Plains rangelands. *Rangelands* 30, 18–22.

Okin, G.S., Gillette, D.A. and Herrick, J.E. (2006) Multi-scale controls on and consequences of aeolian processes in landscape change in arid and semi-arid environments. *Journal of Arid Environments* 65, 253–275.

Peters, D.C. and Havstad, K.M. (2006) Nonlinear dynamics in arid and semi-arid systems: interactions among drivers and processes across scales. *Journal of Arid Environments* 65, 196–206.

Radford, I.J., Nicholas, D.M., Brown, J.R. and Kriticos, D.J. (2001) Paddock-scale patterns of seed production and dispersal in the invasive shrub *Acacia nilotica* (Mimosaceae) in northern Australian rangelands. *Austral Ecology* 26, 338–348.

Radosevich, S.R., Holt, J.S. and Ghersa, C.M. (2007) *Ecology of Weeds and Invasive Plants: Relationship to Agriculture and Natural Resource Management,* 3rd edn. John Wiley and Sons, Hoboken, New Jersey.

Rango, A., Tartowski, S.L., Laliberte, A., Wainwright, J. and Parsons, A. (2006) Islands of hydrologically enhanced biotic productivity in natural and managed arid ecosystems. *Journal of Arid Environments* 65, 235–252.

Rumpff, L., Duncan, D.H., Vesk, P.A., Keith, D.A. and Wintle, B.A. (2011) State-and-transition modeling for adaptive management of native woodlands. *Biological Conservation* 144, 1224–1236.

Schlesinger, W.H., Reynolds, J.F., Cunningham, G.L., Huenneke, L.F., Jarrell, W.M., Virginia, R.A. and Whitford, W.G. (1990) Biological feedbacks in global desertification. *Science* 247, 1043–1048.

Sheley, R., James, J., Smith, B. and Vasquez, E. (2010) Applying ecologically based invasive plant management. *Rangeland Ecology and Management* 63, 605–613.

Society for Range Management (2010) Special Issue – Ecological Sites. *Rangelands* 32, 3–66.

Stringham, T.K., Krueger, W.C. and Shaver, P.L. (2003) State and transition modeling, an ecological process approach. *Journal of Range Management* 56, 106–113.

Theoharides, K.A. and Dukes, J.S. (2007) Plant invasion across space and time: factors affecting nonindigenous species success during four stages of invasion. *New Phytologist* 176, 256–273.

USDA Natural Resources Conservation Service (2007) *National Range and Pasture Handbook.* 190-VI, revision 1. Washington, DC.

Washington-Allen, R.A., West, N.E., Ramsey, R.D. and Efroymson, R.A. (2006) A protocol for retrospective remote sensing-based ecological monitoring of rangelands. *Rangeland Ecology and Management* 59, 19–29.

Westoby, M., Walker, B.H. and Noy-Meir, I. (1989) Opportunistic management for rangelands not at equilibrium. *Journal of Range Management* 42, 266–274.

Wu, X.B. and Archer, S.R. (2005) Scale-dependent influence of topography-based hydrologic features on patterns of woody plant encroachment in savanna landscapes. *Landscape Ecology* 20, 733–742.

Young, T.P., Chase, J.M. and Huddleston, R.T. (2001) Community succession and assembly: comparing, contrasting, and combining paradigms in the context of ecological restoration. *Ecological Restoration* 19, 5–18.

2 Linking Disturbance Regimes, Vegetation Dynamics, and Plant Strategies Across Complex Landscapes to Mitigate and Manage Plant Invasions

Samuel D. Fuhlendorf, Brady W. Allred, R. Dwayne Elmore, and David M. Engle

Natural Resource Ecology and Management, Oklahoma State University, USA

Introduction

Composition and structure of ecosystems have long been described as dynamic, but a recent focus on invasive species has revitalized discussion of assembly rules and dynamics of species composition. For rangelands, our understanding of species invasions focused first on the effects of woody plant encroachment on livestock production. Recent research emphasizes intentional and accidental introduction of exotic species that are capable of dramatic changes to ecosystem structure and function. Increased attention to invasive species, including woody plants, has altered our understanding and methods of describing rangeland dynamics. In early 20th-century descriptions of rangeland dynamics, grazing was seen as the primary driver that produced dynamics with predictable, linear, and reversible patterns. Species invasions that can greatly alter dynamics led to the development of such rangeland concepts as thresholds, multiple steady states, and our modern perception of rangeland change.

Following initial introduction and establishment, invasion of both exotic and native species has been associated with many factors, including introduction of livestock, altered fire regimes, short- and long-term fluctuations in climate, and increased atmospheric trace gases. Rates and patterns of species invasions are: (i) often non-linear and characterized by gradual initial changes followed by rapid increases; (ii) accentuated by drought with variable rates over time; (iii) influenced by soil and topographic features; and (iv) often non-reversible over management timeframes (Archer, 1994). Research and observations on rangelands leave little doubt that species invasions have caused major changes over the past 100–200 years, frequently converting ecosystem types (e.g. converting open grasslands to closed canopy woodlands and converting shrublands to annual grasslands).

Species invasions are best understood through studies at multiple scales. The invasion process is a landscape or regional process that has been analogously described as a glacier moving across a region converting the landscape as it progresses. An agronomic approach to research focused on small plots may be useful in understanding small-scale mechanisms of invasion or potential management approaches, but is not effective at understanding broad-scale invasions or landscape/regional management schemes. Most research efforts understandably rely on small plots to focus on the immediate problem of controlling invasive species, but these studies potentially neglect landscape and regional patterns that might be important in understanding the causes and implications of invasions. Understanding and effectively managing invasion processes requires an appreciation of: (i) the ecological relationships that contribute to invasive species success; (ii) the ecological factors that promote efficacy of specific management strategies; and (iii) the most appropriate life stages and techniques for management. Management of these species also requires an ecological approach that incorporates the life history of the invasive species and the ecological structure and function of the ecosystem that is being invaded.

Our goal is to present an overview of relevant ecological considerations for invasive species in general and to suggest some ecological factors to consider before initiating management efforts. We will focus much of our discussion on rangelands, which are disturbance-dependent ecosystems with a long history of fire and grazing, as well as highly variable topo-edaphic and weather patterns. We conceptualize species invasion by focusing on pre-invasion conditions and trying to understand the interplay between changes in disturbance processes and traits of invasive species that would best prohibit or inhibit invasions (Fig. 2.1). From this perspective, the portfolio of factors influencing species assemblages can be viewed as a filter (climate, soils, grazing, fire, etc.) and

species life history traits determines its establishment and abundance in the novel ecosystem. Using this framework we will specifically examine the following questions: (i) What traits allow some species to gain dominance within some ecosystems and not others? (ii) What environmental factors have contributed to the number and extent of invasive species? and (iii) What can we take from our understanding of succession and rangeland dynamics that will contribute to our management of invasive species? Even though causal factors associated with species invasions are variable across species and ecosystems, most invasions can be explained by a few plant adaptations that relate to current and historical environmental conditions.

Evolutionary Disturbance Processes on Rangelands

Rangeland ecosystems are characterized by disturbances, and most species within these landscapes vary in abundance depending on their adaptation to disturbance patterns (Fig. 2.2). Generally disturbance refers to a discrete, punctuated killing, displacement, or damaging of one or more individuals that directly or indirectly creates an opportunity for new individuals to become established (Sousa, 1984). By this definition, it is somewhat circular to argue that diversity is dependent on disturbance because the definition

How do species become invasive?

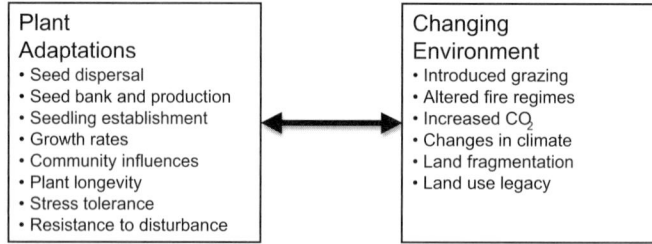

Fig. 2.1. The potential for a species to invade is dependent upon the interaction among the plant traits of the invasive species and the changing environment of the ecosystem. The environment acts as a filter and the traits allow certain plants to be more or less successful.

includes *opportunity* for new species. However, consideration of the pattern of this relationship and analysis of variable responses across different disturbances is warranted. Studies of rangelands suggest the dominant disturbances prior to European settlement were fire and grazing, and their interaction with each other (Fuhlendorf *et al.*, 2009a). Rangeland biotic communities also evolved with many secondary disturbances (e.g. prairie dogs, insect outbreaks) and their interaction with dominant disturbances. Drought also is

sometimes considered a disturbance that is central to understanding rangeland dynamics resulting from evolutionary pressures that shaped rangeland communities. Most processes that are described as disturbances can also be considered as a critical part of the portfolio of factors that contribute to ecosystem structure and function (e.g. climate, soils, fire, grazing, etc.).

Based on the simple definition of disturbance processes, it is circular to predict changes in species diversity and species invasion but it is useful to discuss predicted

(a)

(b)

Fig. 2.2. Fire and grazing are common disturbances on rangelands throughout the world. (a) Fire in a sand sagebrush/mixed prairie ecosystem being invaded by eastern redcedar. (b) Bison grazing on a recently burned patch at the Tallgrass Prairie Preserve in Oklahoma. Both photos by Stephen Winter.

patterns. The diversity/disturbance relationship relates to invasive species because by reducing the dominance of a few species or by widening niche space, exotic or invasive species can capitalize on the opportunity (Elton, 1958; Tilman, 1997; Gurvich *et al.*, 2005) depending on the type, duration, timing, frequency, etc. of the disturbance and other processes (e.g. change in atmospheric deposition of nitrogen). The processes that maintain species diversity can potentially be the dominant contribution to invasion. While we might assume that evolutionary disturbances benefit native species over exotic species, there is no certainty that invasive species will not increase from these disturbances as well. Moreover, invasion can ultimately alter the disturbance regime, increase positive feedbacks at the expense of negative feedbacks, and shift the ecosystem across a threshold to a new system state that may contain reduced species diversity (Fig. 2.3). Finally, it is possible that more species-rich plant communities are more susceptible to invasion than less species-rich plant communities (Ganguli *et al.*, 2008) at least at some scales (Stohlgren *et al.*, 2002).

For at least the 10,000–15,000 years before Europeans discovered the Americas, herbivory in North America was largely controlled by the interaction between fire and grazing. Disturbance by grazing animals on prehistorical landscapes was dependent upon selection of grazing locations and the interaction among landscapes, people, and disturbance processes. For example, recent evaluations of grazing animal behavior suggest that grazing animals are strongly responsive to fires, which were dominant forces ignited by lightning or aboriginal peoples. The net effect was a disturbance-dependent landscape where the interaction of fire and grazing was a dominant feature, hereafter termed *pyric herbivory* (i.e. herbivory shaped by fire) (Fuhlendorf *et al.*, 2009a). From this perspective, grazing and fire may best be viewed as a single disturbance (pyric herbivory) that creates a shifting mosaic of disturbance patches across a complex landscape (e.g. Fuhlendorf and Engle, 2001; Salvatori *et al.*, 2001; Hassan *et al.*, 2008). This results from grazing animals freely selecting between burned and unburned portions of the landscape, and the dependence of fire

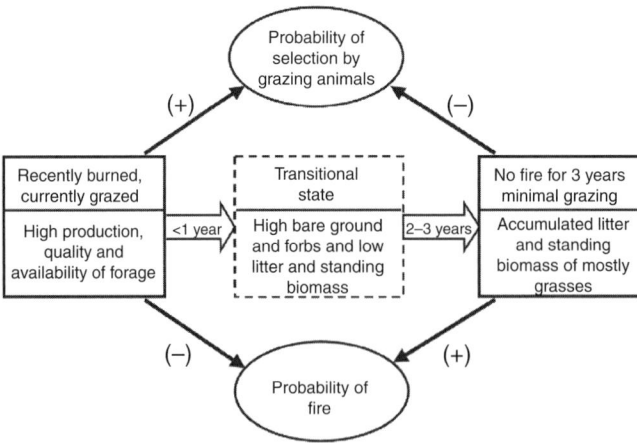

Fig. 2.3. A conceptual model of path dynamics within a shifting mosaic landscape where each patch is experiencing similar but out-of-phase dynamics driven by fire and grazing (Fuhlendorf and Engle, 2004). Ovals represent the primary drivers (fire and grazing) while squares represent the ecosystem states within a single patch as a function of time since focal disturbance. All states have the potential for fire or grazing. Solid arrows indicate positive (+) and negative (−) feedbacks in which plant community structure is influencing the probability of fire and grazing.

occurrence on the removal of fuel by herbivores (e.g. Norton-Griffiths, 1979; Fuhlendorf and Engle, 2001) (Fig. 2.3). That rangeland biodiversity depends largely on the mosaic pattern that results from pyric herbivory reflects the notion that fire and grazing were historically coupled. Over the past century or more these processes have been decoupled, largely to facilitate utilitarian efforts to maximize livestock production (Fuhlendorf and Engle, 2001). While other disturbance processes were also important, the interaction of multiple herbivores with fire across large and complex landscapes was central to the development of the Great Plains of North America prior to settlement. Restoration of these processes and patterns is frequently the focus of conservation of rangeland ecosystems but conversion and fragmentation of these landscapes are serious barriers.

Life Histories and Plant Strategies that Promote Invasiveness

Although life history strategies of invasive species and the conditions of the communities they invade are highly variable, some generalizations are useful in understanding invasion processes and developing ecological management strategies. Some plant traits can enable a species to invade into a relatively open environment while other traits may allow a species to invade under a closed canopy. Likewise, some species are well adapted to specific disturbances, such as fire and grazing, while others may be extremely sensitive to only one or all disturbances. Efficient ecological management requires an understanding of plant traits that best describe the potential strengths and weaknesses species have toward invasion, as well as the conditions of the ecosystem that make it susceptible to invasion.

By identifying these traits and understanding some basic ecological generalizations, a land manager can more efficiently evaluate the need for management and accurately focus on critical phases of vegetation change. Linking successional changes in plant communities to traits or 'vital attributes' of plants has long been proposed (Noble and Slatyer, 1980; Glenn-Lewin, 1992). Research has often focused on important life history traits that can facilitate the establishment of species following disturbance, including propagule persistence or dispersal, age at first reproduction, and longevity or life span. The importance of these traits, as well as the ability to persist or reproduce vegetatively, has been demonstrated for forest community response to fire, and they have been used to predict the species that would most likely increase or decrease immediately following the disturbance.

Other approaches to predicting species invasions from life history traits have focused on a species' ability to persist in resource-rich conditions (competitors), resource-poor conditions (stress tolerators), and with frequent/intense disturbances (ruderals) (Grime, 1977; Blumenthal et al., 2009; Seastedt, 2009). The use of life history traits of functional categories of species is easily transferred to invasive species invasion following disturbance (Horvitz et al., 1998), but considerable debate persists about whether it is productive to categorize species as 'native' or 'exotic' rather than simply focus on the species traits. A recent analysis of species traits (competitor, stress-tolerator, ruderal) highlighted an important consideration when discussing species traits and potential invasive species. Blumenthal et al. (2009) describe a paradox in which resource availability can have a strong influence on plant success but that would hold equally true for native and exotic species. They concluded that life history traits, resource availability, and the structure of the pathogen community were predictive of the potential invasiveness of exotic species, suggesting that invasive species are unique due to rapid growth and limited pathogen burden (Blumenthal et al., 2009; Seastedt, 2009). Pressures from herbivory can be evaluated in similar light to the pathogen burden perspective. If a species is capable of tolerating or avoiding herbivory, it would be similar to low pathogen burdens in the novel ecosystems that they invade.

What has Changed? Altered Disturbance Regimes and Global Change

While plant adaptations are useful in explaining species invasions on rangelands, patterns and rates of invasion have been highly variable in space and time. This suggests that the biotic and abiotic structure and function of these ecosystems, as well as the dynamics over time, may be useful in explaining the invasion process. Since we are interested in the invasion of species into a matrix that was dominated by historic disturbances and other processes that acted as assembly factors for historical plant communities, it is useful to think about how structure and function of these eco-systems have been altered since European settlement. Major alterations of landscapes that contribute to species invasions include: (i) altered fire regimes; (ii) introduction of domestic livestock; (iii) increases in atmospheric trace gases; (iv) climatic influences; and (v) land use legacies (Archer, 1994; Fuhlendorf *et al.*, 1996, 2002; Polley *et al.*, 1997).

Altered fire regimes

Prior to European settlement, periodic fires burned across many landscapes maintaining fairly open grasslands with woody plants limited to discrete landscape units where fire frequency and intensity were insufficient to control their dominance (Axelrod, 1985). These patterns were highly variable depending on regional differences in climate, soils, vegetation, and people. For centuries, the majority of fires were set by aboriginal people for myriad of reasons, but the resulting fire regimes were as critical to the evolution of species and the development of ecosystems as climate and other abiotic factors. At the time of settlement, the use of fire to clear the land of woody vegetation and provide more

open areas for grazing and crops was commonplace, possibly resulting in a short-lived increase in fires (Stambaugh and Guyette, 2006; DeSantis *et al.*, 2010). Not long after settlement, however, natural-occurring and anthropogenic fires were reduced in most places because of fear of fire, which made wildfire suppression a high priority. Also, heavy stocking of rangelands with livestock reduced the probability of natural-occurring fires by reducing the standing biomass available for fuel (Belsky and Blumenthal, 1997; Fuhlendorf *et al.*, 2008). The result is an invasion of species (both native and exotic) that are intolerant of fire.

In some regions, disturbances such as grazing greatly alter the response of ecosystems to fire, resulting in both increased and decreased invasion following fires (Cummings *et al.*, 2007; Davies *et al.*, 2009). Grazing alters the fuel load and ultimately fire intensity, which can strongly influence fire severity and the rate and pattern of invasion (Twidwell *et al.*, 2009). Also, long-term grazing can greatly alter composition and seed bank of plant communities limiting the rate and pattern of recovery following fire (Fuhlendorf and Smeins, 1997), leaving sites available for invasive species establishment. Alternatively, patch fires can result in focused intense grazing pressure limiting the stature of all species and reducing the importance of traits that may promote tolerance or avoidance of grazing (Cummings *et al.*, 2007).

Invasion of species to novel ecosystems can be associated with a feedback where initial invasion can alter fire regimes, leading to further invasion. The common example of this is the invasion of sagebrush eco-systems in the western USA by cheatgrass, which promotes fire, reducing sagebrush dominance, and promoting further invasion and additional fires. This occurs due to life history, as cheatgrass grows early in the growing season, rapidly maturing and providing fuel for subsequent fires through-out the summer (the predominant fire season for this region; Brooks *et al.*, 2004). These frequent fires remove non-sprouting shrubs, such as *Artemsia* spp., shifting the plant community from shrub to annual grass. An alternative example that is less well described is the invasion of tall fescue (*Lolium arundinaceum* (Schreb.) S.J. Darbyshire) in central tallgrass prairies, the most abundant plant in the eastern USA (Rudgers and Clay, 2007). This is a cool-season grass that is invading warm-season grasslands. Once invaded, the species is active during much of the season when fire can be used, in some cases limiting the ability to conduct pre-scribed burns, which can further promote invasion of woody plants and shift the plant community to woodland. These two examples demonstrate that while we have greatly altered fire regimes, invasion by some species can result in positive feedbacks that further alter fire regimes and these changes can either increase or decrease fire-return interval and fire intensity.

Introduction of domestic livestock

Over the past 20,000 years, animal influences on North American rangelands have gone from a high diversity of now extinct large herbivores (mammoths, sloths, giant bison, camels, etc.) to vast herds of smaller herbivores (primarily bison) that were largely interactive with fire and large complex landscapes at the time of settlement (Burkhardt, 1996; Fuhlendorf *et al.*, 2009a). Many rangelands developed for many years with high numbers of herbivores while others were only grazed periodically or by small resident herds of animals. It is difficult to determine the evolutionary influence of grazing on current rangelands of North America, but it is likely that the introduction and ultimately the confinement of livestock 100–150 years ago has considerably changed the structure and function of rangeland ecosystems, influencing the invasion processes that we are observing today.

Most of the settlers were inexperienced in managing the relatively dry rangelands of North America. Consequently, initial stock-ing rates were often extremely high,

resulting in high mortality to livestock during droughts before 1900 (Lehmann, 1969). Livestock mortality rates were reported as high as 85% in the southwestern USA before the turn of the century, suggesting that the resource had been over-used soon after settlement. These severe stocking rates likely resulted in complete removal of vegetation cover and may have led to significant erosion (Smeins *et al.*, 1997). These influences should be considered when evaluating vegetation change over the past 100–200 years, particularly when we base our ecological discussion and manage-ment decisions on plant community descriptions of pre-settlement conditions. The abiotic environment associated with those descriptions frequently no longer exists (Fuhlendorf and Smeins, 1997).

Invasion of native and exotic woody plants has been presumed to be facilitated by the introduction of livestock (Walker *et al.*, 1981; Belsky and Blumenthal, 1997). In some cases, similar relationships could be drawn to herbaceous invasive species, although such relationships can be merely correlative with many changes (livestock, reduced fire, species introductions, etc.) that have occurred for the past century or more. However, several studies have suggested that the presence of dense grassland vegetation, as in ungrazed or moderately grazed conditions, may not restrict the establish-ment of woody species or more recently invasive species, but instead may actually improve conditions for seedling survival and development (Schultz *et al.*, 1955; Archer, 1995; O'Connor, 1995; Cummings *et al.*, 2007). Woody plant or invasive species dominance could be inhibited by harsher micro-environmental conditions in a heavily grazed community that may limit seedling establishment, or invasion could be inhibited by consumption of seedlings by livestock, the level of which depends on the specific species of herbivore and invader. Many plant seedlings are frequently more susceptible to mortality from grazing than mature plants as immature plants can be more palatable with less defense mechanisms, such as secondary chemicals or thorns, and may be less tolerant of defoliation (Taylor *et al.*, 1997; Cummings *et al.*, 2007).

Grazing also has a major influence on other disturbance processes. For example, grazing alters fire regimes through the removal of fuel required to ignite and carry a fire (Fuhlendorf and Smeins, 1997; Fuhlendorf *et al.*, 2008). For fire-sensitive species, reduction of fuel from grazing may be the greatest effect of grazing on the invasion process. Similarly, pre-settlement grazing was largely interactive with fire, creating a shifting mosaic where grazing followed a fire, which would result in a continuum of grazing intensity and thus variability in plant composition and structure. Modern range management has decoupled fire and grazing and developed technologies to promote spatially uniform utilization so that current grazing patterns have little resemblance to evolutionary patterns. Evolutionary patterns of disturbance are understood to be critically important to biodiversity, but evolutionary patterns of disturbance may also be critically important to invasion ecology if the same processes that promote diversity also promote or retard invasion.

Increase in atmospheric trace gases

Increases in atmospheric trace gases such as CO_2 can alter the relationship between woody and herbaceous plants to favor woody plant dominance (Polley *et al.*, 1997). Similar arguments can be made for some non-woody invasive plants (Smith *et al.*, 2000). These arguments are based on observations that indicate: (i) many invasive species possess the C_3 photosynthetic pathway, whereas some rangelands are dominated primarily by plants that possess the C_4 photosynthetic pathway; (ii) increased atmospheric CO_2 may confer a significant advantage to C_3 species relative to C_4 species with respect to physiological activity, growth, and competitive ability; (iii) C_4 grasslands appear to have evolved at CO_2 concentrations below 200 ppm and therefore at low CO_2/O_2 ratios; and (iv) invasion of many of these species has been accompanied by a 30% increase in atmospheric CO_2 over the past 200 years (Archer *et al.*, 1995; Smith *et al.*, 2000).

While the relationship between C_3 plants and CO_2 is theoretically sound and experimentally demonstrated, the ultimate role of these changes in the invasion processes remains largely unstudied (Smith *et al.*, 2000). We know that many aspects of plant physiology are altered by elevated levels of CO_2 and many of these factors have the potential to alter ecosystems and landscapes, but studying the effect on large and complex areas is challenging. Direct comparisons of native and invasive species in controlled environments have demonstrated that invasive species benefited more than native species from elevated CO_2 levels, which could promote invasion (Song *et al.*, 2009). Elevated CO_2 has been shown to alter species and potentially communities and landscapes, but it remains difficult to develop accurate predictions of future change. However, this suggests that focusing management on 'pristine' conditions at the time of settlement may not return these plant communities to a desired composition.

Climatic influences

Many long-standing ecological principles are based on the premise that the dominant vegetation on a landscape is largely dependent upon the climate (Clements, 1916). The importance of short-term and long-term weather patterns on the increase in woody plants over the past 100–200 years has been debated in the ecological literature (Miller and Wigand, 1994; Belsky, 1996). A series of studies in the deserts of the southwestern USA suggest a variety of climate-related causal factors are associated with the increase in woody plants, including short-term droughts, long-term downward trends in precipitation, and long-term warming and/or cooling trends (Neilson, 1986; Conley *et al.*, 1992; Archer, 1994). However, grazing has been identified as a confounding factor that may be at least as important as climatic patterns as a causal factor of woody plant increase (Conley *et al.*, 1992).

Analyses of fossil pollen, pack-rat middens, and stable isotopes of organic matter suggest that populations of woody plants varied with fluctuations in climate over the past 20,000 years (Van Devender and Spaulding, 1979; Nordt *et al.*, 1994; Hall and Valastro, 1995). For instance, mean annual precipitation and proportion of winter precipitation explained 62% of the variability in the relative abundance of shrubs across temperate grasslands and savannas of North America (Paruelo and Lauenroth, 1996). From about 600 to 150 years ago, the earth experienced a cooling period known as the Little Ice Age when conditions were cooler and moister than they were both before and after this period. It has been suggested that this climatic anomaly favored the establishment of grasslands and the warming since has been a significant factor associated with the increase in woody plants (Neilson, 1986). Obviously, dynamics in climate patterns would result in ecosystem changes that may include shifts in general physiognomy.

Climate-driven vegetation change often lags behind the change in climate because vegetation established under one climate regime can persist for tens to hundreds of years until disturbed (Prentice, 1986). The pristine grassland described by many settlers may have been established under very different climatic conditions than those that exist today. Change in climate over the past 1000 years, particularly when considered with other changes (i.e. altered grazing and fire regimes, and carbon dioxide), suggests that the increase in woody plants could have potentially been initiated before settlement (Smeins, 1984). Change in climate to conditions more appropriate for woody plants may have occurred before settlement, and human disturbances could have acted as the catalyst that facilitated the directional increase in woody plants. This is another indication that management should not focus on 'pristine' historical conditions

by assuming that removal of human-caused disturbance will cause native plant communities to become more similar to historical accounts, as humans were shaping plant communities for thousands of years prior to European settlement.

Land use legacies: cultivation, development, and fragmentation

Over 100 years ago, most of North America was settled and society focused on developing and improving landscapes through the introduction of agriculture. Government policies required homesteaders to cultivate portions of their land to demonstrate improvements (see Morris, Chapter 3, this volume). In many cases these lands were unsuited for cultivation and nestled in a large landscape dominated by rangelands. Much of this cropland has been abandoned or restored leaving plant communities that are considerably different from untilled prairies in terms of soil properties, vegetation productivity, and plant composition.

Studies of restoration or recovery of Great Plains grasslands in North America, that were cultivated in the early 1900s, have resulted in highly variable conclusions. Many of these 'restored' croplands have not recovered after 30–70 years (Burke *et al.*, 1995; Fuhlendorf *et al.*, 2002) with higher bare ground and lower production, soil carbon, and soil nitrogen than uncultivated prairies. Some evidence suggests that these marginal lands that have been abandoned or restored serve as a source of invasive species for the larger landscape. At a minimum, these lands need to be considered independent of uncultivated lands and in some regions (e.g. Southern Great Plains of North America) previously cultivated land may be a dominant landscape feature.

Habitat fragmentation and other human alterations of ecosystems and landscapes are deemed important factors as environmental degradation. Human facilitated plant invasions are one consequence of these alterations (Lonsdale, 1999; With,

2002). Among the human activities and alterations that promote plant invasions are roads (Trombulak and Frissell, 2000; Christen and Matlack, 2009), housing in rural (Gavier-Pizarro et al., 2010), exurban (Lenth et al., 2006), and urban areas (Song et al., 2005), and cropland abandonment (Vilà et al., 2003). Processes directly linking plant invasions to human activities include intentional introductions of ornamental plants for landscaping (Mack and Erneberg, 2002), soil and vegetation alteration associated with site development (Hobbs and Huenneke, 1992; Wania et al., 2006), and road construction and maintenance activities that create fertile safe sites for invasive plants to germinate and establish (Greenberg et al., 1997; Trombulak and Frissell, 2000; Barbosa et al., 2010).

Landscape structure, land-use pattern, and land ownership pattern are key determinates of the extent and severity of exotic species invasions. Severity and extent of encroachment of woody plants into grassland is greater in landscapes containing large numbers of smaller, intermingled patch types by providing isolated patches juxtaposed with pockets of seed sources (Coppedge et al., 2001). When only a few seed-producing woody plants become established with avian seed dispersers (Holthuijzen and Sharik, 1985) and under an altered fire regime in which fire spread is interrupted by fragmentation, a feedback cycle results in dense stands of woody vegetation (Fuhlendorf et al., 1996). This same pattern holds true with invasive non-native plants with cattle as seed vectors (Brown and Carter, 1998). Fragmented ownerships in a landscape can result in severe invasions because the incentive to control the invasion on a single ownership is relatively small (i.e. individual managers are responsible for small portions of total damage caused by the invasive species) whereas escalating cost of control is passed to neighbors by a manager who chooses not to control the invasion and whose land serves as a propagule source (Epanchin-Niell et al., 2010).

Land use legacies: improved forages gone bad

For over a century, North American agronomists have imported plants from many environments in an attempt to increase forage and food production (Ball et al., 2002; Barnes et al., 2003). For the most part, these introductions have contributed positively to agriculture, and most species introductions do not create unforeseen problems. However, forage species are often selected for their ability to establish and persist in diverse environments, elevating the potential for invasion into unintended ecosystems. Forage species are often planted and managed as monocultures because they respond favorably and uniformly to fertilizer, are persistent in monoculture, and are often less palatable than native plants. Primarily grown in monocultures for livestock grazing, exotic forage species are selected for traits that promote persistence under intense grazing. Rapid maturation, secondary chemicals, and fungal associations are among the traits that enhance persistence when these species are grown in grazed monocultures, but these same persistence traits can facilitate dominance in plant communities with diverse plant species in which native and domestic herbivores preferentially select among multiple plant species.

Forage plant invasions may in part be explained by the plant's evolutionary history as shaped by fire and grazing. Native plants from the Great Plains evolved under a fire–grazing interaction (Fuhlendorf et al., 2009a) whereas many introduced forages evolved with heavy grazing pressure and may be less dependent on fire. Therefore, when grazing and fire are decoupled and/or fire is removed, the new environment does not resemble the evolutionary environment of native species, and exotic forages gain competitive advantage because grazers preferentially select the more palatable native species. Grazers shift their preference from individual plants to patches (i.e. burned patches over unburned patches) in landscapes that include burned and unburned patches (Fuhlendorf and Engle, 2004). This suggests that individual plant traits that promote persistence and dominance under grazing alone (i.e. when the grazing animal is allowed full selectivity among plants) would become less important as patch-level selectivity increases among burned and unburned patches. So, the fire–grazing interaction that functioned in the evolution of Great Plains grasslands resulted in selection patterns at the patch level depending on time since fire. This contrasts with the traditional agroecosystem model that minimizes preferential selectivity by using monocultures of plants that possess traits that promote persistence. When introduced forage species possess these traits and escape into diverse native communities, grazing animals select native species and the introduced forage species is competitively favored and increases in abundance. Some highly palatable species (i.e. lucerne) do not become dominant in grazed ecosystems because they lack many of the persistence traits of other introduced forages.

Conclusion

The long-standing emphases placed on vegetation dynamics, life history traits, disturbance ecology, population ecology, and landscape ecology are strongly informative to the study and management of invasive species. Evaluation of literature on species invasion patterns indicates that generalized conclusions are difficult because they are highly scale dependent like most other ecological processes (Fuhlendorf and Smeins, 1999; Fridley et al., 2007; Fuhlendorf et al., 2009a). To more easily interpret results, ecologists often simplify their approach in an experiment or observation by focusing on one factor while purposefully excluding the effects of other factors or ecological interactions. Although the simplifying approach eases analysis and interpretation, it also excludes inherently complex yet critically important processes or traits, often resulting in conclusions that may not scale up. The effect of grazing, for example, is commonly viewed as a simple binary response (i.e. often either yes or no regarding defoliation) and in

---

Box 2.1. Case study: *Lespedeza cuneata*

Lespedeza cuneata (Dum.-Cours.) G. Don. is a perennial legume, introduced from eastern and central Asia for forage production, erosion control, and land reclamation. Intentionally introduced, *L. cuneata* is now considered invasive throughout the eastern and midwestern USA. Invasion into disturbed habitat and rangeland is rapid, displacing native species and forming dense patches strongly dominated by *L. cuneata* (Eddy and Morre, 1998; Brandon *et al.*, 2004; Cummings *et al.*, 2007). Dominance by *L. cuneata* often reduces biodiversity and biomass production of other species, as well as important ecosystem processes, such as altering nitrogen cycling (Price and Weltzin, 2003; Garten *et al.*, 2008).

Several important life history and plant traits of *L. cuneata* contribute to its successful establishment and invasion in competitive and stressful environments. A primary mechanism of *L. cuneata* invasion may be its ability to suppress native vegetation by intercepting solar radiation and shading neighboring plants (Brandon *et al.*, 2004). Large amounts of leaf area contribute to dense, light-limited understory herbaceous canopies. Once established, monocultures of *L. cuneata* will limit light energy available to other plant species. Greater leaf area than native species also increases acquisition of CO_2, compensating for reduced photosynthetic rates compared to other native species (Allred *et al.*, 2010). The efficient use of biomass allocated to foliage within *L. cuneata*, compared to native species, can also aid in stressed environments or adverse conditions. Higher specific leaf area (more leaf area for less biomass investment) in *L. cuneata* may sustain growth and resource acquisition if resources become limiting (Allred *et al.*, 2010).

Continued

Box 2.1. – Continued

Introduced primarily as a forage species, *L. cuneata* is highly nutritious and palatable in early growth stages, but grazers shift to other forage plants when palatability and digestibility decline with rapid plant maturation (Donnelly, 1954; Schutzenhofer and Knight, 2007). Condensed tannins also decrease digestibility and can cause gastrointestinal discontent in ruminants (Mosjidis, 1990). These herbivory avoidance traits reduce herbivore pressure on *L. cuneata* thereby providing it competitive advantage in species mixes, which assists in successful invasion. Simplified disturbance regimes that promote uniform grazing distribution have aided in the spread of *L. cuneata* by providing an environment where grazing avoidance traits are a benefit. Cummings *et al.* (2007) found that the invasion rate of *L. cuneata* was four times greater on grasslands in which the fire–grazing interaction was not present than those in which it was present (Fig. 2.4). Under the interaction animals selected for fresh growth in burned patches and were able to maintain plants, including *L. cuneata*, in palatable regrowth stages. Herbivory avoidance and competition traits characteristics of *L. cuneata* are not as beneficial under intense focal grazing created by the fire–grazing interaction.

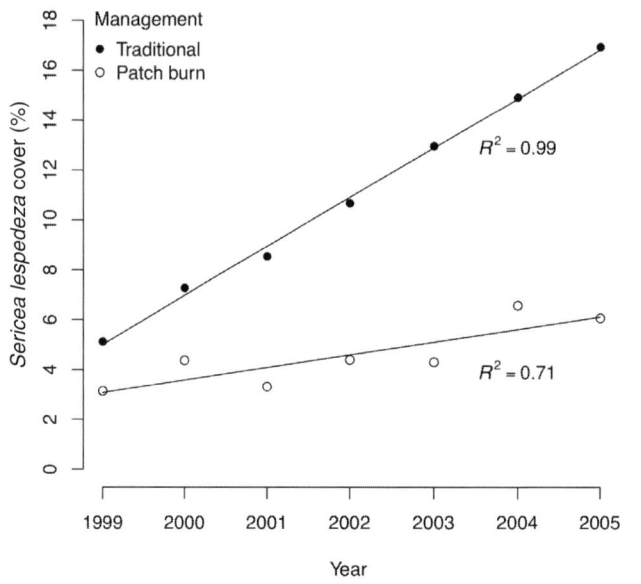

Fig. 2.4. *Sericea lespedeza* invasion over time in the traditional management and patch-burn treatments (Cummings *et al.*, 2007). In 1999, the treatments had similar levels (p>0.05) of invasion. By 2000, the *Sericea* invasion in the traditional management treatment was higher than in the patch-burn treatment. In the spring of 2003, the traditional treatment pastures were burned in entirety, so at that point both treatments had the same amount of fire over the pasture. After this point, invasion in the traditional management treatment followed the same linear increase of 2% per year, while the patch-burn treatment displayed fluctuating invasions. Modified from Cummings *et al.* (2007).

Lespedeza cuneata is now widespread throughout the eastern portion of the Southern Great Plains where uniform cattle grazing is an objective. While control with herbicides is common (Koger *et al.*, 2002), the use of new management strategies based on life history traits and historical disturbances can be more effective (Cummings *et al.*, 2007) and economically and ecologically advantageous. In this case, restoring the pattern of historical disturbances for a site benefited native species and limited an invasive species.

isolation of other disturbance effects. However, the effect of grazing is more complex and is influenced by a myriad of factors (e.g. number, distribution, and species of herbivore; climate; resources; disturbances) beyond the simple removal of biomass. Assuming that grazing and fire patterns on small plots are similar to large complex landscapes is obviously erroneous. Incorporating other factors, as well as interactions among factors, leads to a more appropriate, and often different, understanding of the system (Fuhlendorf and Engle, 2001). Evaluation of invasive species literature leads to similar conclusions where small-scale experimental studies may do little to explain large-scale landscape patterns. Scientific understanding and management of invasive plants on landscapes requires some innovative and alternative experimental approaches.

In general, the approach used to study interactions, processes, and traits related to invasive species is frequently simplified to understand current invasions and to predict future invasions. This simplification is often employed by studying single species in small and controlled environments (e.g. small plots) over short time periods. Although studying species traits and processes at fine spatial and temporal scales informs our understanding of plant invasions, eliminating natural complexity and variation limits the application to managing invasions. Simple studies of plot-level herbicide applications can lead to conclusions that cannot be extrapolated to large scales (Fuhlendorf *et al.*, 2009b). Although studying vegetation dynamics, ecosystem processes, and invasions at broader spatial and temporal scales will inherently increase complexity and variation, only actively restoring complexity will complete our understanding of system dynamics.

While species invasions are complex and extrapolation across scales and regions should be done with caution, some generalizations are possible. First of all, species invasions are best understood by focusing on the species traits, such as dispersal, establishment, and response and resistance to disturbance. Understanding the invasion of species requires an understanding of the interplay between these traits and the environment in which the species is invading. Second, the environment is constantly changing and many of these changes can enhance or inhibit the invasion process. We often use historical landscapes and disturbance regimes as our baseline and frequently restoration of disturbance processes can minimize invasions. However, sometimes, invasions are facilitated by historical disturbance and critical feedbacks may occur that promote even more invasions. Finally, studies of vegetation dynamics and assembly rules of plant communities are relevant to invasive species ecology and management. The concepts of thresholds, non-linear dynamics, and ecosystem resilience are central to the dynamics of invasive species. Prediction of invasion and management of these processes requires innovative approaches that focus on multiple spatial and temporal scales.

References

Allred, B., Fuhlendorf, S., Monaco, T. and Will, R. (2010) Morphological and physiological traits in the success of the invasive plant *Lespedeza cuneata*. *Biological Invasions* 12, 739–749.

Archer, S. (1995) Herbivore mediation of grass-woody plant interactions. *Tropical Grasslands* 29, 218–235.

Archer, S.R. (1994) Woody plant encroachment into southwestern grasslands and savannas: rates, patterns and proximate causes. In: Vavra, M.P., Laycock, W.A. and Pieper, R.D. (eds) *Ecological Implications of Livestock Herbivory in the West*. Society for Range Management, Denver, Colorado, pp. 13–68.

Archer, S., Schimel, D.S. and Holland, E.A. (1995) Mechanisms of shrubland expansion: land use, climate, or CO_2? *Climatic Change* 29, 91–99.

Axelrod, D.I. (1985) Rise of the grassland biome, central North-America. *Botanical Review* 51, 163–201.

Ball, D.M., Hoveland, C.S. and Lacefield, G.D. (2002) *Southern Forages. Modern Concepts for Forage Crop Management*. Potash & Phosphate

Institute and the Foundation for Agronomic Research, Norcross, Georgia.

Barbosa, N.P.U., Wilson Fernandes, G., Carneiro, M.A.A. and Júnior, L.A.C. (2010) Distribution of non-native invasive species and soil properties in proximity to paved roads and unpaved roads in a quartzitic mountainous grassland of southeastern Brazil (rupestrian fields). *Biological Invasions* 12, 3745–3755.

Barnes, R.F., Nelson, C.J., Collins, M. and Moore, K.J. (2003) *Forages: an Introduction to Grassland Agriculture.* Iowa State Press, Ames, Iowa.

Belsky, A.J. (1996) Viewpoint: Western juniper expansion: is it a threat to arid northwestern ecosystems? *Journal of Range Management* 49, 53–59.

Belsky, A.J. and Blumenthal, D.M. (1997) Effects of livestock grazing on stand dynamics and soils in upland forests of the interior west. *Conservation Biology* 11, 315–327.

Blumenthal, D., Mitchell, C.E., Pyšek, P. and Jarošík, V. (2009) Synergy between pathogen release and resource availability in plant invasion. *Proceedings of the National Academy of Sciences* 106, 7899–7904.

Brandon, A.L., Gibson, D.J. and Middleton, B.A. (2004) Mechanisms for dominance in an early successional old field by the invasive non-native *Lespedeza cuneata* (Dum. Cours.) G. Don. *Biological Invasions* 6, 483–493.

Briske, D.D., Fuhlendorf, S.D. and Smeins, F.E. (2006) A unified framework for assessment and application of ecological thresholds. *Rangeland Ecology & Management* 59, 225–236.

Brooks, M.L., D'Antonio, C.M., Richardson, D.M., Grace, J.B., Keeley, J.E., DiTomaso, J.M., Hobbs, R.J., Pellant, M. and Pyke, D. (2004) Effects of invasive alien plants on fire regimes. *BioScience* 54, 677.

Brown, J. and Carter, J. (1998) Spatial and temporal patterns of exotic shrub invasion in an Australian tropical grassland. *Landscape Ecology* 13, 93–102.

Burke, I.C., Lauenroth, W.K. and Coffin, D.P. (1995) Soil organic matter recovery in semiarid grasslands: implications for the conservation reserve program. *Ecological Applications* 5, 793–801.

Burkhardt, J.W. (1996) Herbivory in the Intermountain West: An overview of evolutionary history, historic cultural impacts and lessons from the past. *Idaho Station Bulletin 58.* Idaho Forest, Wildlife and Range Experiment Station, Moscow, Idaho.

Christen, D.C. and Matlack, G.R. (2009) The habitat and conduit functions of roads in the spread of three invasive plant species. *Biological Invasions* 11, 453–465.

Clements, F.E. (1916) *Plant Succession: An analysis of the development of vegetation.* Carnegie Institution of Washington, Washington, DC.

Conley, W., Conley, M.R. and Karl, T.R. (1992). A computational study of episodic events and historical context in long-term ecological processes: climate and grazing in the Northern Chihuahuan desert. *Coenoses* 7, 55–60.

Coppedge, B.R., Engle, D.M., Fuhlendorf, S.D., Masters, R.E. and Gregory, M.S. (2001) Landscape cover type and pattern dynamics in fragmented southern Great Plains grasslands, USA. *Landscape Ecology* 16, 677–690.

Cummings, D.C., Fuhlendorf, S.D. and Engle, D.M. (2007) Is altering grazing selectivity of invasive forage species with patch burning more effective than herbicide treatments? *Rangeland Ecology & Management* 60, 253–260.

Davies, K.F., Holyoak, M., Preston, K.A., Offeman, V.A. and Lum, Q. (2009) Factors controlling community structure in heterogeneous metacommunities. *Journal of Animal Ecology* 78, 937–944.

DeSantis, R.D., Hallgren, S.W., Lynch, T.B., Burton, J.A. and Palmer, M.W. (2010) Long-term directional changes in upland Quercus forests throughout Oklahoma, USA. *Journal of Vegetation Science* 21, 606–618.

Donnelly, E.D. (1954) Some factors that affect palatability in sericea lespedeza, *L. cuneata. Agronomy Journal* 46, 96–97.

Eddy, T.A. and Moore, C.M. (1998) Effects of sericea lespedeza (*Lespedeza cuneata* (Dumont) G. Don) invasion on oak savannas in Kansas. *Transactions of the Wisconsin Academy of Sciences, Arts, and Letters* 86, 57–62.

Elton, C.S. (1958) *The Ecology of Invasions by Animals and Plants.* University of Chicago Press, Chicago, Illinois.

Epanchin-Niell, R.S., Hufford, M.B., Aslan, C.E., Sexton, J.P., Port, J.D. and Waring, T.M. (2010) Controlling invasive species in complex social landscapes. *Frontiers in Ecology and the Environment* 8, 210–216.

Fridley, J.D., Stachowicz, J.J., Naeem, S., Sax, D.F., Seabloom, E.W., Smith, M.D., Stohlgren, T.J., Tilman, D. and Holle, B.V. (2007) The invasion paradox: reconciling pattern and process in species invasions. *Ecology* 88, 3–17.

Fuhlendorf, S.D. and Engle, D.M. (2001) Restoring heterogeneity on rangelands: ecosystem management based on evolutionary grazing patterns. *Bioscience* 51, 625–632.

Fuhlendorf, S.D. and Engle, D.M. (2004) Application of the fire-grazing interaction to restore a shifting mosaic on tallgrass prairie. *Journal of Applied Ecology* 41, 604–614.

Fuhlendorf, S.D. and Smeins, F.E. (1997) Long-term vegetation dynamics mediated by herbivores, weather and fire in a *Juniperus-Quercus* savanna. *Journal of Vegetation Science* 8, 819–828.

Fuhlendorf, S.D. and Smeins, F.E. (1999) Scaling effects of grazing in a semi-arid grassland. *Journal of Vegetation Science* 10, 731–738.

Fuhlendorf, S.D., Smeins, F.E. and Grant, W.E. (1996) Simulation of a fire-sensitive ecological threshold: a case study of Ashe juniper on the Edwards plateau of Texas, USA. *Ecological Modelling* 90, 245–255.

Fuhlendorf, S.D., Zhang, H., Tunnell, T.R., Engle, D.M. and Cross, A.F. (2002) Effects of grazing on restoration of southern mixed prairie soils. *Restoration Ecology* 10, 401–407.

Fuhlendorf, S.D., Archer, S.R., Smeins, F.E. and Engle, D.M. (2008) From the Dust Bowl to the Green Glacier: Human Activity and Environmental Change in Great Plains Grasslands. In: Van Auken, O.W. (ed.) *Western North American Juniperus Communities: A Dynamic Vegetation Type*. Springer, New York, pp. 219–238.

Fuhlendorf, S.D., Engle, D.M., Kerby, J.D. and Hamilton, R.G. (2009a) Pyric herbivory: rewilding landscapes through the recoupling of fire and grazing. *Conservation Biology* 23, 588–598.

Fuhlendorf, S.D., Engle, D.M., O'Meilia, C.M., Weir, J.R. and Cummings, D.C. (2009b) Does herbicide weed control increase livestock production on non-equilibrium rangeland? *Agriculture, Ecosystems & Environment* 132, 1–6.

Ganguli, A.C., Engle, D.M., Mayer, P.M. and Hellgren, E.C. (2008) Plant community diversity and composition provide little resistance to *Juniperus* encroachment. *Botany* 86, 1416–1426.

Garten, C.T., Classen, A.T., Norby, R.J., Brice, D.J., Weltzin, J.F. and Souza, L. (2008) Role of N2-fixation in constructed old-field communities under different regimes of CO_2, temperature, and water availability. *Ecosystems* 11, 125–137.

Gavier-Pizarro, G.I., Radeloff, V.C., Stewart, S.I., Huebner, C.D. and Keuler, N.S. (2010) Rural housing is related to plant invasions in forests of southern Wisconsin, USA. *Landscape Ecology* 25, 1505–1518.

Glenn-Lewin, D.C.V. (1992) *Plant Succession:*

Theory and Prediction. Chapman and Hall, London.

Greenberg, C., Crownover, S. and Gordon, D. (1997) Roadside soils: a corridor for invasion of xeric scrub by nonindigenous plants. *Natural Areas Journal* 17, 99–109.

Grime, J.P. (1977) Evidence for existence of three primary strategies in plants and its relevance to ecological and evolutionary theory. *American Naturalist* 111, 1169–1194.

Gurvich, D.E., Tecco, P.A. and Díaz, S. (2005) Plant invasions in undisturbed ecosystems: the triggering attribute approach. *Journal of Vegetation Science* 16, 723.

Hall, S.A. and Valastro, S. (1995) Grassland vegetation in the southern Great Plains during the last glacial maximum. *Quaternary Research* 44, 237–245.

Hassan, S.N., Rusch, G.M., Hytteborn, H., Skarpe, C. and Kikula, I. (2008) Effects of fire on sward structure and grazing in western Serengeti, Tanzania. *African Journal of Ecology* 46, 174–185.

Hobbs, R.J. and Huenneke, L.F. (1992) Disturbance, diversity, and invasion: implications for conservation. *Conservation Biology* 6, 324–337.

Holthuijzen, A. and Sharik, T. (1985) The avian seed dispersal system of eastern red cedar (*Juniperus virginiana*). *Canadian Journal of Botany-Revue Canadienne de Botanique* 63, 1508–1515.

Horvitz, C.C., Pascarella, J.B., McMann, S., Freedman, A. and Hofstetter, R.H. (1998) Functional roles of invasive non-indigenous plants in hurricane-affected subtropical hardwood forests. *Ecological Applications* 8, 947–974.

Koger, C.H., Stritzke, J.E. and Cummings, D.C. (2002) Control of sericea lespedeza (*Lespedeza cuneata*) with triclopyr, fluroxypyr, and metsulfuron. *Weed Technology* 16, 893–900.

Lehmann, V.W. (1969) *Forgotten Legions: sheep in the Rio Grande Plain of Texas*. Texas Western Press, El Paso, Texas.

Lenth, B.A., Knight, R.L. and Gilgert, W.C. (2006) Conservation value of clustered housing developments. *Conservation Biology* 20, 1445–1456.

Lonsdale, W.M. (1999) Global patterns of plant invasions and the concept of invasibility. *Ecology* 80, 1522–1536.

Mack, R.N. and Erneberg, M. (2002) The United States naturalized flora: largely the product of deliberate introductions. *Annals of the Missouri Botanical Garden* 89, 176–189.

Miller, R.F. and Wigand, P.E. (1994) Holocene

changes in semiarid Pinyon-Juniper woodlands. *BioScience* 44, 465–474.

Mosjidis, J.A. (1990) Daylength and temperature effects on emergence and early growth of sericea lespedeza. *Agronomy Journal* 82, 923–926.

Neilson, R.P. (1986) High resolution climatic analysis and southwest biogeography. *Science* 232, 27–34.

Noble, I.R. and Slatyer, R.O. (1980) The use of vital attributes to predict successional changes in plant communities subject to recurrent disturbances. *Vegetatio* 43, 5–21.

Nordt, L.C., Boutton, T.W., Hallmark, C.T. and Waters, M.R. (1994) Late quaternary vegetation and climate changes in central Texas based on the isotopic composition of organic carbon. *Quaternary Research* 41, 109–120.

Norton-Griffiths, M. (1979) The influence of grazing, browsing, and fire on the vegetation dynamics of the Serengeti. In: Sinclair, A.R.E. and Norton-Griffiths, M. (eds) *Serengeti: Dynamics of an Ecosystem.* University of Chicago Press, Chicago, Illinois, pp. 378–392.

O'Connor, T.G. (1995) Acacia Karroo invasion of grassland: environmental and biotic effects influencing seedling emergence and establishment. *Oecologia* 103, 214–223.

Paruelo, J.M. and Lauenroth, W.K. (1996) Relative abundance of plant functional types in grasslands and shrublands of North America. *Ecological Applications* 6, 1212–1224.

Polley, H.W., Mayeux, H.S., Johnson, H.B. and Tischler, C.R. (1997) Viewpoint: Atmospheric CO_2, soil water, and shrub/grass ratios on rangelands. *Journal of Range Management* 50, 278–284.

Prentice, I.C. (1986) Vegetation responses to past climatic variation. *Vegetatio* 67, 131–141.

Price, C.A. and Weltzin, J.F. (2003) Managing non-native plant populations through intensive community restoration in Cades Cove, Great Smoky Mountains National Park, USA. *Restoration Ecology* 11, 351–358.

Rudgers, J.A. and Clay, K. (2007) Endophyte symbiosis with tall fescue: how strong are the impacts on communities and ecosystems? *Fungal Biology Reviews* 21, 107–124.

Salvatori, R., Egunyu, F., Skidmore, A.K., de Leeuw, J. and van Gils, H.A.M. (2001) The effects of fire and grazing pressure on vegetation cover and small mammal populations in the Maasai Mara National Reserve. *African Journal of Ecology*, 39, 200–204.

Schultz, A.M., Launchbaugh, J.L. and Biswell, H.H. (1955) Relationship between grass density and brush seedling survival. *Ecology* 36, 226–238.

Schutzenhofer, M.R. and Knight, T.M. (2007) Population-level effects of augmented herbivory on *Lespedeza cuneata*: implications for biological control. *Ecological Applications* 17, 965–971.

Seastedt, T. (2009) Ecology: traits of plant invaders. *Nature* 459, 783–784.

Smeins, F.E. (1984) Origin of the brush problem: a geological and ecological perspective of contemporary distributions. In: McDaniel, K.W. (ed.) *Proceedings of Brush Management Symposium.* Texas Tech Press, Lubbock, Texas, pp. 5–16.

Smeins, F.E., Fuhlendorf, S.D. and Taylor, C.A. (1997) Environmental and land use changes: a long-term perspective. In: Ansley, R.J. (ed.) *Juniper Symposium Proceedings.* Texas A&M University Press, San Angelo, Texas, pp. 1.3–1.21.

Smith, S.D., Huxman, T.E., Zitzer, S.F., Charlet, T.N., Housman, D.C., Coleman, J.S., Fenstermaker, L.K., Seemann, J.R. and Nowak, R.S. (2000) Elevated CO_2 increases productivity and invasive species success in an arid ecosystem. *Nature* 408, 79–82.

Song, I., Hong, S., Kim, H., Byun, B. and Gin, Y. (2005) The pattern of landscape patches and invasion of naturalized plants in developed areas of urban Seoul. *Landscape and Urban Planning* 70, 205–219.

Song, L., Jinrong, W., Changhan, L., Li, F., Peng, S. and Chen, B. (2009) Different responses of invasive and native species to elevated CO_2 concentration. *Acta Oecologica* 35, 128–135.

Sousa, W.P. (1984) The role of disturbance in natural communities. *Annual Review of Ecology and Systematics* 15, 353–391.

Stambaugh, M.C. and Guyette, R.P. (2006) Fire regime of an ozark wilderness area, Arkansas. *The American Midland Naturalist* 156, 237–251.

Stohlgren, T.J., Chong, G.W., Schell, L.D., Rimar, K.A., Otsuki, Y., Lee, M., Kalkhan, M.A. and Villa, C.A. (2002) Assessing vulnerability to invasion by nonnative plant species at multiple spatial scales. *Environmental Management* 29, 566–577.

Taylor, C., Launchbaugh, K., Huston, E., Straka, E. and Pritz, R. (1997) Improving the efficacy of goating for biological juniper management. In: Ansley, R.J. (ed.) *Juniper Symposium Proceedings.* Texas A&M University Press, San Angelo, Texas, pp. 2.10–2.27.

Tilman, D. (1997) Community invasibility, recruitment limitation, and grassland biodiversity. *Ecology* 78, 81–92.

Trombulak, S.C. and Frissell, C.A. (2000) Review of

ecological effects of roads on terrestrial and aquatic communities. *Conservation Biology* 14, 18–30.

Twidwell, D., Fuhlendorf, S.D., Engle, D.M. and Taylor, C.A. (2009) Surface fuel sampling strategies: linking fuel measurements and fire effects. *Rangeland Ecology and Management* 62, 223–229.

Van Devender, T.R. and Spaulding, W.G. (1979) Development of vegetation and climate in the southwestern United States. *Science* 204, 701–710.

Vilà, M., Burriel, J.A., Pino, J., Chamizo, J., Llach, E., Porterias, M. and Vives, M. (2003) Association between Opuntia species invasion and changes in land-cover in the Mediterranean region. *Global Change Biology* 9, 1234–1239.

Walker, B.H., Ludwig, D., Holling, C. and Peterman, R.M. (1981) Stability of semi-arid savanna grazing systems. *Journal of Ecology* 69, 473–498.

Wania, A., Kuhn, I. and Klotz, S. (2006) Plant richness patterns in agricultural and urban landscapes in Central Germany – spatial gradients of species richness. *Landscape and Urban Planning* 75, 97–110.

With, K.A. (2002) The landscape ecology of invasive spread. *Conservation Biology* 16, 1192–1203.

3 Land-use Legacy Effects of Cultivation on Ecological Processes

Lesley R. Morris

US Department of Agriculture, Agricultural Research Service, Utah State University, USA

Introduction

Land-use legacies are long-lasting changes in ecosystems following human utilization of resources. Cultivation for crop production is known worldwide for creating land-use legacies that can persist for decades, centuries, and even millennia (Foster *et al.*, 2003; McLauchlan, 2006). These legacies are important because they represent fundamental changes in basic ecosystem processes such as plant species reproduction and colonization, soil nutrient cycling and availability, and soil water movement and availability. When these basic processes are interrupted, old fields can become more difficult to manage and potentially impossible to restore to the pre-disturbance native plant community. Cultivation (plowing, seeding, and harvesting a crop annually) involves both soil disturbance and the sowing and harvesting of the intended crop. When cultivation ceases, the abandoned fields ('old fields' hereafter) are commonly dominated by exotic and invasive plant species. Among the temperate grasslands of the world, for instance, the introduction and spread of exotic species is consistently linked to the cultivation of crops such as cereal grains, legumes, and forage grasses from western Europe (Mack, 1989). The crop species can become invasive in their new habitats as *Sorghum halepense* did in Argentina and *Elettaria cardamomum* did in Sri Lanka and south India (Heywood, 1989; Mack, 1989). Cultivation also brings the unintended introduction of exotic plant species that accompany crop seeds as contaminants, thereby creating a wide species pool of potentially invasive plants (Mack, 1989). The connection between cultivation, land abandonment, and exotic species invasion is increasingly relevant because the extent of old fields is increasing worldwide (McLauchlan, 2006; Hobbs and Cramer, 2007).

Certainly, not all exotic species in old fields are detrimental and some even assist in the re-establishment of native species following disturbance, called 'secondary succession' (Lugo, 2004). However, exotic species can also arrest secondary succession on old fields and dominate these sites for decades to over half a century (Stylinski and Allen, 1999; Elmore *et al.*, 2006; Standish *et al.*, 2007). Exotic annual grasses, in particular, are known to arrest secondary succession of old fields around the world (Cramer *et al.*, 2008). For example, *Saccharum spontaneum* prevents tropical forest regeneration in Panama, *Avena barbata* limits the reestablishment of eucalypt woodlands in southwestern Australia, and *Bromus tectorum* restricts shrubland regeneration in the western USA (Mack, 1989; Hooper *et al.*, 2005; Standish *et al.*, 2006). Exotic shrubs and forbs can also dominate old fields even after attempts to establish more desirable species through seeding (Doren *et al.*, 1991; Stylinski and Allen, 1999; Banerjee *et al.*, 2006). Because cultivation involves several anthropogenic disturbances coupled with introductions of exotic plant species, cultivation is

© CAB International 2012. *Invasive Plant Ecology and Management: Linking Processes to Practice* (eds T.A. Monaco and R.L. Sheley)

inextricably linked to invasive species management. In this chapter, I will identify how the anthropogenic disturbances associated with cultivation differ from natural ones, how these exotic disturbances create land-use legacies through alteration of biotic and abiotic processes, and what managers might do to help repair these processes.

Linkages Between Cultivation and Exotic Plant Invasion

Anthropogenic disturbance versus natural disturbance

Ecologists recognize that disturbance (e.g. fire, flood, and herbivory) is a natural ecosystem function (Pickett and White, 1985; Hobbs and Huenneke, 1992). Natural disturbance regimes are characterized by their distribution, frequency, rotation period, predictability, area disturbed, and magnitude (or severity) (Pickett and White, 1985). There is growing recognition that anthropogenic disturbance is different than natural disturbance because it can involve both modifications to these natural disturbance regimes and/or the introduction of new disturbances (e.g. soil excavation and plowing) (Stylinski and Allen, 1999). In comparison to the natural disturbance regimes under which ecosystems evolve, anthropogenic disturbances are introduced to the ecosystem and accompany changing human activities (McIntyre and Hobbs, 1999). Therefore, anthropogenic disturbance regimes are unpredictably linked to economic and social factors, rather than any environmental cycle (Lugo, 2004). Cultivation involves several anthropogenic disturbances including plowing, fertilization, and irrigation. These cultivation disturbances are outside of the magnitude and frequency of natural disturbance regimes and can result in modification of the ecosystem and irreversible legacy effects, such as species loss, because the environment can become mismatched with the species' traits (Byers, 2002).

Disturbance, whether natural or anthro-pogenic, varies in magnitude, size, and frequency (Pickett and White, 1985; Myster, 2008). However, unlike natural disturbance, cultivation also varies locally and regionally with human history. In fact, the current composition of plant species on old fields may have more connection to historical land use and management than current climate or resource conditions (Motzkin et al., 1999; Buisson and Dutoit, 2004). Thus, under-standing old fields requires a holistic approach that encompasses ecological, historical, and spatial variables in a new way (Benjamin et al., 2005). Disturbance magnitude includes both intensity (amount of biomass destroyed and degree of soil disruption) and severity (biological effect) (Pickett and White, 1985). Cultivation can have different disturbance magnitude on adjacent old fields in the same soil type due to historical differences in equipment and practices between farmers (Coffin et al., 1996; Buisson and Dutoit, 2004). Disturbance magnitude can also vary over time because cultivation methods vary over time. For example, cultivation intensity was greater with early conventional tillage methods (continuous plowing) until conservation tillage (reduced plowing) was developed in the western USA (Schillinger and Papendick, 2008). Conversely, disturbance severity can be greater in more recent times due to technological advances in plowing, fertilization, irrigation, and subsoil drainage practices (Doren et al., 1991; Benjamin et al., 2005; Standish et al., 2006).

Likewise, the size and frequency of cultivation disturbance is affected by human history (Benjamin et al., 2005). For example, innovations in fertilization led to clearing larger areas for cultivation in Australia (Standish et al., 2006). The size of an old field is important for native species because colonization tends to be higher at the edges and decrease into the center of fields (Standish et al., 2007). Disturbance frequency depends on the historical duration of cultivation (how long it was in use) and the age since abandonment (time since cultivation ceased). Longer duration in cultivation can produce more depleted native seed banks, more need for control of exotic species, more soil compaction, more erosion, and more organic matter loss (Burke et al., 1995; Jenkins and Parker, 2000; Standish et al., 2007). Although exceptions have been reported, species richness and similarity to the pre-disturbance plant community generally increase with time since abandonment (Motzkin et al., 1999; Öster et al., 2009).

Ecosystem properties (e.g. system structure, resource base, and landscape characteristics) and plant community properties (e.g. life history and competitive ability of resident plant species) can make some ecosystems more susceptible than others to land-use legacies (Pickett and White, 1985; Kruger et al., 1989; Cramer et al., 2008). The severity of cultivation disturbances can be greater in resource limited systems versus resource rich ones. Invasion of exotic species is enhanced by disturbance when it increases the availability of a limiting resource (Hobbs, 1989). For example, in light-limited forest systems, clearing land for cultivation elevates the limited resource (i.e. solar radiation at the soil surface) that assists invasion of shade-intolerant exotic species (Meiners et al., 2002). Finally, landscape position plays a role in the severity of cultivation disturbance (Fig. 3.1). For example, species diversity is often higher in old fields when they are adjacent to a native plant community (Lawson et al., 1999; Hooper et al., 2004). In contrast, old fields that are adjacent to currently cultivated fields are subject to a persistent species pool of potentially invasive plants and additional disturbances (e.g. pesticide and herbicide drift or fire) (Cramer et al., 2008).

Plant community properties include life history and competitive ability of the resident species. Life history refers to an organism's rate of growth, allocation of assimilates and structure, and timing of life cycle events (Pickett and White, 1985). How a plant community responds to disturbance depends on life history traits and competitive ability of the resident plant species (see, James, Chapter 8, this volume). The presence of exotic species may not greatly alter plant community development in old fields if vegetation is composed of native species with life histories that overlap, and thus can compete with them for space and resources after abandonment (Meiners et al., 2007). For example, in tropical forests where exotic species are not shade tolerant, they can dominate an old field until native trees close the canopy (Fine, 2002). However, ecological processes like nutrient cycling or soil water movement may be altered by cultivation to an extent that old fields become foreign to the adaptations that native species formed over their evolutionary history (Byers, 2002). Native species, therefore, may no longer have a prior-resident advantage over exotic species (Byers, 2002). These altered ecological processes may now favor exotic species that evolved with frequent cultivation disturbance (Byers, 2002). Thus, exotic species from habitats with a long history of cultivation can become successful dominants in disturbed land where cultivation has a shorter history (Hobbs and Huenneke, 1992). For example, exotic annuals from Mediterranean regions that were historically cultivated for centuries have successfully dominated the newly disturbed grasslands of California, USA (Hobbs and Huenneke, 1992). The plant community composition of an old field will largely depend upon the suite of life histories and adaptations found in the resident native vegetation (Hobbs and Huenneke, 1992). Therefore, in old fields, generalizations about the alteration of ecological processes and the ecological principles that govern their management are always relative to human history and inherent ecosystem and plant community properties in which cultivation takes place (Kruger et al., 1989; Cramer and Hobbs, 2007).

How Land-use Legacies are Created

To explain how land-use legacies are created, Cramer et al. (2008) proposed a stepwise model of degradation in ecological processes created by cultivation (Fig. 3.2). As the duration, intensity, and extent of cultivation disturbance increases, biotic, and then

(a)

(b)

Fig. 3.1. Aerial photographs depicting different landscape positions of old fields. (a) Old fields surrounded by native vegetation (photo from United States Geological Survey, 1957). (b) Old fields surrounded by active agricultural fields (photo from Google Earth, 2010).

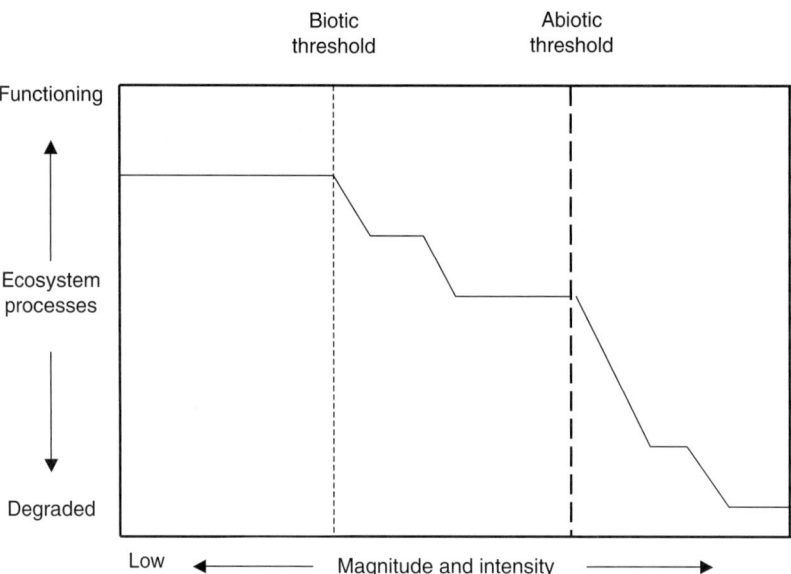

Fig. 3.2. Model of stepwise degradation in old fields (adapted from Cramer *et al.*, 2008 with permission of Elsevier). As the duration, magnitude and intensity of cultivation increases, ecosystem function will become more degraded. Crossing biotic thresholds will require less intervention to repair function. Crossing abiotic thresholds will require greater intervention and investment, and may not return to pre-disturbance function.

abiotic thresholds are crossed. Thresholds are defined by the ability of the plant community to recover without human intervention. When a threshold is crossed, the ecological processes cannot repair themselves, and human intervention is required (Whisenant, 1999; Cramer *et al.*, 2008). Biotic processes involve reproduction, seed dispersal, seed banking, and competition; while abiotic processes involve physical characteristics of soils, soil fertility, and hydrology. When a biotic threshold is crossed, secondary succession of the plant community may be altered and novel assemblages of exotic species become established. With biotic thresholds, there is potential to return to the original plant community, but the rate of return is slowed. When the duration, intensity, and severity of cultivation cross both the biotic and the abiotic thresholds, return to the historical plant community is further stalled and the changes may be irreversible. In these cases, succession is arrested and old fields can remain with a persistent exotic plant com-

munity, which remains in a degraded state beyond the timeframe of a single human lifetime (Cramer *et al.*, 2008). Any attempt at restoration of old fields, therefore, requires identification of the key biotic and abiotic processes that have been compromised and evaluation of the potential for their repair (Hobbs and Cramer, 2007). Likewise, the magnitude, frequency, and size of cultivation disturbances are all important for assessing the extent of damage to ecological processes, and thereby, defining expectations and goals for how much energy and resources are applied to repair them.

Biotic processes and thresholds

Thresholds controlled by biotic processes require manipulation of vegetation for recovery (Whisenant, 1999). In old fields, biotic thresholds are governed by biological processes such as reproduction, seed dispersal, seed banking, and resource competition (Cramer *et al.*, 2008). When thresholds for these biotic processes are crossed, the old fields can become reproductively limited, become dispersal limited, and have altered competitive interactions for both resource exploitation and establishment space. Crossing thresholds associated with any or all of these biotic processes has the potential to favor exotic over native plant species.

Reproductive limitations can favor exotic plants

Both sexual and asexual reproduction of native plant communities can be restricted due to historical cultivation. For example, pollinator populations may have declined under historical pesticide use, drift from nearby agricultural fields, or be inhibited by the level of fragmentation in the landscape around old fields (Hobbs and Yates, 2003; Krug and Krug, 2007). If exotic plant species are wind pollinated, this may only become a limiting factor to the native plant community. Further, because cultivation removes entire plants, species that reproduce primarily vegetatively are less likely to occupy old fields (Fernández *et al.*, 2004; Standish *et al.*, 2007; Dyer, 2010; Morris *et al.*, 2011). Old fields with longer duration and higher intensity of cultivation tend to have few if any vegetatively reproducing species because soil cultivation destroys most of the root system (Jenkins and Parker, 2000). This life history trait may make these species less able to compete for establishment space with exotic species that reproduce primarily by seed. Therefore, plant composition can be altered from a diverse community of sexual and asexually reproducing species to a less-diverse one dominated by species that rely primarily on seed for establishment (Fernández *et al.*, 2004; Morris *et al.*, 2011). Exotic species typically rely upon and produce more seeds than native plant species (Mason *et al.*, 2008). However, size and landscape position can also interact with this pattern. For example, smaller old fields in the Grazalema mountains of southern Spain, that are surrounded by forests and shrublands, contain more species that reproduce vegetatively than larger old fields (Fernández *et al.*, 2004).

Seed limitations can favor exotic plants

Seed banking and dispersal are two processes that contribute to seed limitation on old fields. In order to form a store in the soil, seeds must survive beyond the duration and intensity of cultivation (Standish *et al.*, 2007). Native species of environments without frequent disturbance do not typically form long-lasting seed banks, or their seed banks are depleted by longer and higher cultivation duration and intensity, respectively (Cramer *et al.*, 2008). Exotic crop and pasture weeds, on the other hand, form persistent soil seed banks that are likely to dominate the seed pool long after fields are abandoned (Ellery and Chapman, 2000; Buisson and Dutoit, 2004; Cramer *et al.*, 2007). Therefore, seed banks in old fields have often changed from predominantly native to exotic species. Furthermore, some crop weeds remain in the seed bank even though they are not present in the aboveground vegetation because the conditions are no longer conducive for their growth (e.g. plowing), but their seed remains available for germination and emergence once soil is disturbed again (Buisson and Dutoit, 2004; Banerjee *et al.*, 2006). Ultimately, altered seed bank composition and the specific germination requirements of exotic species can arrest secondary succession on old fields to pre-disturbance plant community composition (Hobbs and Huenneke, 1992).

When seed banks and vegetative reproductive parts have been destroyed, the secondary succession of old fields to native species is heavily dependent upon native seed dispersal (Hooper *et al.*, 2004). However, seed dispersal of native species is often limited in old fields. Wind-dispersed species are sometimes the few native plants to consistently reestablish in old fields (Standish *et al.*, 2006; Morris *et al.*, 2011). Wind-dispersed seeds are typically more likely to be carried into old fields than those reliant on animals, whose populations may have declined due to habitat fragmentation, land clearing, or historical pesticide use and pest control (Hobbs and Yates, 2003; Dyer, 2010). Plants with generalized seed dispersal mechanisms are more likely to provide greater seed rain into an old field (Fine, 2002). However, plant species abundance in the surrounding landscape can be equally as important as dispersal mode in secondary succession (Lawson *et al.*, 1999; Ruprecht, 2006). Native species seed dispersal into old fields usually declines with distance from intact vegetation despite the diversity of dispersal mechanisms (Hooper *et al.*, 2004; Cramer *et al.*, 2008). The exceptions are when animals assist in long distance seed dispersal or when remnant trees are found in the old fields (Hooper *et al.*, 2004; Ganade, 2007). Dispersal limitation is more likely to affect late successional species, whose seeds tend to be larger and disperse via animals, than early successional species, which have smaller seeds that spread easily (Bonet and Pausas, 2004; Santana *et al.*, 2010). Large-seeded species can also become dispersal limited because the majority of animals in old fields are smaller (e.g. birds or bats) and carry smaller seeds (Hooper *et al.*, 2005). Consequently, many herbaceous species with limited or unassisted animal dispersal can be under-represented in old fields (Dyer, 2010). Seed dispersal plays an important role in determining the trajectory of the plant community within old fields. Plant communities are more likely to remain in a persistent degraded state where native seed dispersal is limited and there are few dispersal mechanisms (Cramer *et al.*, 2008). Conversely, exotic species have an increased likelihood of long distance dispersal over native species (Mason *et al.*, 2008).

Altered competitive interactions can favor exotic plants

As described above, many of the exotic species that invade old fields evolved under harsh conditions with frequent anthropogenic disturbances, like cultivation, which means they have adapted life history traits for disturbance tolerance (Mack, 1989). This often confers a competitive advantage for resource exploitation and establishment space to exotic species over the native species that have evolved life history traits under conditions without frequent and intense disturbance (Hobbs and Huenneke, 1992). The competitive effects of exotic grasses and forbs create an important barrier to native plant reestablishment in old fields (Hartnett and Bazzaz, 1983; Holl *et al.*, 2000; Riege and del Moral, 2004; Standish *et al.*, 2007). Exotic grasses compete with tree seedlings for moisture and nutrients and decrease the performance of native woody species (Hooper *et al.*, 2002). Exotic grasses can also differ from the native community in root and shoot morphology, phenology, and water uptake due to different photosynthetic pathways (e.g. C_3 versus C_4 grasses) and other traits traits associated with life form (e.g. annual versus perennial) (Mack, 1989; Davis *et al.*, 2005). All of these factors enable exotic grasses to successfully compete with native species and arrest secondary succession in many ecosystems. For example, the exotic perennial grass *Saccharum spontaneum* ssp. *spontaneum* inhibits forest regeneration on old fields in Panama (Hooper *et al.*, 2005). Similarly, competition with the exotic rhizomatous grass *Agrostis giganta* was the primary barrier to native tree colonization of old fields in rainforests of the Pacific Northwest, USA (Riege and del Moral, 2004). In the arid western USA, the exotic annual grass *Bromus tectorum* has replaced the native perennial bunchgrasses on old fields where it can persist for over half a century after cultivation ceased (Daubenmire, 1975; Elmore *et al.*, 2006). In old fields in Australia, invasive annual grasses like *Avena barbata* have remained dominant for up to 45 years after cultivation (Standish *et al.*, 2007) (Fig. 3.3).

Cultivation can also alter competitive interaction for resource exploitation by altering the soil microflora and fauna. Soil microbes are fundamental in decomposition, nutrient cycling, and nutrient availability for plants, which can all influence plant secondary succession by aiding plants to acquire soil nutrients (Biondini *et al.*, 1985; Paul and Clark, 1989). Cultivation and

(a)

(b)

Fig. 3.3. Photographs of old fields and exotic species. (a) Old field in southern Florida (photo by Joy Brunk, Everglades National Park), USA, where *Schinus terebinthifolius* (on the left) has excluded the native sawgrasses (in foreground). (b) Old field in Western Australia where *Avena barbata* (in foreground) has arrested succession of the eucalypt woodland (on the right, foreground) (reproduced from Cramer *et al.*, 2008 with permission from Elsevier).

fertilization can reduce abundance and diversity of soil microbial communities, which also require decades to centuries to recover (Lovell *et al.*, 1995; Buckley and Schmidt, 2001; Steenwerth *et al.*, 2002). For example, Burke *et al.* (1995) reported that soil microbial biomass between cultivated and noncultivated soils required at least 53 years to recover in the Central Plains of the USA. However, recovery of microbial populations may be heavily influenced by plant species composition and abundance as well as time since abandonment. For example, while some have found decreasing microbial abundance with time since abandonment, others found old fields with dominance of exotic annual grasses to have similar soil microbial biomass regardless of time since abandonment (Steenwerth et al., 2002; Kulmatiski and Beard, 2008). One study suggested that cultivation increased non-native plant establishment because it decreased microbial abundance, not because it increased resource conditions (Kulmatiski and Beard, 2008). Although the mechanism was unclear, presumably the advantage from plant–microbe feedbacks was benefitting the exotic plants over the native species (Klironomos, 2002; Kulmatiski and Beard, 2008).

Cultivation management practices (e.g. pesticides, herbicides, and fertilizers) have also been shown to reduce mycorrhizal populations and community structure in soils around the world (Douds and Millner, 1999; Buckley and Schmidt, 2001; Oehl et al., 2003). Because mycorrhizal associations can improve mineral nutrition, water absorption, and drought tolerance in plants, their reduction can influence secondary succession in old fields (Janos, 1980; Allen and Allen, 1984; Auge, 2001). Plant–mycorrhizal relationships can confer competitive advantages to some plant species in low-nutrient environments where they are adapted to acquire nutrients through mycorrhizal associations (Allen and Allen, 1984). However, invasive and native ruderal species that colonize old fields tend to be non-mycorrhizal (Reeves et al., 1979; Janos, 1980; Kulmatiski and Beard, 2008). A legacy of high nutrient concentrations from fertilization in old fields, particularly with phosphate, can reverse the competitive advantage to favor non-mycorrhizal exotic species (Standish et al., 2006). There is increasing interest in understanding and using mycorrhizal associations in restoration; however, higher levels of infection with mycorrhizal fungi are not always linked with higher levels of native vegetation recovery on old fields (Richter et al., 2002).

Recovery of biotic processes

When only a biotic threshold has been crossed, the potential to return to the original plant community is greater than when there is additional degradation of abiotic processes, but the rate of recovery is slowed (Cramer et al., 2008). Whether or not to intervene when a biotic threshold is crossed depends upon the amount of change from the pre-disturbance plant community that society will accept, and how much time is needed for its return (Cramer et al., 2008). For example, even without dominance by exotic species, unassisted secondary succession to pre-disturbance calcareous grassland communities on old fields in Europe is estimated to take over 50 years (Fagan et al., 2008; Öster et al., 2009).

Although it may seem that seed limitation can be overcome simply by adding more seeds of the native species, there are many other environmental and biological factors that influence their establishment. Sown species can establish in old fields, but recruitment from the seedings is generally lower (Öster et al., 2009; L.R. Morris and T.A. Monaco, unpublished results). Arid areas, in particular, have had limited success in establishing native plant communities on old fields with reseeding (Roundy et al., 2001; Banerjee et al., 2006). Many of the failures are blamed on low precipitation (see Hardegree et al., Chapter 6, this volume). Successful revegetation of old fields in extremely arid areas of the Intermountain Region, USA, have been improved by using irrigation along with seeding (Roundy et al., 2001). However, seeding is still considered an unreliable method in arid regions, especially when seed banks of exotic species compete with seed banks of native species at all times of the year, regardless of irrigation (Banerjee et al., 2006). Furthermore, granivore preference for native seeds over exotic ones as well as preferential herbivory of seedlings has direct and indirect effects on native species recruitment (Lawson et al., 1999; Holl et al., 2000; Riege and del Moral, 2004; MacDougall and Wilson, 2007; Standish et al., 2008).

In an effort to reestablish native vegetation on exotic-plant dominated landscapes (including old fields), repair strategies are often employed that include additional exotic disturbances (e.g. plowing, disking, drill seeding) (Whisenant, 1999). Although considerably less research has examined the legacy of the exotic disturbances associated with these repair strategies on ecological processes, they may have similar consequences as cultivation (DePuit and Redente, 1988; Kettle et al., 2000). For example, because exotic plants can create persistent seed banks, the most effective way to manage the exotic plant population may be to manage seed banks (Marushia and Allen, 2011). A typical method to manage an exotic species seed bank is through annual disking (or plowing), which repeatedly mixes the soil and breaks down its stability and structure (Whisenant, 1999). This repair strategy may further degrade the abiotic processes in the old field past abiotic thresholds.

Therefore, many researchers have begun to call for more innovative methods that take advantage of the facilitative interactions among plants and emphasize using species that stabilize and modulate ecosystem function, called 'foundation species' (Buisson and Dutoit, 2004; Ellison et al., 2005; Banerjee et al., 2006; Prevéy et al., 2010). Planting patches of woody native species in old fields, for example, can facilitate seed dispersal and recruitment via animals as well as serve as catchments for wind-dispersed seed (Holl et al., 2000; Buisson and Dutoit, 2004; Hooper et al., 2005). This strategy enables foundational species to facilitate secondary succession by ameliorating harsh climatic conditions while concentrating nutrients under their canopies (Allen, 1988; Holl et al., 2000; Banerjee et al., 2006; Prevéy et al., 2010). Planting desirable species directly under the canopy edge of trees and shrubs has also been shown to enhance their growth due to higher mycorrhizal infection, less competition from exotic grasses under shaded conditions, and increased moisture and nutrients (Holl et al., 2000). Some suggest that only a few foundational species may need to be planted to initiate recovery of species diversity in old fields because of their critical role in structuring the plant community (Buisson and Dutoit, 2004).

Abiotic processes and thresholds

Thresholds controlled by abiotic processes require physical manipulation of the soils for recovery (Whisenant, 1999). Cultivation practices often involve creating a uniform soil environment both physically (e.g. land leveling) and in nutrients (e.g. fertilization) to maximize crop production (Homburg and Sandor, 2011). In old fields, these practices impact abiotic processes associated with soil structure, surface microtopography, nutrient distribution, nutrient content, and hydrology. As with biotic processes, crossing a threshold associated with any of these abiotic variables also has the potential to favor exotic over native plant species.

Soil structure and microtopography can favor exotic plants

Where an abiotic threshold has been crossed, the establishment niche of the native species pool may no longer exist (Cramer et al., 2008). Cultivation can increase soil compaction (i.e. bulk density) and soil surface penetration resistance that create physical barriers to seedling emergence and impede root growth (Unger and Kaspar, 1994). For example, cultivation can create a plow pan (a compacted layer 20–25 cm beneath the soil surface) that restricts root growth (Buschbacher et al., 1988; Uhl et al., 1988). A plow pan may be less encumbering on exotic annuals with shallow root systems. Soil compaction has been shown to favor exotic species in old fields in eastern US forests, possibly because the exotic species were horticultural introductions for ground cover on highly disturbed sites (e.g. rock pathways) with traits allowing them to thrive in compacted soils (Parker et al., 2010). In contrast, Kyle et al. (2007) reported soil compaction on old fields in western US steppe was negatively correlated with exotic species while native species were unaffected. This advantage could be related to enhanced root growth of the exotic species in loosened soil, which allows more access to soil resources earlier in its life cycle (Kyle et al., 2007).

Leveling of the soil surface and reducing small-scale heterogeneity also has been implicated in reducing seedling establishment by decreasing the occurrence of depressions for water capture (Whisenant et al., 1995). Cultivation also can reduce the number of safe sites for seed germination and establishment. There may be as many as 15-fold more wind-dispersed seeds than vertebrate-dispersed seeds in an old field, but the density of established wind-dispersed seed can be lower due to microsite limitations (Ingle, 2003). Flinn (2007) showed that small scale heterogeneity of post-cultivation forest floors had limited suitable sites for native fern establishment, and sowing the seed had little effect. The loss of microtopographic heterogeneity may represent a larger limitation to seedling establishment or plant survival than soil quality in some systems (Flinn, 2007). However, there are fewer studies on establishment limitation connected to abiotic properties of the soil, perhaps because so many old fields are already known to be seed limited, or because individual species interact differently depending upon their life history traits (Bakker and Berendse, 1999; Flinn and Vellend, 2005; Standish et al., 2007).

Soil nutrient distribution can favor exotic plants

Cultivation practices not only homogenize the soil physically, but also homogenize soil resources, and patterns of nutrient cycling across old fields that last for decades (Robertson et al., 1993; Standish et al., 2006). Homogenization is more profound in nutrient-limited ecosystems where soil nutrients tend to be patchy in the environment and are concentrated under shrubs in zones known as 'islands of fertility' (Bochet et al., 1999; Whisenant, 1999). These islands of fertility are mechanically homogenized and diffused across the whole soil surface during cultivation (Charley and West, 1975). On a landscape scale, processes like nitrogen mineralization may be similar between old fields and noncultivated land but their distribution is altered because the spatial aggregation has been removed (Bolton et

al., 1993; Robertson et al., 1993). This effect is further enhanced with the application of fertilizers (Standish et al., 2006). Soil homogenization on old fields due to cultivation can contribute to the dominance of early successional species, like exotic annual species (Robertson et al., 1993; Standish et al., 2006). Homogeneous soil resources can favor exotic plant species that tolerate regularly high competition for resources across the landscape as opposed to native species that may be more competitive when resources are patchy (Hutchings et al., 2003; Standish et al., 2006).

Soil nutrient availability can favor exotic plants

Old fields often have altered soil organic carbon and soil nutrients in comparison to similar noncultivated sites (McLauchlan, 2006). Even prehistoric old fields in southwestern North America have been found to contain lower soil organic carbon 1000 years after cultivation ceased (Sandor et al., 1986). The loss of soil organic matter is linked to further degradation in soil structure; namely increased compaction that restricts soil water movement, nutrient cycling, and microbial activity (Whisenant, 1999; McLauchlan, 2006; Homburg and Sandor, 2011). Soil nutrients can be lost through annual crop harvesting, reduction of soil organic material, and loss of topsoil from wind and water erosion. In contrast, soil organic carbon and nutrients can be increased in old fields through the application of either organic (e.g. manure and ash) or inorganic fertilizers (McLauchlan, 2006). The introduction and use of synthetic fertilizers played a role in expanding cultivation agriculture into areas where it would not have been possible otherwise, and where legacies are often greater (Li and Norland, 2001; Standish et al., 2006; Cramer et al., 2008). Still, without fertilization, long-term differences in soil nutrients can remain for decades to centuries (McLauchlan, 2006; Homburg and Sandor, 2011). Even when ecosystems are gaining soil organic matter, the rate of recovery does not always match the rate of loss (Ihori et al.,

1995). Remarkably, the recorded losses in organic matter, nitrogen, and phosphorus on a mass concentration basis can be similar between soils of old fields abandoned for centuries and modern old fields (Homburg and Sandor, 2011). In addition, the agricultural crop itself can influence soil-nitrogen levels because some nitrogen-fixing plants (e.g. legumes) can add nitrogen to the soil (McLauchlan, 2006). Likewise, the vegetation composition following abandonment, native or exotic, can influence both the retention and loss of nutrients in the soil (Lugo, 2004).

Fertilization can elevate soil nutrients for decades to centuries (Dupouey et al., 2002; McLauchlan, 2006; Standish et al., 2008). Fertilization of old fields usually elevates the soil nutrients that are typically limiting in the native system, most typically nitrogen (N), phosphorus (P), and potassium (K). However, the potential for P retention is much higher than N because it has a greater sorption in many soils, and fewer loss pathways (e.g. leaching) than N (McLauchlan, 2006; Grossman and Mladenoff, 2008). Indeed, elevated P legacies have been found in Roman agricultural terraces from AD 50 to 250 (Dupouey et al., 2002). Elevated nutrient content can make it easier for exotic species to colonize and compete with native species (Li and Norland, 2001; Zimmerman et al., 2007; see Grant and Paschke, Chapter 5, this volume). For example, some old fields in Puerto Rico have maintained exotic species dominance for 22 years due to altered soils from liming practices (Lugo and Brandeis, 2005; Zimmerman et al., 2007). Nutrient addition in combination with soil disturbance can increase the establishment and growth of exotic species more than nutrient addition alone (Hobbs and Atkins, 1988; Li and Norland, 2001). For example, in the southern coastal state of Florida in the USA, areas were opened for cultivation by the introduction of synthetic fertilizers and rock-plowing, which crushed the natural oolitic limestone bedrock to create a growth medium and increase soil depth for vegetable production (Doren et al., 1991; Li and

Norland, 2001). The exotic and invasive shrub, Schinus terebinthifolius, responds more favorably to the additional P found in abandoned rock-plowed fields where it competitively excludes the native sawgrass species, Cladium jamaicense, which is more adapted to low nutrient soils (Li and Norland, 2001). Many exotic species competitively exclude native species when soil nutrients are elevated, but increased fertility does not always confer a competitive advantage to exotic species. For example, on old fields in Australia with residually elevated P levels, reduction of P was not enough to manage the exotic Avena barbata because it could still obtain more P than the native species at all levels (Standish et al., 2008).

These alterations in soil fertility can change nutrient cycling and nutrient uptake by plants, conditions which can be created and perpetuated through feedbacks by the exotic plants themselves (see Eviner and Hawkes, Chapter 7, this volume). There is debate over whether the plant–soil relationships found in exotic-dominated old fields are due to land-use legacies or plant-growth influences (Kulmatiski et al., 2006). In other words, are exotic species passengers or drivers of the increases in soil fertility (MacDougall and Turkington, 2005)? It can be difficult if not impossible to separate these feedbacks in scientific studies (Woods, 1997). However, there is some literature that highlights the role of the cultivation disturbance in altering soil fertility, and thereby community composition, because soil structural and nutrient modifications are also found in old fields that are not dominated by exotic species (Motzkin et al., 1996; McLauchlan, 2006). Physical soil disturbance without nutrient addition may or may not increase invasion potential (Hobbs, 1989). Hobbs (1989) found that soil disturbance promoted the establishment of exotic species, but their performance increased more when soil disturbance was combined with nutrient addition. As discussed above, ecosystem and plant community properties may determine which has greater influence (Pickett and White, 1985; Hobbs, 1989).

Altered hydrology can favor exotic plants

The land-use legacies of cultivation also include changes in hydrological processes such as soil moisture avaliability, soil water-holding capacity, runoff and infiltration, and groundwater flows. Cultivation legacies can have a greater effect on differences in soil water movement between plowed and unplowed sites than the differences in soil water movement between two different soil series (Schwartz et al., 2003). In fact, soil hydraulic conductivity can remain affected for well over 25 years after cultivation ceases, and such alterations may be very difficult to restore to pre-disturbance function (Fuentes et al., 2004). Water availability can also be reduced by soil compaction in old fields (Standish et al., 2006). Plowing has been shown to reduce infiltration rates (Gifford, 1972; Tromble, 1980). Recovery potential, for infiltration rates on old fields with livestock grazing, is lower than is predicted for livestock grazing alone (Gifford, 1982). Finally, irrigation is implicated in both raising water tables and lowering them during intensive pumping (Whisenant, 1999; Elmore et al., 2006; Cramer et al., 2007).

Altered hydrology can impact secondary plant community succession on old fields at the local and the landscape scale. Ground-water pumping can lower the water table under which the native plant community was dependent, inviting exotic plant species to colonize and dominate old fields that are more dependent upon precipitation, and therefore modify the entire hydrologic cycle in soils (Elmore et al., 2006). Compacted soils in old fields have been blamed for poor plant germination and establishment due to less pore space for water movement, root growth, and water storage (Cramer et al., 2007; Hillhouse, 2008). At the landscape scale, irrigation can also alter soil hydrology. Irrigation can cause soil damage through anthropogenic salinization (Whisenant, 1999; McLauchlan, 2006). A dramatic example of this is the wheatbelt region of western Australia where expeditious land clearing for cultivation raised the water tables and increased surface soil salinity

(Cramer and Hobbs, 2002; Standish et al., 2006). The replacement of deep-rooted plants with shallow-rooted annual crops increased groundwater recharge and raised the water table, bringing salts to the surface (Whisenant, 1999). Concentrations of salt on soil surfaces reduced infiltration by dispersing clays, which further reduced porosity of the soil (Whisenant, 1999). This legacy not only threatens the agricultural output and restoration potential of the fields, but also the remnant vegetation in the surrounding landscape (Cramer and Hobbs, 2002). Similarly, the soil surfaces of abandoned sugarcane fields in Puerto Rico, that were created out of drained wetlands, are at a lower elevation than the surrounding forest due to burning off of the peat for sugarcane harvests. This has led to longer periods of flooding and higher risk of saltwater intrusion (Zimmerman et al., 2007). As I mentioned previously, when exotic disturbances are outside of the magnitude and frequency of natural disturbance regimes, they can result in modifications with irreversible legacy effects, such as native species loss, because the environment becomes mismatched with the species' traits (Byers, 2002). Sometimes exotic species are the only plants in the community that are adapted to these new environments.

Recovery of abiotic processes

Once both the biotic and abiotic thresholds have been crossed, restoration involves remediation of the physical environment, which requires intense effort and increased capital (Whisenant, 1999; Cramer et al., 2008). For example, where soil structure and fertility had been altered in the rock-plowed areas of southern Florida, USA, the most effective way to manage the invasive shrub *Schinus* was to completely remove topsoil (Li and Norland, 2001). Restoration that prevents reinvasion by exotic species is a lengthy process with uncertain outcomes and risks. Even with intense efforts, return to the pre-cultivation plant community and ecosystem state may not be feasible

(Standish *et al.*, 2008). For example, when wetlands or groundwater have been drained, it may not be realistic, nor is it always sufficient to re-water these areas because soil properties (e.g. pH, salinity, and nutrient levels) have also been altered (Bakker and Berendse, 1999; Cramer *et al.*, 2007). Clearly, abiotic and biotic processes are linked through feedbacks, but there must be a baseline of primary abiotic structure and function in order for biotic process to be reestablished (Whisenant, 1999; King and Hobbs, 2006).

Attempts to directly repair altered soil nutrient availability have met varied success (Blumenthal *et al.*, 2003; Corbin and D'Antonio, 2004). When soil nitrogen is elevated, strategies to reduce it have included manipulations of the C:N ratio of soils via organic carbon additions (e.g. sawdust) to promote microbial immobilization of N (Blumenthal *et al.*, 2003). Similarly, where systems have elevated P, N addition has been used to lower the P:N ratio (Gough and Marrs, 1990; Fagan *et al.*, 2008). Others have found that simply seeding desired species into areas with elevated soil fertility is more effective than attempting to reduce soil fertility with straw mulch (Kardol *et al.*, 2008). In some cases, plant species not included in the pre-disturbance community may be more suited to the new conditions at the site, such as plants with functional traits and life histories more similar to the dominant exotic plants (Whisenant, 1999).

Another approach to repairing abiotic processes involves using biotic intervention with foundational species that exert control over abiotic processes (Buisson and Dutoit, 2004; Ellison *et al.*, 2005; Banerjee *et al.*, 2006; King and Hobbs, 2006; Prevéy *et al.*, 2010). Both carbon and nitrogen pools can recover, although very slowly, in cultivated soils once perennial grasses replace the annual species that dominate in initial stages of abandonment (Burke *et al.*, 1995). However, the traditional agronomic approaches to native plant restoration (e.g. landscape-wide herbicide application, plowing, and seeding) are unlikely to work in exotic plant-dominated communities where abiotic processes are altered and may

be maintained by exotic plant–soil feedbacks (Kulmatiski, 2006; Whisenant, 1999). Therefore, in order to take advantage of foundational species influence, it may be more effective to plant patches containing foundational species and other native seedlings (Holl *et al.*, 2000; Hooper *et al.*, 2005; Banerjee *et al.*, 2006; King and Hobbs, 2006). In addition to ameliorating the environment for other desired plants, foundational species may improve the function of compromised abiotic processes like water infiltration, nutrient cycling, and soil structure (King and Hobbs, 2006).

Few studies have examined how land-use legacies in soil affect recovery in ecosystem processes during restoration (Anderson, 2008). More research, and explicit consideration of *a priori* legacies in soils, is needed and will help improve restoration efforts (Callaham *et al.*, 2008). Understanding the initial soil conditions at time of abandonment (e.g. nutrient levels and mycorrhizal abundance) is an important first step for increasing restoration success because these initial conditions can influence the trajectory of secondary succession more than the restoration treatment (Anderson, 2008).

The Debate about When and How to Intervene

It is important to acknowledge the global debate about whether old fields dominated by exotic species should be restored back to native plant communities, back to cultural agricultural landscapes, or if there should be intervention at all (Foster *et al.*, 2003; Marty *et al.*, 2007; Hobbs *et al.*, 2009). Intervention depends upon the length of delay in recovery and amount of divergence of the old field from the historical plant community that is deemed acceptable by society (Cramer *et al.*, 2008). There are also considerations about 'if the cure is worse than the disease' when it comes to removing exotic species, especially when the management techniques involve additional exotic disturbance (Zavaleta *et al.*, 2001; Hobbs *et al.*, 2009). For example, removal of unwanted exotic species may not necessarily bring the ecosystem back to its pre-disturbance condition, and may move it toward a less desirable one (Pierson et al., 2007; Seastedt et al., 2008). In fact, there is a risk that compounded disturbances will lead to longer-term alterations and new ecological surprises (Paine et al., 1998). Unfortunately, the legacies of management and repair strategies have seldom been studied explicitly, even though some of them involve further cultivation dis- turbances that can also alter the same biotic and abiotic processes discussed in this chapter (DePuit and Redente, 1988; Kettle et al., 2000). Still, even when there is a reintroduction of historically natural disturbances, the altered systems on old fields can remain relatively unchanged (Motzkin et al., 1996).

However, not intervening in secondary succession is not strongly favored either (Standish et al., 2007). Furthermore, the reasons for and the goals of restoration efforts are not always the same worldwide and will depend upon social values as well as economic needs (Hobbs et al., 2009). For example, some may seek to restore old fields to the pre-disturbance plant community for historical conservation and biodiversity while others may want more palatable forage than the exotic species offer so the land can be used for grazing (Marty et al., 2007; Hobbs et al., 2009). In some places, like the tropics, reestablishing trees in old fields plays an important role in providing fast-growing carbon sinks for carbon sequestration while in other places it is more about human survival (Hobbs and Cramer, 2007; Hobbs et al., 2009). Therefore, intervention in secondary succession on old fields can have a wide variety of goals, which also influences how and what types of repair strategies will be used.

Conclusion

Cultivation involves exotic disturbances that leave land-use legacies for decades, centuries, or millennia (Foster et al., 2003; McLauchlan, 2006). Exotic species can become a long-term problem and arrest secondary succession on old fields when they are better adapted to cultivation disturbances than native species (Mack, 1989; Hobbs and Huenneke, 1992). Cultivation legacies can affect both biotic and abiotic processes of ecosystems and can, therefore, influence every aspect of management and restoration. Management and restoration of old fields requires indentifying the suite of ecological processes that have been compromised – biotic, abiotic, or both (Hobbs and Cramer, 2007). Biotic processes, such as reproduction and seed dispersal, will be difficult to repair in old fields, but less difficult than abiotic ones. Old fields that are relatively small and surrounded by native plant communities may have a better chance for unassisted succession to take place if an abiotic threshold has not been crossed, although it will likely proceed slowly. Abiotic thresholds, such as changes in soils that influence nutrient uptake in plants, may not be reversible with repair strategies, especially at large scales (Cramer et al., 2008). It may be necessary to consider using plants with traits that are more adapted to the new conditions at the site, even if they are not a part of the pre-disturbance community (Whisenant, 1999). The level of required intervention increases when more than one process is compromised and when several processes interact to create feedbacks (Hobbs and Cramer, 2007). Because different processes may be interacting simultaneously at different sites, effective strategies will have to be site specific (Cramer et al., 2008).

Old fields containing exotic species can be hybrid systems where some of the plant species and ecological processes of the pre-cultivation system remain; or they can be completely novel systems with funda- mentally new species assemblages and altered biotic and abiotic processes (Seastedt et al., 2008; Hobbs et al., 2009). Hybrid systems, where both biotic and abiotic thresholds have not been crossed, may have more potential for restoration to the pre-disturbance system, or at least its structure and function (Hobbs et al., 2009). Novel systems, where both biotic and abiotic thresholds have been crossed, have less potential for return to pre-cultivation function. Such novel systems may require creative and innovative restoration alternatives that do not include a return to the historical ecosystem (Hobbs *et al.*, 2009).

Acknowledgements

I wish to extend my thanks to Tom Monaco, Roger Sheley, and Rachel Standish for their helpful comments on early drafts of this chapter. I also want to thank Rachel Standish for her help with the photo in Fig. 3.3. And, thanks to Jonathan E. Taylor and Alice Clarke at the Everglades National Park, USA, for their assistance in locating an image of the Hole-in-the-Donut used in Fig. 3.3.

References

Allen, E.B. and Allen, M.F. (1984) Competition between plants of different successional stages: mycorrhizae as regulators. *Canadian Journal of Botany* 62, 2625–2629.

Allen, M.F. (1988) Belowground structure: a key to reconstructing a productive arid ecosystem. In: Allen, E.B. (ed.) *Reconstruction of Disturbed Arid Lands: An ecological approach*. Westview Press, Boulder, Colorado, pp. 113–135.

Anderson, R.C. (2008) Growth and arbuscular mycorrizal fungal colonization of two prairie grasses grown in soil from restoration of three ages. *Restoration Ecology* 16, 650–656.

Auge, R.M. (2001) Water relations, drought and vesicular-arbuscular mycorrhizal symbiosis. *Mycorrhiza* 11, 3–42.

Bakker, J.P. and Berendse, F. (1999) Constraints in the restoration of ecological diversity in grassland and heathland communities. *Trends in Ecology and Evolution* 14, 63–68.

Banerjee, M.J., Gerhart, V.J. and Glenn, E.P. (2006) Native plant regeneration on abandoned desert farmland: effects of irrigation, soil preparation, and amendments on seedling establishment. *Restoration Ecology* 14, 339–348.

Benjamin, K., Domon, G. and Bouchard, A. (2005) Vegetation composition and succession of abandoned farmland: effects of ecological, historical and spatial factors. *Landscape Ecology* 20, 627–647.

Biondini, M.E., Bonham, C.D. and Redente, E.F. (1985) Secondary successional patterns in a sagebrush (*Artemisia tridentata*) community as

they relate to soil disturbance and soil biological activity. *Vegetatio* 60, 25–36.

Blumenthal, D.M., Jordan, N.R. and Russelle, M.P. (2003) Soil carbon addition controls weeds and facilitates prairie restoration. *Ecological Applications* 13, 605–615.

Bochet, E., Rubio, J.L. and Poessen, J. (1999) Modified topsoil islands within patchy Mediterranean vegetation in SE Spain. *Catena* 38, 23–44.

Bolton, H., Smith, J.L. and Link, S.O. (1993) Soil microbial biomass and activity of a disturbed and undisturbed shrub-steppe ecosystem. *Soil Biology and Biochemistry* 25, 545–552.

Bonet, A. and Pausas, J.G. (2004) Species richness and cover along a 60-year chronosequence in old-fields of southeastern Spain. *Plant Ecology* 174, 257–270.

Buckley, D.H. and Schmidt, T.M. (2001) The structure of microbial communities in soil and the lasting impact of cultivation. *Microbial Ecology* 42, 11–21.

Buisson, E. and Dutoit, T. (2004) Colonisation by native species of abandoned farmland adjacent to a remnant patch of Mediterranean steppe. *Plant Ecology* 174, 371–384.

Burke, I.C., Lauenroth, W.K. and Coffin, D.P. (1995) Soil organic matter recovery in semiarid grasslands: implications for the Conservation Reserve Program. *Ecological Applications* 5, 793–801.

Buschbacher, R., Uhl, C. and Serrao, E.A.S. (1988) Abandoned pastures in eastern Amazonia. 2. Nutrient stocks in the soil and vegetation. *Journal of Ecology* 76, 682–699.

Byers, J.E. (2002) Impact of non-indigenous species on natives enhanced by anthropogenic alteration of selection regimes. *Oikos* 97, 449–458.

Callaham Jr, M.A., Rhoades, C.C. and Heneghan, L. (2008) A striking profile: soil ecological knowledge in restoration management and science. *Restoration Ecology* 16, 604–607.

Charley, J.L. and West, N.E. (1975) Plant-induced soil chemical patterns in some shrub-dominated semi-desert ecosystems of Utah. *Journal of Ecology* 63, 945–963.

Coffin, D.P., Lauenroth, W.K. and Burke, I.C. (1996) Recovery of vegetation in semiarid grassland 53 years after disturbance. *Ecological Applications* 6, 538–555.

Corbin, J.D. and D'Antonio, C.M. (2004) Can carbon addition increase competitiveness of native grasses? A case study from California. *Restoration Ecology* 12, 36–43.

Cramer, V.A. and Hobbs, R.J. (2002) Ecological consequences of altered hydrological regimes in fragmented ecosystems in southern Australia: impacts and possible management responses. *Austral Ecology* 27, 546–564.

Cramer, V.A. and Hobbs, R.J. (2007) *Old Fields: Dynamics and Restoration of Abandoned Farmland.* Island Press, Washington, DC.

Cramer, V.A., Standish, R.J. and Hobbs, R.J. (2007) Prospects for the recovery of native vegetation in western Australian old fields. In: Cramer, V.A. and Hobbs, R.J. (eds) *Old Fields: Dynamics and Restoration of Abandoned Farmland.* Island Press, Washington, DC, pp. 286–306.

Cramer, V.A., Hobbs, R.J. and Standish, R.J. (2008) What's new about old fields? Land abandonment and ecosystem assembly. *Trends in Ecology and Evolution* 23, 104–112.

Daubenmire, R. (1975) Plant succession on abandoned fields, and fire influences, in a steppe area of southeastern Washington. *Northwest Science* 49, 36–48.

Davis, M.A., Bier, L., Bushelle, E., Diegel, C., Johnson, A. and Kujala, B. (2005) Non-indigenous grasses impede woody succession. *Plant Ecology* 178, 249–264.

DePuit, E.J. and Redente, E.G. (1988) Manipulation of ecosystem dynamics on reconstructed semiarid lands. In: Allen, E.B. (ed.) *Reconstruction of Disturbed Arid Lands: An ecological approach.* Westview Press, Boulder, Colorado, pp. 162–204.

Doren, R.F., Whiteaker, L.D. and LaRosa, A.M. (1991) Evaluation of fire as a management tool for controlling *Schinus terebinthifolius* as secondary successional growth and abandoned agricultural land. *Environmental Management* 15, 121–129.

Douds, D.D. and Millner, P.A. (1999) Biodiversity of arbuscular mycorrhizal fungi in agroecosystems. *Agriculture Ecosystems and the Environment* 74, 77–93.

Dupouey, J.L., Dambrine, E., Laffite, J.D. and Moares, C. (2002) Irreversible impact of past land use on forest soils and biodiversity. *Ecology* 83, 2978–2984.

Dyer, J.M. (2010) Land-use legacies in a central Appalachian forest: differential response of trees and herbs to historic agricultural practices. *Applied Vegetation Science* 13, 195–206.

Ellery, A.J. and Chapman, R. (2000) Embryo and seed coat factors produce seed dormancy in capeweed (*Arctotheca calendula*). *Australian Journal of Agricultural Research* 51, 849–854.

Ellison, A.M., Bank, M.S., Clinton, B.D., Colburn, E.A., Elliott, K., Ford, C.R., Foster, D.R., Kloeppel, B.D., Knoepp, J.D., Lovett, G.M., Mohan, J., Orwig, D.A., Rodenhouse, N.L., Sobczak, W.V., Stinson, K.A., Stone, J.K.,

Swan, C.M., Thompson, J., Von Holle, B. and Webster, J.R. (2005) Loss of foundation species: consequences for the structure and dynamics of forested ecosystems. *Frontiers in Ecology and the Environment* 3, 479–486.

Elmore, A.J., Mustard, J.F., Hamburg, S.P. and Manning, S.J. (2006) Agricultural legacies in the Great Basin alter vegetation cover, composition, and response to precipitation. *Ecosystems* 9, 1231–1241.

Fagan, K.C., Pywell, R.F., Bullock, J.M. and Marrs, R.H. (2008) Do restored calcareous grasslands on former arable fields resemble ancient targets? The effect of time, methods and environment on outcomes. *Journal of Applied Ecology* 45, 1293–1303.

Fernández, J.B., García Mora, M.R. and García Novo, F. (2004) Vegetation dynamics of Mediterranean shrublands in former cultural landscape at Grazalema Mountains, South Spain. *Plant Ecology* 172, 83–94.

Fine, P.V.A. (2002) Invasibility of tropical forests by exotic plants. *Journal of Tropical Ecology* 18, 687–705.

Flinn, K.M. (2007) Microsite-limited recruitment controls fern colonization of post-agricultural forests. *Ecology* 88, 3103–3114.

Flinn, K.M. and Vellend, M. (2005) Recovery of forest plant communities in post-agricultural landscapes. *Frontiers in Ecology and the Environment* 3, 243–250.

Foster, D., Swanson, F., Aber, J., Burke, I., Brokaw, N., Tilman, D. and Knapp, A. (2003) The importance of land-use legacies to ecology and conservation. *Bioscience* 53, 77–87.

Fuentes, J.P., Flury, M. and Bezdicek, D.F. (2004) Hydraulic properties in a silt loam soil under natural prairie, conventional till and no till. *Soil Science Society of America Journal* 68, 1679–1688.

Ganade, G. (2007) Processes affecting succession in old fields of Brazilian Amazonia. In: Cramer, V.A. and Hobbs, R.J. (eds) *Old Fields: Dynamics and Restoration of Abandoned Farmland.* Island Press, Washington, DC, pp. 75–92.

Gifford, G.F. (1972) Infiltration rate and sediment production trends on a plowed big sagebrush site. *Journal of Range Management* 25, 53–55.

Gifford, G.F. (1982) A long-term infiltrometer study in southern Idaho, USA. *Journal of Hydrology* 58, 367–374.

Gough, M.W. and Marrs, R.H. (1990) A comparison of soil fertility between semi-natural and agricultural plant communities: implications for the creation of species-rich grassland on abandoned agricultural land. *Biological Conservation* 51, 83–96.

Grossman, E.B. and Mladenoff, D.J. (2008) Farms, fires, and forestry: disturbance legacies in the soils of the Northwest Wisconsin (USA) Sand Plain. *Forest Ecology and Management* 256, 827–836.

Hartnett, D.C. and Bazzaz, F.A. (1983) Physiological integration among intraclonal ramets in *Solidago canadensis. Ecology* 64, 779–788.

Heywood, V.H. (1989) Patterns, extents and modes of invasions by terrestrial plants. In: Drake, J.A., Mooney, H.A., di Castri, F., Groves, R.H., Kruger, F.J, Rejmánek, M. and Wiliamson, M. (eds) *Biological invasions: A global perspective.* John Wiley and Sons, Chichester, UK, pp. 31–51.

Hillhouse, H.L. (2008) Plant establishment in tallgrass prairie plantings. Dissertation, University of Wisconsin-Madison, Wisconsin.

Hobbs, R.J. (1989) The nature and effects of disturbance relative to invasions. In: Drake, J.A., Mooney, H.A., di Castri, F., Groves, R.H., Kruger, F.J., Rejmánek, M. and Wiliamson, M. (eds) *Biological Invasions: A global perspective.* John Wiley and Sons, Chichester, UK, pp. 389–401.

Hobbs, R.J. and Atkins, L. (1988) Effect of disturbance and nutrient addition on native and introduced annuals in plant communities in the Western Australian wheatbelt. *Australian Journal of Ecology* 13, 171–179.

Hobbs, R.J. and Cramer, V.A. (2007) Why old fields? Socioeconomic and ecological causes and consequences of land abandonment. In: Cramer, V.A. and Hobbs, R.J. (eds) *Old Fields: Dynamics and Restoration of Abandoned Farmland.* Island Press, Washington, DC, pp. 309–318.

Hobbs, R.J. and Huenneke, L.F. (1992) Disturbance, diversity, and invasion: implications for conservation. *Conservation Biology* 6, 324–337.

Hobbs, R.J. and Yates, C.J. (2003) Impacts of ecosystem fragmentation on plant populations: generalizing the idiosyncratic. *Australian Journal of Botany* 51, 471–488.

Hobbs, R.J., Higgs, E. and Harris, J.A. (2009) Novel ecosystems: implications for conservation and restoration. *Trends in Ecology and Evolution* 24, 599–605.

Holl, K.D., Loik, M.E., Lin, E.H. and Samuels, I.A. (2000) Tropical montaine forest restoration in Costa Rica: overcoming barriers to dispersal and establishment. *Restoration Ecology* 8, 339–349.

Homburg, J.A. and Sandor, J.A. (2011) Anthropogenic effects on soil quality of ancient agricultural systems of the American Southwest. *Catena* 85(2), 144–154.

Hooper, E., Condit, R. and Legendre, P. (2002) Responses of 20 native tree species to reforestation strategies for abandoned farmland in Panama. *Ecological Applications* 12, 1626–1641.

Hooper, E.R., Legendre, P. and Condit, R. (2004) Factors affecting community composition of forest regeneration in deforested, abandoned land in Panama. *Ecology* 85, 3313–3326.

Hooper, E., Legendre, P. and Condit, R. (2005) Barriers to forest regeneration of deforested and abandoned land in Panama. *Journal of Applied Ecology* 42, 1165–1174.

Hutchings, M.J., John, E.A. and Wijesinghe, D.K. (2003) Toward understanding the consequences of soil heterogeneity for plant populations and communities. *Ecology* 84, 2322–2334.

Ihori, T., Burke, I.C. and Hook, P.B. (1995) Nitrogen mineralization in native cultivated and abandoned fields in shortgrass steppe. *Plant and Soil* 171, 203–208.

Ingle, N.R. (2003) Seed dispersal by wind, birds, and bats between Philippine montane rainforest and successional vegetation. *Oecologia* 134, 251–261.

Janos, D.P. (1980) Mycorrhizae influence tropical succession. *Biotropica* 12, 56–64.

Jenkins, M.A. and Parker, G.R. (2000) The response of herbaceous-layer vegetation to anthropogenic disturbance in intermittent stream bottomland forests of southern Indiana, USA. *Plant Ecology* 151, 223–237.

Kardol, P., Van der Wal, A., Bezemer, T.M., deBoer, W., Dyts, H., Holtkamp, R. and Van der Putten, W.H. (2008) Restoration of species-rich grasslands on ex-arable land: seed addition outweighs soil fertility reduction. *Biological Conservation* 141, 2208–2217.

Kettle, W.D., Rich, P.M., Kindscher, K., Pittman, G.L. and Fu, P. (2000) Land-use history in ecosystem restoration: a 40-year study in the prairie-forest ecotone. *Restoration Ecology* 8, 307–317.

King, E.G. and Hobbs, R.J. (2006) Identifying linkages among conceptual models of ecosystem degradation and restoration: towards an integrative framework. *Restoration Ecology* 14, 369–378.

Klironomos, J.N. (2002) Feedback with soil biotia contributes to plant rarity and invasiveness in communities. *Nature* 417, 67–70.

Krug, C.B. and Krug, R.M. (2007) Restoration of old fields in Renosterveld: a case study in a Mediterranean-type shrubland of South Africa. In: Cramer, V.A. and Hobbs, R.J. (eds) *Old Fields: Dynamics and Restoration of Abandoned Farmland.* Island Press, Washington, DC, pp. 265–285.

Kruger, F.J., Breytenbach, G.J., MacDonald, I.A.W. and Richardson, D.M. (1989) The characteristics of invaded Mediterranean-climate regions. In: Drake, J.A., Mooney, H.A., di Castri, F., Groves, R.H., Kruger, F.J, Rejmánek, M. and Wiliamson, M. (eds) *Biological Invasions: A global perspective*. John Wiley and Sons, Chichester, UK, pp. 181–204.

Kulmatiski, A. (2006) Exotic plants establish persistent communities. *Plant Ecology* 187, 261–275.

Kulmatiski, A. and Beard, K.H. (2008) Decoupling plant-growth from land-use legacies in soil microbial communities. *Soil Biology and Biochemistry* 40, 1059–1068.

Kulmatiski, A., Beard, K.H. and Stark, J.M. (2006) Soil history as a primary control on plant invasion in abandoned agricultural fields. *Journal of Applied Ecology* 43, 868–876.

Kyle, G.P., Beard, K.H. and Kulmatiski, A. (2007) Reduced soil compaction enhances establishment of non-native plant species. *Plant Ecology* 193, 223–232.

Lawson, D., Inouye, R.S., Huntly, N. and Carson, W.P. (1999) Patterns of woody plant abundance, recruitment, mortality, and growth in a 65 year chronosequence of old-fields. *Plant Ecology* 145, 267–279.

Li, Y. and Norland, M. (2001) The role of soil fertility in invasion of Brazilian pepper (*Schinus terebinthifolius*) in Everglades National Park, Florida. *Soil Science* 166, 400–405.

Lovell, R.D., Jarvis, S.C. and Bardgett, R.D. (1995) Soil microbial biomass and activity in long-term grassland: effects of management changes. *Soil Biology and Biochemistry* 27, 969–975.

Lugo, A.E. (2004) The outcome of alien tree invasions in Puerto Rico. *Frontiers in Ecology and the Environment* 2, 265–273.

Lugo, A.E. and Brandeis, T.J. (2005) A new mix of alien and native species coexist in Puerto Rico's landscapes. In: Burslem, D.F.R.P., Pinard, M.A. and Hartley, S.E. (eds) *Biotic Interactions in the Tropics: Their role in the maintenance of species diversity*. Cambridge University Press, Cambridge, UK, pp. 484–509.

MacDougall, A.S. and Turkington, R. (2005) Are invasive species the drivers or passengers of change in degraded ecosystems? *Ecology* 86, 42–55.

MacDougall, A.S. and Wilson, S.D. (2007) Herbivory limits recruitment in an old field seed addition experiment. *Ecology* 88, 1105–1111.

Mack, R.N. (1989) Temperate grasslands vulnerable to plant invasions: characteristics and consequences. In: Drake, J.A., Mooney, H.A., di Castri, F., Groves, R.H., Kruger, F.J., Rejmánek, M. and Wiliamson, M. (eds) *Biological Invasions: A global perspective*. John Wiley and Sons, Chichester, UK, pp. 155–173.

Marty, P., Aronson, J. and Lepart, J. (2007) Dynamics and restoration of abandoned farmland and other old fields in Southern France. In: Cramer, V.A. and Hobbs, R.J. (eds) *Old Fields: Dynamics and Restoration of Abandoned Farmland*. Island Press, Washington, DC, pp. 202–224.

Marushia, R.G. and Allen, E.B. (2011) Control of exotic annual grasses to restore native forbs in abandoned agricultural land. *Restoration Ecology* 19, 45–54.

Mason, R.A.B., Cooke, J., Moles, A.T. and Leishman, M.R. (2008) Reproductive output of invasive versus native plants. *Global Ecology and Biogeography* 17, 633–640.

McIntyre, S. and Hobbs, R.J. (1999) A framework for conceptualizing human effects on landscapes and its relevance to management and restoration. *Conservation Biology* 13, 1282–1292.

McLauchlan, K. (2006) The nature and longevity of agricultural impacts on soil carbon and nutrients: a review. *Ecosystems* 9, 1364–1382.

Meiners, S.J., Pickett, S.T.A. and Cadenasso, M.L. (2002) Exotic plant invasions over 40 years of old field successions: community patterns and associations. *Ecography* 25, 215–223.

Meiners, S.J., Cadenasso, M.L. and Pickett, T.A. (2007) Succession on the Piedmont of New Jersey and its implications for ecological restoration. In: Cramer, V.A. and Hobbs, R.J. (eds) *Old Fields: Dynamics and Restoration of Abandoned Farmland*. Island Press, Washington, DC, pp. 145–161.

Morris, L.R., Monaco, T.A. and Sheley, R.L. (2011) Land-use legacies and recovery 90 years after cultivation in sagebrush ecosystems of the Great Basin, USA. *Rangeland Ecology and Management* 64, 488–497.

Motzkin, G., Foster, D., Allen, A., Harrod, J. and Boone, R. (1996) Controlling site to evaluate history: vegetation patterns of a New England sand plain. *Ecological Monographs* 66, 345–365.

Motzkin, G., Wilson, P., Foster, D.R. and Arthur, A. (1999) Vegetation patterns in heterogeneous landscapes: the importance of history and environment. *Journal of Vegetation Science* 10, 903–920.

Myster, R.W. (2008) *Post-Agricultural Succession in the Neotropics*. Springer, New York.

Oehl, F., Sieverding, E., Ineichen, K., Mäder, P., Boller, T. and Wiemken, A. (2003) Impact of land use intensity on the species diversity of arbuscular mycorrhizal fungi in agroecosystems of central Europe. *Applied and Environmental Microbiology* 69, 2816–2824.

Öster, M., Ask, K., Cousins, S.A.O. and Eriksson, O. (2009) Dispersal and establishment limitation reduces the potential for successful restoration of semi-natural grassland communities on former arable fields. *Journal of Applied Ecology* 46, 1266–1274.

Paine, R.T., Tegner, M.J. and Johnson, E.A. (1998) Compounded perturbations yield ecological surprises. *Ecosystems* 1, 535–545.

Parker, J.D., Richie, L.J., Lind, E.M. and Maloney, K.O. (2010) Land use history alters the relationship between native and exotic plants: the rich don't always get richer. *Biological Invasions* 12, 1557–1571.

Paul, E.A. and Clark, F.E. (1989) *Soil Microbiology and Biochemistry*. Academic Press, San Diego, California.

Pickett, S.T.A. and White, P.S. (1985) *The Ecology of Natural Disturbance and Patch Dynamics*. Academic Press, Orlando, Florida.

Pierson, F.B., Blackburn, W.H. and Van Vactor, S.S. (2007) Hydraulic impacts of mechanical seeding treatments on sagebrush rangelands. *Rangeland Ecology and Management* 60, 666–674.

Prevéy, J.S., Germino, M.J. and Huntly, N.J. (2010) Loss of foundation species increases population growth of exotic forbs in sagebrush steppe. *Ecological Applications* 20, 1890–1902.

Reeves, F.B., Wagner, D., Moorman, T. and Kiel, J. (1979) The role of endomycorrhizae in revegetation practices in the semi-arid West. I. A comparison of incidence of mycorrhizae in severely disturbed vs. natural environments. *American Journal of Botany* 66, 6–13.

Richter, B.S., Tiller, R.L. and Stutz, J.S. (2002) Assessment of arbuscular mycorrhizal fungal propagules and colonization from abandoned agricultural fields and semi-arid grasslands in riparian floodplains. *Applied Soil Ecology* 20, 227–238.

Riege, D. and del Moral, R. (2004) Differential tree colonization of old fields in a temperate rain forest. *American Midland Naturalist* 151, 251–264.

Robertson, G.P., Crum, J.R. and Ellis, B.G. (1993) The spatial variability of soil resources following long-term disturbance. *Oecologia* 96, 451–456.

Roundy, B.A., Heydari, H., Watson, C., Smith, S.E., Munda, B. and Pater, M. (2001) Summer establishment of Sonoran Desert species for revegetation of abandoned farmland using line source sprinkler irrigation. *Arid Land Research and Management* 15, 23–39.

Ruprecht, E. (2006) Successfully recovered grassland: a promising example from Romanian old-fields. *Restoration Ecology* 14, 473–480.

Sandor, J.A., Gersper, P.L. and Hawley, J.W. (1986) Prehistoric agricultural terraces and soils in the Mimbres area, New Mexico. *World Archeology* 22, 70–86.

Santana, V.M., Baeza, M.J., Marrs, R.H. and Vallejo, V.R. (2010) Old-field secondary succession in SE Spain: can fire divert it? *Plant Ecology* 211, 337–349.

Schillinger, W.F. and Papendick, R.I. (2008) Then and now: 125 years of dryland wheat farming in the Inland Pacific Northwest. *Agronomy Journal* 100 (Suppl.), S166–S182.

Schwartz, R.C., Evett, S.R. and Unger, P.S. (2003) Soil hydraulic properties of cropland compared with reestablished and native grasslands. *Geoderma* 116, 47–60.

Seastedt, T.R., Hobbs, R.J. and Suding, K.N. (2008) Management of novel ecosystems: are novel approaches required? *Frontiers in Ecology and the Environment* 6, 547–553.

Standish, R.J., Cramer, V.A., Hobbs, R.J. and Kobryn, H.T. (2006) Legacy of land-use evident in soils of western Australia's wheat belt. *Plant and Soil* 280, 189–207.

Standish, R.J., Cramer, V.A., Wild, S.L. and Hobbs, R.J. (2007) Seed dispersal and recruitment limitation are barriers to native recolonization of old-fields in western Australia. *Journal of Applied Ecology* 44, 435–445.

Standish, R.J., Cramer, V.A. and Hobbs, R.J. (2008) Land-use legacy and the persistence of invasive *Avena barbata* on abandoned farmland. *Journal of Applied Ecology* 45, 1576–1583.

Steenwerth, K.L., Jackson, L.E., Calderón, F.J., Stromberg, M.R. and Scow, K.M. (2002) Soil microbial community composition and land use history in cultivated and grassland ecosystems of coastal California. *Soil Biology and Biochemistry* 34, 1599–1611.

Stylinski, C.D. and Allen, E.B. (1999) Lack of native species recovery following severe exotic disturbance in southern Californian shrublands. *Journal of Applied Ecology* 36, 544–554.

Tromble, J.M. (1980) Infiltration rates on rootplowed rangeland. *Journal of Range Management* 33, 423–425.

Uhl, C., Buschbacher, R. and Serrao, E.A.S. (1988) Abandoned pastures in eastern Amazonia. 1. Patterns of plant succession. *Journal of Ecology* 76, 663–681.

Unger, P.W. and Kaspar, T.C. (1994) Soil compaction and root growth: a review. *Agronomy Journal* 86, 759–766.

Whisenant, S.G. (1999) *Repairing Damaged Wildlands: A process-oriented, landscape scale approach.* Cambridge University Press, Cambridge, UK.

Whisenant, S.G., Thurow, T.L. and Maranz, S.J. (1995) Initiating autogenic restoration on shallow semiarid sites. *Restoration Ecology* 3, 61–67.

Woods, K. (1997) Community response to plant invasion. In: Luken, J.O. and Thieret, J.W. (eds) *Assessment and Management of Plant Invasions.* Springer-Verlag, New York, pp. 56–68.

Zavaleta, E.S., Hobbs, R.J. and Mooney, H.A. (2001) Viewing invasive species removal in a whole-ecosystem context. *Trends in Ecology and Evolution* 16, 454–459.

Zimmerman, J.K., Aide, T.M. and Lugo, A.E. (2007) Implication of land use history for natural forest regeneration and restoration strategies in Puerto Rico. In: Cramer, V.A. and Hobbs, R.J. (eds) *Old Fields: Dynamics and Restoration of Abandoned Farmland.* Island Press, Washington, DC, pp. 51–74.

4 Resource Pool Dynamics: Conditions That Regulate Species Interactions and Dominance

A. Joshua Leffler[1] and Ronald J. Ryel[2]

[1] *US Department of Agriculture, Agricultural Research Service, USA*
[2] *Department of Wildland Resources and the Ecology Center, Utah State University, USA*

Introduction

A primary obstacle in the restoration of native plant communities dominated by invasive exotic species is the limited capacity managers have to influence ecological processes that perpetuate the undesired condition. Although much attention has been given to understanding how disturbances lead to invasion (Sher and Hyatt, 1999; Davis *et al.*, 2000; Davis and Pelsor, 2001), the physiological and morphological differences between invasive and non-invasive plant species (Pyšek and Richardson, 2007), the existence of multiple stable ecosystem states and the transitions among them (Sutherland, 1974; Westoby *et al.*, 1989; Briske *et al.*, 2003), and feedbacks that keep invasive plants entrenched on the landscape (Schlesinger *et al.*, 1990; Klironomos, 2002; Sperry *et al.*, 2006), scientists have not linked ecological theory and management (Gurevitch *et al.*, 2011). If scientists cannot mechanistically describe how basic ecological principles directly influence plant invasions, sound management strategies cannot be developed to effectively prevent invasion and restore landscapes to a more desired state. One underlying process known to be critically important to community structure, ecosystem function, and exotic plant invasion is resource-use dynamics (Davis *et al.*, 2000; Davis and Pelsor, 2001). However, much of our current view of resource acquisition by plants has focused on gross availability of resources and does not effectively address the variable timing and spatial distribution of resources that are closely tied to plant requirements and performance (Theoharides and Dukes, 2007). In this chapter we detail resource pools as a common currency for species interactions and demonstrate that understanding how plants modify resource availability for other species can lead to an improved framework for making decisions to manage invasive species.

Understanding resource pool dynamics is fundamental to understanding performance of invasive species and the susceptibility of ecosystems to plant invasion. Plants do not merely use required resources such as light energy, water, or nutrients, but modify the availability of these resources to other plants. Plants are 'ecosystem engineers' (*sensu* Jones *et al.*, 1994) or organisms that modify the supply of resources to other organisms. Plants modify resource availability through their morphology or structure (i.e. shading other plants or intercepting rainfall) and through their physiology (i.e. uptake of nitrogen); and because plant species are unique, the influence each has on resources is also unique (Eviner, 2004). The influence of plants on resource pool dynamics gained recognition in the field of environmental biophysics (Campbell, 1977) and more recently in ecohydrology (Baird and Wilby, 1999; Eamus *et al.*, 2006).

Resource availability is widely recognized as a primary factor controlling species assembly in a plant community and consequently influences invasion. The niche concept (Hutchinson, 1957) and the principle of competitive exclusion (Gause, 1934) suggest communities are composed of species that spatially and temporally segregate resource use and that niche requirement differences are sufficient to maintain species diversity in communities (Adler et al., 2010). While stochastic events (i.e. disturbance and seed arrival) clearly play a role in community assembly (Gleason, 1926; Hubbell, 2001), resource availability determines which species persist (Tilman, 1982) and increase when rare (Crawley et al., 1996; Theoharides and Dukes, 2007; Gurevitch et al., 2011), which is required for invasion. The fluctuating resource availability hypothesis (Davis et al., 2000; Davis and Pelsor, 2001) links disturbance, resources, and plant community invisibility, and is broadly conceptualized as niche opportunity (Shea and Chesson, 2002). Accordingly, invasive species exploit resources that are liberated following a disturbance. We need, however, to link theoretical concepts of resources, communities, and invasive species with practical management. The influence of plants on resources pools and the likelihood of invasion suggest managers can manipulate resource pools to achieve a desired outcome. Understanding the roles that individual species play in a community is the critical link between theory and management.

In this chapter, we first present the concept of a resource pool and the fluxes that contribute to or subtract from the pool; next we describe how plants influence pools and fluxes and the consequences these modifications have for other species; then we examine how species interactions are mediated by resource pools; and finally we describe links between resources pools, invasion, and management. We conclude the chapter by describing three scenarios that illustrate the importance of considering resource pools when making decisions about the management of specific invasive species. Invasion is a multi-step process; here we focus on interactions between plants and resources, which necessarily addresses three early phases of invasion, namely, colonization, establishment, and spread (Theoharides and Dukes, 2007). While numerous conceptual frameworks for invasion have been proposed (Gurevitch et al., 2011), the resource dynamic framework is especially applicable to management.

Resource Pools and Fluxes

Necessary resources for plants, 'substances required for and consumed during growth' (Field et al., 1992), have been recognized for over 200 years. The resource pools concept (Fig. 4.1a) comes from mid-20th-century ecologists who described the transfer of energy and materials among species in ecosystems (Ovington, 1962; Bormann and Likens, 1967; Odum, 1969). Here, we emphasize that a resource pool is a measurable, limited, and temporally and spatially variable quantity of a resource. All pools are influenced by fluxes, rates of addition or subtraction of resources to or from the pool (Chapin et al., 2002). Modulators are physical or chemical properties (i.e. temperature, soil pH) that affect a pool or flux but are neither consumed nor depleted (Field et al., 1992). Modulators can affect plants (e.g. low temperature inhibits nitrogen uptake) but plants can also affect modulators (e.g. root exudates reduce soil pH).

Resource pools are critical for plant community assembly and invasion. Most simply, invasion is a process by which a species increases in abundance when rare (Crawley et al., 1996; Theoharides and Dukes, 2007; Gurevitch et al., 2011). While several processes may be operating to allow an increase in population size such as enemy release or propagule pressure (Gurevitch et al., 2011), the basic resources required for growth and reproduction must be present. Consequently, given adequate seed supply and lack of enemies, invasion will be a resource-driven process (Crawley et al., 1996; Davis et al., 2000; Davis and Pelsor, 2001).

(a)

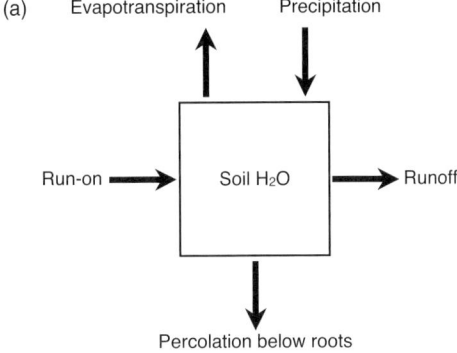

(b)

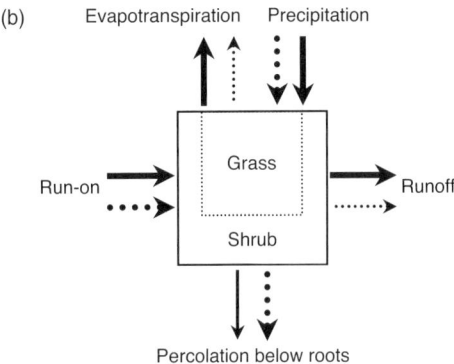

Fig. 4.1. (a) A typical ecosystem 'bucket' model describing a pool of soil water and the fluxes that modify pool size. (b) Pools and fluxes of soil water as 'perceived' by and modified by two species, a shrub (solid lines) and a grass (dotted lines). Shrubs as larger individuals access a larger volume of soil water, transpire more water, and may be more tolerant of low soil water potential and allow less water to percolate below the roots. Grasses have a smaller pool of soil water and are better at slowing water movement over the soil surface. Precipitation and run-on are assumed equal for each species in this model.

Invasion is possible because resource pools, fluxes, and modulators vary considerably in space and time. Some variation is predictable such as seasonal patterns of light, temperature, and precipitation, or the spatial arrangement of soil nitrogen with respect to plants (Burke, 1989; Hooker *et al.*, 2008). Other variation is difficult to predict such as tree-fall gaps in a forest canopy (Hubbell *et al.*, 1999) or discrete patches of nitrogen derived from animal waste

(Peek and Forseth, 2003). Predictability is important to community assembly and plant invasion because invasive species often take advantage of high resource availability that is outside the normal range of variation experienced by the species in the undisturbed community (Sher and Hyatt, 1999).

Pools and fluxes of resources are not independent. Soil pools of water and nitrogen are inherently linked because plants must take up nitrogen in solution (Nye and Tinker, 1977; Barber, 2001). Hence N supply to plants is strongly affected by diffusion rate and plant growth may become N-limited when soils remain wet (Ryel *et al.*, 2008). Linkages among resources suggest that changes in pool size or flux of one resource can influence the pool or flux of another resource. Consequently, even minor disturbances can influence a suite of resources and have a cascading effect on the ecosystem. These types of changes likely cause abrupt transitions among community states on the landscape (Sutherland, 1974; Westoby *et al.*, 1989; Briske *et al.*, 2003).

Critical to linking resources and community assembly to invasive species management is our assertion that resource pools can be species specific (Fig. 4.1b). Two species, which differ in their morphology and physiological tolerance, essentially 'perceive' different sizes of the same resource pools. Consequently, modification of a resource pool by each of these species will differentially influence resource availability for other species and ultimately coexistence. Management actions can be taken to effect the same result; intentional removal or addition of a resource can influence community structure (Perry *et al.*, 2010; Dickson and Foster, 2011).

Plant Modification of Resource Pools

Many organisms can be considered ecosystem engineers (*sensu* Jones *et al.*, 1994). Ecosystem engineers are organisms that modify the supply of a resource for other plants in the community. While the classic example of an animal engineer is the beaver (*Castor canadensis*), plant species that profoundly modify the influx and efflux of resources, as well as microclimate and disturbance regimes, can also be considered ecosystem engineers (Fig. 4.2).

Plants considerably modify the influx of resources. The canopy of a plant can intercept precipitation preventing it from entering the pool of soil water (Bosch and Hewlett, 1982; Brown et al., 2005; LaMalfa and Ryel, 2008), or promote condensation of fog (Azevedo and Morgan, 1974; Ewing et al., 2009), essentially creating precipitation when it would not have occurred. Plants slow the flow of water over soil surfaces preventing erosion but also increasing local infiltration (Schlesinger et al., 1990; Bhark and Small, 2003; Ludwig et al., 2005). Roots can encourage deep recharge of water through the creation of preferential flow

paths along the surface of roots and through soil macropores created by roots (Beven and Germann, 1982; Li and Ghodrati, 1994), or through the process of hydraulic redistribution among soils of different water potential (Richard and Caldwell, 1987; Caldwell et al., 1998; Leffler et al., 2005). Plants influence nutrient influx through the deposition of leaves or roots and organic molecules from root exudates (Haynes, 1986; Porazinska et al., 2003; Zhang et al., 2008). Plants reduce the flux and modify the spectrum of light between above- and below-canopy environments (Fahnestock and Knapp, 1994; Hubbell et al., 1999; Kobe, 1999).

Plants modify the efflux of resources. Even in ecosystems with considerable precipitation, water use by plants will substantially dry soils (Duff et al., 1997). At the scale of a watershed, plant transpiration

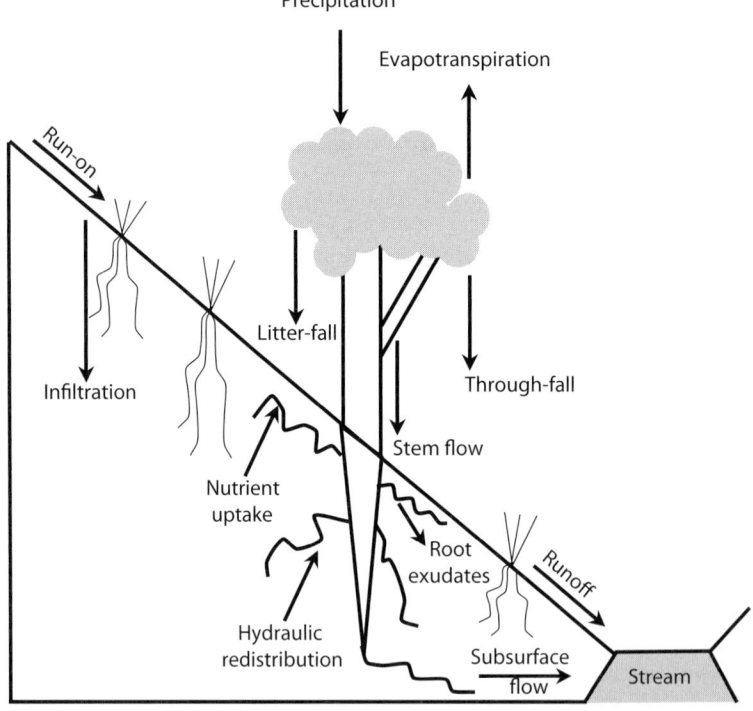

Fig. 4.2. Plants as ecosystem engineers. Arrows indicate fluxes associated with pools of soil water and nutrients. Plants modify precipitation by intercepting it, channeling it along stems, or slowing overland flow of water. Plants modify soil water through transpiration and hydraulic redistribution. Plants modify soil nutrient pools through deposition of litter, nutrient uptake, and root exudates.

can be observed in diminished stream flow during daylight hours (Bond *et al.*, 2002). Plants cast shadows, which reduce evaporation (LeMaitre *et al.*, 1999; Zhang and Schilling, 2006) and cool the soil surface (D'Odorico *et al.*, 2010; He *et al.*, 2010). Harvest of trees in semi-arid Australia causes upward movement of groundwater and soil salinization because plants are no longer using the water resource (Peck, 1978; McFarlane and Williamson, 2002). Plants use considerable mineral nutrient resources; when plants senesce in late summer, their lack of use results in an accumulation of inorganic nitrogen in the soil (Booth *et al.*, 2003b; Walvoord *et al.*, 2003; Sperry *et al.*, 2006; Hooker *et al.*, 2008; Adair and Burke, 2010). Fire is an important disturbance in many ecosystems; the density and size of plants influences frequency and intensity of fires and consequently the loss of carbon and nitrogen during fires (Cleary *et al.*, 2010; Wan *et al.*, 2001).

Plant modification of resource influx and efflux alters fluxes among resource pools. Rate and depth of water infiltration is a function of soil water content (Brooks *et al.*, 1991), which depends on plant transpiration. The slowing of surface water movement by plants (Schlesinger *et al.*, 1990; Bhark and Small, 2003; Ludwig *et al.*, 2005) reduces export of carbon and nitrogen from the point of precipitation to downstream eco-systems. Plants influence nutrient cycling by altering the C:N ratio of soils (Hooper and Vitousek, 1998; Eviner *et al.*, 2006) and consequently the production of nitrate, which is highly mobile and easily leached (Miller and Cramer, 2004).

Finally, land use and management can influence resource pools. Livestock grazing and timber harvest remove nutrients stored in plant tissue (Thompson Tew *et al.*, 1986; Milchunas and Lauenroth, 1993). Grazing can either increase or decrease soil pools of water, carbon, and nitrogen (Milchunas and Lauenroth, 1993; Naeth and Chanasyk, 1995; Schuman *et al.*, 1999; Donkor *et al.*, 2006; Bagchi and Ritchie, 2010), and increases light availability (Fahnestock and Knapp, 1994); harvest of forests reduces interception of precipitation (Brown *et al.*,

2005; LaMalfa and Ryel, 2008). Shrub-reduction treatments in semi-arid lands (Mueggler and Blaisdell, 1958; Wambolt and Payne, 1986) slow water use (Sturges, 1973, 1993; Inouye, 2006; Prevéy *et al.*, 2010a, b), but result in carbon and nitrogen loss (Tiedemann and Klemmedson, 1986; McClaran *et al.*, 2008).

Modification of a Resource Pool by One Species has Consequences for Others

Species are linked by common resource pools in an ecosystem (Bormann and Likens, 1967; Odum, 1969). Consequently, when a resource pool is modified, other species in the community will have access to a larger or smaller pool (Jones *et al.*, 1994). For example, it is often difficult for seedlings to establish when other plants have prior access to soil water (Gross and Werner, 1982; Prevéy *et al.*, 2010b); these seedlings essentially experience microsite drought, even though soil moisture conditions at the landscape scale are favorable for germination and establishment. Conversely, hydraulic redistribution allows deep-soil water, which is accessed by shrubs and trees, to be used by shallow-rooted grasses (Richard and Caldwell, 1987; Dawson, 1993; Caldwell *et al.*, 1998; Ludwig *et al.*, 2004).

The failure of a species to use a resource pool creates a pool that can be used by other species. Shrub removal results in an increase in soil water availability (Sturges, 1973, 1993) but also leads to the establishment of invasive annual species such as cheatgrass (*Bromus tectorum*) and western salsify (*Tragopogon dubius*), which require increased soil water for successful establishment (Prevéy *et al.*, 2010a, b). Soil NO_3^- concentration increases following plant senescence in autumn or removal of plants (Booth *et al.*, 2003b; Walvoord *et al.*, 2003; Sperry *et al.*, 2006; Hooker *et al.*, 2008; Adair and Burke, 2010), which favors species or genotypes most capable of using the excessive resource (Rosenzweig, 1971; Maron and Connors, 1996). For example, nitrogen influx is enhancing invasion by non-native *Phragmites* in the northeastern USA (Holdredge *et al.*, 2010).

Modification of resource pools by plants at small spatial and temporal scales can supersede large-scale influences. For example, soil water content can vary with plant communities (Caldwell *et al.*, 2008). Replacement of a conifer forest with a deciduous forest type may increase water yield by 30%, locally equivalent to 300 mm of additional precipitation (LaMalfa and Ryel, 2008). In southwestern Australia, waterlogging of soils following conversion of savanna to cropland is so profound that crop yields are negatively correlated with precipitation (McFarlane and Williamson, 2002). Tree-fall gap studies indicate the profound community change that results following sudden increases in solar flux (Hubbell *et al.*, 1999).

Plants can also influence physical and chemical properties to the benefit of other individuals of the same species. So called positive feedbacks can promote invasion by plants with apparently different life histories. In the Chihuahuan desert of North America, encroachment by creosote bush (*Larrea tridentata*) into grasslands increases minimum soil temperature and facilitates further shrub establishment (D'Odorico *et al.*, 2010; He *et al.*, 2010). In much of the Great Basin, frequent fires in cheatgrass-dominated sites not only kill young shrubs, they create a near-surface supply of NO_3^- that promotes rapid growth of this invasive grass (Booth *et al.*, 2003b; Sperry *et al.*, 2006).

Management for one species can enhance resource pool size for other species in a community. Intentionally planting the nitrogen-fixing shrub bush lupine (*Lupinus arboreus*) doubled soil NH_4^+ and altered the species composition of a coastal plant community (Maron and Connors, 1996). The exotic nitrogen-fixing tree Russian olive (*Elaeagnus angustifolia*), initially planted for stream-bank stabilization, increases soil N in riparian areas in the southwestern USA (DeCant, 2008) and fixed nitrogen has been observed in cottonwood (*Populus angustifolia*) trees near Russian olive in southern Idaho (Leffler,

unpublished data). Because these nitrogen-enhanced soils are riparian, this excess N can be exported to downstream ecosystems (Binkley *et al.*, 1992; Stein *et al.*, 2010), altering N pools in other communities.

All Species in a Community do not Rely on Identical Resource Pools

Species coexist when they are not limited by the same niche dimension (Gause, 1934; Hutchinson, 1957; Hardin, 1960; Chesson, 2000). This basic tenet of ecology theory indicates that species differ in how they use resources or perceive resource pools. These different resource requirements led Elton (1958) to propose that diverse communities are more resistant to invasion because there are no niche opportunities (Shea and Chesson, 2002); all resources are exhausted by the species in the community.

The soil water pool is commonly viewed as divided among plant functional types. Walter (1971) proposed a two-layer model of water availability, near-surface water and water at depth. Ehleringer *et al.* (1991) expanded this concept, demonstrating that different life forms (i.e. grasses and trees) draw water from different depths. Some functional groups, such as shrubs, relied on water from a range of soil depths while other species were closely tied to near-surface water (annual species), or only used surface water if it was reliably available (Williams and Ehleringer, 2000).

The recent concept of growth and maintenance pool water (Ryel *et al.*, 2008, 2010) is especially relevant to management. Rather than viewing water among different soil depths as divided among functional groups, soil water is separated into pools for different functions and explicitly links water and nutrient availability. The growth pool of water exists when soil water potential is high enough to allow diffusion of nutrients to the root surface. This water is used by all species and growth will cease when nutrients can no longer be extracted. The maintenance pool is the water that remains in the soil after growth has stopped. Species will differ in the timing of their requirements of these pools (Fig. 4.3). Annual plants do not require a maintenance pool; they senesce when the growth pool is exhausted. Perennial plants need a maintenance pool to persist through the growing season and the amount of water in this pool is a function of soil water content and root depth (Fig. 4.3). This concept can be extended further. A winter annual plant requires a 'germination' pool of water during the autumn. If this resource is limited or absent, population growth of winter annual species will suffer, yet perennial and summer annual species in the community would be

unaffected. Consequently, a land manager can examine the resource pool requirements of desired and undesired species and take action to influence appropriate resource pools.

The soil nitrogen pool may also be perceived differently by each species in a community. While soil nitrogen is typically concentrated in the uppermost soil layers (Barber, 2001; Hooker et al., 2008), it exists in multiple organic and inorganic forms. Species differ in their ability to acquire organic nitrogen (mostly as amino acids),

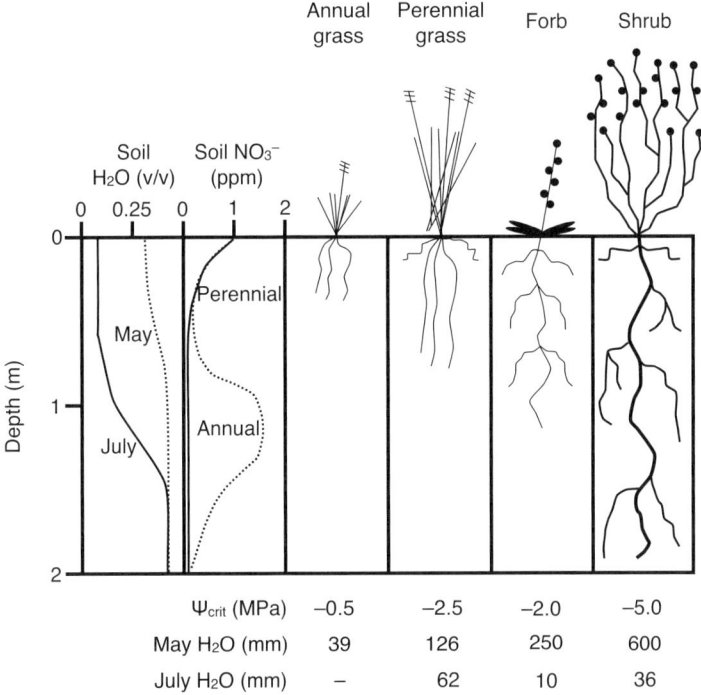

Fig. 4.3. Soil water and NO_3^- pools available to different life forms in a semi-arid ecosystem. The growth pool, water at high potential available in the near-surface soil, is similar for each species but maintenance pools differ considerably based on root depth and critical water potential (Ψ_{crit}, Ψ which causes uncontrolled xylem cavitation). For soil water content in annual grass, perennial grass, and shrub communities: values represent the difference between soil water content in the rooting zone during May, the soil water content derived from Ψ_{crit}, and depth of water extraction from Peek et al. (2005), Ryel et al. (2010), and Leffler et al. (2005), respectively. For May, we assumed water content of 30% to 40% throughout the soil profile in calculations. Calculations of Ψ_{crit} and soil water content in July based on moisture release curve in Leffler et al. (2002) and data in Ryel et al. (2010). Forb data are based on Prevéy et al. (2010a) and an assumed root depth of 1 m. For soil NO_3^-: profiles in annual and perennial communities to 1 m derived from Hooker et al. (2008); below 1 m values are hypothetical but NO_3^- reservoirs have been observed previously (Walvoord et al., 2003; Sperry et al., 2006; Graham et al., 2008). Soil NO_3^- values are appropriate for spring and early summer. Drawings not to scale.

NH_4^+, and NO_3^- (Weigelt *et al.*, 2005), potentially allowing specialization on these different forms (McKane *et al.*, 2002; Ashton *et al.*, 2010). McKane *et al.* (2002) observed species to differ in timing, depth, and form of uptake, and Aanderud and Bledsoe (2009) observed partial separation of N-form uptake between native and invasive grasses in an oak woodland in California, USA.

Finally, solar flux is a critical resource divided among species based on intensity. Early successional species, grasses, forest canopy emergent trees, and others require high intensity light while other species cannot tolerate direct sun (Hubbell *et al.*, 1999; Kobe, 1999). Consequently, high-light species create an environment only for low-light species (Messier *et al.*, 1998), essentially preempting this resource from other species requiring high light. Grass cover reduces available light and limits forb biomass in tallgrass prairie (Turner and Knapp, 1996), and MacDougall and Turkington (2005) observed subordinate species to respond to increased light rather than increased N in an oak savanna.

Interactions Among Species are Mediated by Resource Pools

The competition paradigm has dominated ecological theory since inception of ecology as a discipline (Gause, 1934; Clements, 1936; Tilman, 1982; Goldberg and Barton, 1992). Accordingly, species compete directly or indirectly for a set of resources (Pianka, 1987) and the theory implies that the likelihood of competitive exclusion increases with increasing niche overlap (Gause, 1934; Hardin, 1960; Johansson and Keddy, 1991). While competition exists (Goldberg and Barton, 1992), many authors question the importance of competition in structuring plant communities (Hutchinson, 1957; Welden and Slauson, 1986; Rees *et al.*, 1996; Damgaard and Fayolle, 2010). The crux of the argument against the importance of competition is that other factors such as abiotic stress, predators, disease, or dispersal can play a relatively stronger role (Grace,

1991; Brooker and Kikvidze, 2008) because they influence resource pools. Drought directly limits water availability and indirectly limits nutrient availability by reducing diffusion (Nye and Tinker, 1977; Barber, 2001); herbivory or disease can slow plant water use or increase light for sub-canopy species (Fahnestock and Knapp, 1994; Naeth and Chanasyk, 1995; Heil *et al.*, 2000); predation can have the opposite effect by removing herbivores that defoliate plants (McClaren *et al.*, 2008). Thus, even apparently top-down processes (Schmitz, 2008) can ultimately have bottom-up effects through their alteration of resource pools or fluxes. This resource-based framework eliminates the distinction between abiotic stress such as drought caused by physical processes (i.e. lack of rain) or by other individuals (i.e. available water was used) and focuses on the mechanistic interaction among individuals mediated by resource dynamics (Grime, 1977; Welden and Slauson, 1986).

We present this perspective because direct competition for a resource by two individuals is rare, especially for below-ground resources. First, many apparently competitive interactions are separated in time. Individuals may directly compete for soil water if they germinate together and grow at similar rates, but only after reaching sizes where resource acquisition is limited by other individuals (i.e. density dependence, Goldberg *et al.*, 2001). Commonly, however, germination is separated by several days in these individuals. Consequently, one individual germinates and begins growing into an environment previously modified (i.e. drier or shaded) by the species that germinated earlier. Annual and perennial species are often described as in competition (Corbin and D'Antonio, 2004; Blank, 2010), but perennial species are biologically active long after annual species have senesced and therefore they no longer modify water, nitrogen, or light availability for perennial species. Perennial plant species, however, may experience an environment previously modified by an annual species (e.g. low soil N or soil moisture) to the detriment of perennial species performance and subsequent fitness. Second, other competitive interactions are only temporary. For example, trees or shrubs and grasses are often described as in competition (Facelli and Tembly, 2002), but can only compete for water until the woody plants have access to deep soil water (Ehleringer *et al.*, 1991), and grasses cannot effectively limit light to mature trees. Taller statured trees, however, create a shade environment for the grasses; thus the distribution of grass species across the landscape may be affected by the shade patterns of trees. Finally, interactions among species can shift from generally negative (i.e. competition) through neutral to generally positive (facilitative) depending on abiotic conditions (Callaway and Walker, 1997; Maestre *et al.*, 2003). Individual aridland plants growing close together is often interpreted as competition for water (Welden and Slauson, 1986; Keddy, 1989), but when water was scarce, shading of grasses by shrubs reduced leaf temperature and transpiration rate, and promoted grass survival (Maestre *et al.*, 2003).

The processes of community assembly and succession are often thought to depend on availability of species, availability of sites for establishment, and performance of species at those sites (Pickett *et al.*, 1987; Sheley *et al.*, 2006). For a species to enter a community it must be present in the region and be capable of dispersing to the site. This requirement may depend on availability of plants to disperse propagules across the landscape, or the presence of vectors (Theoharides and Dukes, 2007). Sites must contain the resources necessary for germination and establishment, and resource availability must coincide with plant phenology for germination and establishment. A site previously occupied by another seed that germinated likely lacks the necessary water, nitrogen, or light required for the newly germinating individual. Finally, persistence requires different resource pools than establishment. A shrub may establish in a favorable site, but will not persist in the community if deep water is not available. An ecosystem that has a largely untapped resource such as deep-soil water will support a species that requires that resource if those

species are in the regional pool, capable of dispersing to the site, and establishing. Alternatively, if established plant species are using that resource, other species will often be inhibited.

Because resources are dynamic, communities are constantly changing. Modern ecology recognizes that communities are neither completely deterministic (Clements, 1936) nor stochastic (Gleason, 1926) collections of species (Sutherland, 1974; May, 1977). Rather, a community is constantly changing within environmental constraints. These changes are generally small such as a series of wet summers increasing grass abundance and/or biomass in a shrub/ grass community. Occasionally, changes in resources are large, shifting communities to different states (Sutherland, 1974; Westoby *et al.*, 1989; Briske *et al.*, 2003) such as fire removing the shrub component of an ecosystem and making resources available for the establishment of grasses and forbs. Whenever resource pools change through natural processes or management, changes in the community will likely follow (Davis and Pelsor, 2001). Predictable and repeatable patterns of resource dynamics result in stability of plant assemblages, while unpredictable or directional trends in resources will alter species assemblages. A key factor in community change is resource flux outside the typical range of variation (Sher and Hyatt, 1999).

Resource Pool Dynamics and Invasive Species

Changing resource availability, whether caused by natural processes or management, will ultimately result in changes in plant community composition. There is a growing body of evidence that altered resource availability is a key factor influencing colonization of a site by an invasive species (Theoharides and Dukes, 2007). In many cases, the trigger for invasion is a discrete disturbance that alters resource availability either directly, or disrupts plant communities to such an extent that supply and use of a single or multiple resources is altered (White and Pickett, 1985; Hobbs and Huenneke, 1992; Sher and Hyatt, 1999; Adair *et al.*, 2008).

Disturbance is often a prerequisite for invasion (Hobbs and Huenneke, 1992; Davis *et al.*, 2000), and it can occur through natural or anthropogenic processes. Studies often attribute plant invasion to land uses such as livestock grazing (Young *et al.*, 1972; Chambers *et al.*, 2007), cultivation for crops (Williamson and Fitter, 1996; see Morris, Chapter 3, this volume), and recreation (Vilá and Pujadas, 2001). While these land uses have profoundly changed plant communities, from the perspective of altered resource availability, natural processes such as fire (Zedler and Scheid, 1988; Chambers *et al.*, 2007) or drought (Allen and Breshears, 1998) can also promote long-term vegetation change. The nature of the disturbance is less important to plants than the consequences that the disturbance has for resource pools.

All disturbances are not equivalent; some have large influence on resource pools, while others have only minor consequences. Drought can influence landscapes, and if persistent, can alter the nature of a plant community for decades to centuries. The 1950s drought in the southwestern USA shifted the woodland-forest ecotone by over 2 km in distance and 200 m in elevation (Allen and Breshears, 1998). In addition, the influence of livestock grazing can range from minimal to intense. Light grazing likely removes only a small portion of plant biomass and has minimal effect on soil water, while heavy grazing reduces transpiration and water infiltration (Naeth and Chanasyk, 1995; Chanasyk *et al.*, 2004) and increases light availability (Fahnestock and Knapp, 1994). Even light grazing, however, can create bare patches or depressions in soils where water may accumulate, providing opportunities for exotic plants to establish if seed sources are available (Hobbs and Huenneke, 1992). Old-field studies illustrate the influence of former land cultivation practices on soil structure, water-holding capacity, and nutrient content. In many cases, the influence of agriculture on subsequent plant communities is evident for over 100 years (Elmore *et al.*, 2006; see Morris, Chapter 3, this volume).

Plant invasion following disturbance and a change in resource availability is a natural process that would occur even if exotic species were absent. Succession following disturbance can proceed linearly from short-lived, rapidly growing species that can quickly colonize bare soil to larger, slow-growing species, tolerant of high plant density (Tansley, 1935; Clements, 1936; Connell and Slayter, 1977); or non-linearly through multiple stable states (Sutherland, 1974; Westoby *et al.*, 1989; Briske *et al.*, 2003). These previous invasions, however, were by local or regional species; contemporary invasion is often by exotic species, creating novel communities with resource pool dynamics that do not have historical or evolutionary precedent at the site of invasion. The same basic ecological processes operate in native and exotic plant communities (Gurevitch *et al.*, 2011), but there has been insufficient time for an evolutionary response by native species to exotics (Crawley *et al.*, 1996).

Invasion involves a process of matching changes in resource pools with species that can take advantage of those resources. Only a small fraction of exotic species are invasive (Williamson and Fitter, 1996; Levine *et al.*, 2003). Many of these invasive exotic species can be broadly described as having an r-selected life history (MacArthur and Wilson, 1967) or as ruderals (Grime, 1977). Many invasive exotic species can be categorized as 'acquisitive' on the leaf economics spectrum (Wright *et al.*, 2004) and traits such as rapid growth, early reproduction, and high fecundity are almost universally found in invasive species (Pyšek and Richardson, 2007), but other traits such as germination timing, leaf area, and biomass differ little between native and invasive exotic species (Pyšek and Richardson, 2007), and a universal complement of invasive traits has yet to be found (Tecco *et al.*, 2010). The key to understanding if an exotic species will become invasive is not the traits it possesses alone, but if those traits allow the exotic species to exploit an available resource pool (Heger and Trepl,

2003; Roscher *et al.*, 2009). Hence recent studies link invasion to increased water availability following shrub removal (Prevéy *et al.*, 2010a, b) and increased N availability following herbivory by insects (Brown, 1994).

The resource pool perspective for plant invasion has several consequences for management. In addition to taking advantage of favorable resource pools during establishment, an invasive species will further modify these pools, promoting subsequent dominance, impact, and spread. Consequently, invasion by few individuals of one species may promote spread (D'Odorico *et al.*, 2010; He *et al.*, 2010) or secondary invasion by other species. Removal of shrubs creates a deep-soil pool of water that can be tapped by exotic annual and perennial forbs, including *Tragapogon dubius* or *Centaurea maculosa* (Hill *et al.*, 2006; Kulmatiski *et al.*, 2006), which are invasive in western North America. Also, because transitions among stable community states can be influenced by resource pools (i.e. abiotic mechanisms, Briske *et al.*, 2006), opportunities for restoration of native species may be more likely when favorable resource conditions exist for the establishment of the desirable species. For example, abundance of cheatgrass in the Intermountain West, USA can vary considerably among years at the same site (Griffith and Loik, 2010), and cheatgrass population die-offs can occur, creating establishment opportunities for other species, including the native shrub big sagebrush.

Managing Resource Pools to Influence Invasive Plant Abundance

Manipulation of resource pools can be a powerful tool to manage invasive plant species. Resource availability can be managed by either directly manipulating the resource pools or their flux, or indirectly by managing plants and animals to affect a pool or flux. A resource pool approach to management of invasive species is especially powerful because it is based on fundamental ecological principles (Sheley *et al.*, 2006; Gurevitch *et al.*, 2011) that are responsible for community assembly and plant invasion.

Resource pools or fluxes can be directly managed. Manipulation of nitrogen availability through the addition of labile carbon such as sucrose, sawdust, or wood chips can favor the slow growth of native species, a trait more important when nitrogen is not abundant (Perry *et al.*, 2010). Supplemental water can be applied when desirable species are active and invasive annual plants have already senesced, or supplemental water can be injected at soil depths that native species can access but invasive species cannot. These efforts, however, would be costly and indirect management of resource pools may be a better option.

Resource pools and fluxes can be managed indirectly through the manipulation of plants and animals. Grazing, mowing, or burning can be used to remove nitrogen from soils (Marrs, 1993; Augustine, 2003; Härdtle *et al.*, 2006), especially if plants high in nitrogen can be targeted for removal. Plants that contain a high C:N ratio can be established, which will ultimately increase the C:N ratio of the soil and further promote immobilization of N by soil microbes (Zink and Allen, 1998; Perry *et al.*, 2010). Reducing the biomass of shrubs, grasses, and forbs can increase near-surface soil water, while reducing shrub and tree biomass can increase availability of deep-soil water (Sturges, 1993; Leffler *et al.*, 2005; Prevéy *et al.*, 2010a, b). Alternatively, managers can increase near-surface soil water availability in the early summer by reducing the biomass of annual species that rapidly use water in the spring (Harris, 1967; Booth *et al.*, 2003a).

Successful management of invasive species through resource manipulation will require careful identification of the critical resource pools. In most instances, the resources required for plant establishment are different from the resources required for long-term persistence (Holt, 2009). However, it is long-term resource modification by desired species that will reduce the probability of invasion. Hence, adopting a systems approach (Odum, 1994) to resource pool management will require separate identification of the resource pools that are necessary for establishment and persistence of desired species. For example, native grasses in Oregon, USA have high germination rate, low survival immediately following germination, but high survival once established (James *et al.*, 2011). Consequently, the critical resource pool for these native grasses may be moisture at the site of germination in the spring. Moisture status can be improved by removal of competing vegetation or ensuring seed contact with the soil.

Management of a community for long-term resistance to invasion will also require a systems approach. A resistant community will have few available resources that an invasive species can exploit (Elton, 1958; Davis *et al.*, 2000; Shea and Chesson, 2002). This low available-resource state arises not from a diversity of species, but from a diversity of resource-use patterns that coincide with spatial or temporal aspects of the establishment and growth of potential invasive plant species. Rather than approaching an invasive species problem with the goal of establishing plant species that are matched for a specific use, managers need to consider the resource pools present on the landscape and establish desired species that will fully use the resource pools. While this approach may not maximize short-term gain, it will enhance probability of long-term sustainability.

Management Scenarios

We present three scenarios to illustrate the use of resource pools in ecosystem management. Scenarios are based on conceptual models, developed from relevant studies, which are the core tools for understanding resource pools dynamics as a function of vegetation composition, climate, and land management practices. The scenarios describe specific situations, but the reasoning and approach can be applied to other situations.

Soil water pools in the sagebrush steppe ecosystem

Native sagebrush steppe ecosystems in western North America contain a mixture of perennial tussock grasses and big sagebrush (*Artemisia tridentata*), with a variety of less dominant perennial forbs. Native annual species are relatively rare in this ecosystem (West and Young, 2000). Much of this ecosystem type is now dominated by nearly monotypic stands of big sagebrush or cheatgrass (*Bromus tectorum*), an invasive winter annual species. Perennial grasses have largely been lost from this ecosystem in many areas; a change partially attributed to limited grazing tolerance and altered fire regimes (Chambers *et al.*, 2007).

Soil water recharge in these ecosystems is dominated by autumn, winter, and spring precipitation (Caldwell, 1985) that results in accumulation of water during non-growth periods, top-down recharge of a shallow growth pool (surface to 30 cm), and a deeper maintenance pool (to 1–1.5 m). When not intensively disturbed, the intact perennial grass/forb/shrub community rapidly uses the growth pool in the spring, draws the maintenance pool to low water potentials during summer, and little soil water remains in the autumn (Seyfried *et al.*, 2005). Regeneration of sagebrush or grasses occurs in small gaps where resources are available in the spring, and may be enhanced by larger gaps formed by recently deceased tussock grasses or shrubs.

While all species use water from the growth pool, the deeper water is used primarily by sagebrush and can be drawn to water potentials well below the wilting point of grasses and forbs. When the water potential of the growth pool is insufficient for nutrient diffusion, the grasses and forbs produce seed and become dormant (Ryel *et al.*, 2010). Sagebrush, however, relies on the maintenance pool for limited physiological activity throughout the summer and into the autumn. Consequently, the maintenance pool is largely depleted as well. Invasion and dominance by exotic species is difficult because few resources remain for them to exploit (Chambers *et al.*, 2007; Prevéy *et al.*, 2010a, b).

When perennial grasses and forbs are removed, growth of the remaining sagebrush is temporarily enhanced because it has exclusive access to the growth pool (Ryel *et al.*, 2010). Higher sagebrush biomass increases demand on the maintenance pool, which is unaffected by the loss of the grasses. Consequently, the maintenance pool cannot supply adequate water to allow sagebrush to persist through the summer; xylem cavitation (Sperry and Hacke, 2002) and death of branches occurs. Plants in this condition are experiencing severe late-summer drought, but not from lack of precipitation; rather water stress arises from excessive use of the maintenance pool due to greater shrub biomass. In subsequent years, use of the growth pool begins to diminish due to declining shrub condition.

At this point, the stands are very susceptible to invasion from annual species such as cheatgrass that can easily exploit the underutilized growth pool. Once cheatgrass biomass is sufficient to carry fire the remaining mature sagebrush will be killed; a dense monoculture of cheatgrass often follows. These dense, contiguous stands of cheatgrass prevent establishment of sagebrush seedlings, as resource-rich gaps are no longer available. Cheatgrass very effectively depletes the growth water pool, but uses little of the maintenance pool. As such, this water can accumulate in deeper soil layers and provide resources to subsequent invaders such as deeper-rooted biennial species (Prevéy *et al.*, 2010a, b), or may eventually connect to deep ground-water pools where water quality may be low (e.g. saline) and the resulting mixed water inhibits plant growth (Ryel *et al.*, 2010).

Sagebrush steppe as a model system indicates the importance of managing the growth soil water pool. When this pool becomes available on a large scale, invasion by annual grasses is likely if propagules are available in the regional species pool. The simplest management option is to ensure that perennial grasses and forbs are not removed; grazing should be carefully managed to limit impact on growth and seed production, and sites should be allowed enough time to recover from use. If perennial grasses are removed, seeding should be undertaken very quickly, especially in shrub interspaces where growth pool water may be more available, or seed could be added regularly as a precaution. Large-scale treatments to reduce shrub density should not be undertaken if the growth pool is or can be dominated by annual grasses. If treatments are necessary, perennial grasses should be established first.

Soil N in annual and perennial grass communities

Grasslands of the Great Plains are dense communities composed of cool and warm season species. Cool-season grasses can remain green under snow, are physiologically active in the spring, and senesce during late spring and early summer; warm-season grasses are physiologically active in response to summer rains. In these systems, there are active species throughout the growing season and soil-N fluctuations are minimal (McCulley *et al.*, 2009).

When plant growth ceases, soil NO_3^- increases because plants are no longer using this resource (Booth *et al.*, 2003b; Sperry *et al.*, 2006; Adair and Burke, 2010), but nitrification can continue at low soil water potential (Low *et al.*, 1997). Although grazing can have long-term positive or negative effects on soil N (Milchunas and Lauenroth, 1993), an immediate effect of excessive grazing is reduced uptake of N by perennial grasses, which can result in high soil NO_3^- availability (Booth *et al.*, 2003b; Sperry *et al.*, 2006; Adair and Burke, 2010). These systems become invasible by a species that can respond to high N availability with vigorous growth.

Short-lived annual species typically have high relative growth rates (Wright *et al.*, 2004; James, 2008) and can grow rapidly in high-N environments (Maron and Connors, 1996). In many cases, annual grasses will respond to increased N availability to a greater extent than perennial grasses. If an invasive annual species is in the regional species pool, it can more effectively take advantage of the increased N availability following perennial species loss. In subsequent years, high soil NO_3^- in the late summer and autumn will become common and serve to perpetuate annual grass dominance.

This model of annual grass invasion illustrates the importance of careful management of soil N pools. In an intact system, N fluctuations are minimal, but the N resource fluctuates in space and time in a disturbed system, creating windows of opportunity for invasive species (Davis et al., 2000; Davis and Pelsor, 2001). Of course, the best management option is to prevent loss of perennial species in the first place, ensuring that soil N remains low, a difficult proposition in regions with high N deposition (Matson et al., 2002). If a system is already in a poor state, however, other options need to be considered to reduce soil N.

Several methods have been used, each intending to increase the ratio of carbon to nitrogen in soils degraded by annual grasses. Most commonly, addition of carbon-rich compounds such as sugar or sawdust stimulates immobilization of mineral nitrogen by bacteria (Zink and Allen, 1998; Klironomos, 2002; Mazzola et al., 2011). Carbon addition can be effective at limiting N availability to annual grasses, but it must be combined with seeding of perennial grasses to ensure annual species simply do not recolonize and dominate the limited soil N pool. More importantly, however, would be the timing of carbon addition. This treatment should be applied to specifically deny N to annual species when they require it for establishment in the autumn. Favoring perennial grasses over annual species will naturally begin to stabilize soil NO_3^- as well. It is also pos-sible to remove nutrients by removing biomass with fire, mowing, or grazing (Perry et al., 2010). These efforts, however, will likely remove only a small amount of the excess N.

Establishment and maintenance water pool in riparian forests

Alteration of stream flow has led to dramatic declines in riparian forests of the western USA (Webb and Leake, 2006). These groundwater-dependent ecosystems, historically composed of a cottonwood (Populus sp.) overstory and an understory of willow (Salix sp.), have been replaced by the invasive shrub salt cedar (Tamarix chinensis) and the invasive tree Russian olive (Elaeagnus angustifolia). Gallery forests, which once extended for hundreds of meters from the main channel of large rivers such as the Rio Grande in North America, are now restricted to narrow bands less than 50 m wide (Howe and Knopf, 1991). These forests have been lost because of upstream dams, channelization, and water diversion for culinary and agriculture use (Webb and Leake, 2006).

Water in riparian forests is supplied by both local and upstream sources (Leffler and Evans, 1999). The growth pool can be recharged by local precipitation or flooding of the forest during spring snowmelt several hundred kilometers away. The growth pool also may be maintained well into the summer where monsoon precipitation is substantial. The maintenance pool, however, is supplied almost exclusively by ground-water derived from off-site sources. Large trees such as older cottonwoods require this maintenance pool and are highly vulnerable to xylem cavitation when groundwater is limited (Tyree et al., 1994; Leffler et al., 2000). Thus, older individuals may die from drought stress. New recruits, however, face additional challenges because they require the disturbance of a scouring flood, essentially an establishment pool of water, to coincide with a short window of seed viability (Bradley and Smith, 1986). In the absence of these floods, largely eliminated by dams, cottonwoods fail to establish. If they do emerge, root extension must keep pace with accelerated groundwater decline (Scott et al., 1997; Stella et al., 2010) or the seedlings will die.

Salt cedar and Russian olive modify the growth pool to the detriment of cottonwoods and willows. Salt cedar is tolerant of increasingly saline soils caused in part by runoff from agriculture (Shafroth *et al.*, 1995). This species deposits salt on the soil surface when it sheds leaves in the autumn (Shafroth *et al.*, 1995). Because salt effectively lowers the soil water potential, cottonwood and willow seedlings experience drought even when water is abundant, i.e. the growth pool shrinks for the native species. Not only are water pools altered, Russian olive is capable of nitrogen fixation (Katz and Shafroth, 2003) and its growth is not limited by soil nitrogen pools. Consequently, diffusion of N in low water potential soils does not limit its access to the growth pool and it can promote rates of water table decline in excess of that which native, non-nitrogen fixing species can survive.

The key to managing riparian forests for native cottonwood and willow species is periodic flooding to allow establishment and occasional maintenance of a high water table to allow persistence. This requires management primarily of off-site water to enhance the maintenance pool. Consequently, region-wide water management plans are essential because so many stakeholders compete for access to this limited resource. Active removal of invasive species is also necessary for successful restoration because salt cedar and Russian olive have a substantial influence on the growth pool of water.

Summary

While the assertion that resource availability influences invasion (Davis *et al.*, 2000; Davis and Pelsor, 2001) is not new, we articulate a view of resource dynamics that is critical for management of invasive-dominated communities and for initial prevention of invasion. The construct of resource pools allows for assessing functionally important elements of necessary and limiting resources for native species and exotic species that either are, or may become invasive. From this view, it is apparent that the invasion process requires resources to become available and the presence of an exotic species capable of using the resource. The concept of resource pool dynamics allows managers to better understand the implications of management decisions on invasion by exotic species, assess the role of factors outside management influence such as climate and weather on invasion, and comprehend the challenges in restoration of a degraded community. Resource availability will always vary in space and time, but invasion is likely to occur when these dynamics are abruptly disrupted or changed directionally (i.e. nitrogen becomes more available over time). Moreover, management to transition invaded communities to a more desirable state should be taken when resource availability favors that transition.

Relevant management decisions can be made when resource pools are considered rather than merely applying land management treatments to reduce the abundance of certain species or functional types, or modifying land uses. For each decision, a manager needs to ask, 'How do I need to change water and nutrient pools to achieve my desired objective and how will other management activities affect pools of water and nutrients?' Management tools can therefore be more varied and targeted, potentially giving managers additional options and more effective treatments.

Acknowledgements

We thank T.A. Monaco and R.L. Sheley for conceiving of this project and their valuable editorial input. Previous versions of this manuscript were improved with the help of J.J. James and E. Espeland.

References

Aanderud, Z.T. and Bledsoe, C.S. (2009) Preferences for N-15-ammonium, N-15-nitrate, and N-15-glycine differ among dominant exotic and subordinate native grasses from a California oak woodland. *Environmental and Experimental Botany* 65, 205–209.

Adair, E.C. and Burke, I.C. (2010) Plant phenology and life span influence soil pool dynamics: *Bromus tectorum* invasion of perennial C3–C4 grass communities. *Plant and Soil* 335, 255–269.

Adair, E.C., Burke, I.C. and Lauenroth, W.K. (2008) Contrasting effects of resource availability and plant mortality on plant community invasion by *Bromus tectorum* L. *Plant and Soil* 304, 103–115.

Adler, P.B., Ellner, S.P. and Levine, J.M. (2010) Coexistence of perennial plants: an embarrassment of niches. *Ecology Letters* 13, 1019–1029.

Allen, C.D. and Breshears, D.D. (1998) Drought-induced shift of a forest-woodland ecotone: rapid landscape response to climate variation. *Proceedings of the National Academy of Science* 95, 14839–14842.

Ashton, I.W., Miller, A.E., Bowman, W.D. and Suding, K.N. (2010) Niche complementarity due to plasticity in resource use: plant partitioning of chemical N forms. *Ecology* 91, 3252–3260.

Augustine, D.J. (2003) Long-term, livestock-mediated redistribution of nitrogen and phosphorus in an East African savanna. *Journal of Applied Ecology* 40, 137–149.

Azevedo, J. and Morgan, D.L. (1974) Fog precipitation in coastal California forests. *Ecology* 55, 1135–1141.

Bagchi, S. and Ritchie, M.E. (2010) Introduced grazers can restrict potential soil carbon sequestration through impacts on plant community composition. *Ecology Letters* 13, 959–968.

Baird, A.J. and Wilby, R.L. (1999) *Eco-hydrology: plants and water in terrestrial and aquatic environments*. Routledge, London.

Barber, S.A. (2001) *Soil Nutrient Bioavailability: A Mechanistic Approach*. John Wiley and Sons, New York.

Beven, K. and Germann, P. (1982) Macropores and water flows in soils. *Water Resources Research* 18, 1311–1325.

Bhark, E.W. and Small, E.E. (2003) Association between plant canopies and the spatial patterns of infiltration in shrubland and grassland of the Chihuahuan Desert, New Mexico. *Ecosystems* 6, 185–196.

Binkley, D., Sollins, P., Bell, R., Sachs, D. and Myrold, D. (1992) Biogeochemistry of adjacent conifer and alder-conifer stands. *Ecology* 73, 2022–2033.

Blank, R.R. (2010) Intraspecific and interspecific pair-wise seedling competition between exotic annual grasses and native perennials: plant-soil relationships. *Plant and Soil* 326, 331–343.

Bond, B.J., Jones, J.A., Moore, G., Phillips, N., Post, D. and McDonnell, J.J. (2002) The zone of vegetation influence on baseflow revealed by diel patterns of streamflow and vegetation water use in a headwater basin. *Hydrological Processes* 16, 1671–1677.

Booth, M.S., Caldwell, M.M. and Stark, J.M. (2003a) Overlapping resource use in three Great Basin species: implications for community invasibility and vegetation dynamics. *Journal of Ecology* 91, 36–48.

Booth, M.S., Stark, J.M. and Caldwell, M.M. (2003b) Inorganic N turnover and availability in annual- and perennial-dominated soils in a northern Utah shrub-steppe ecosystem. *Biogeochemistry* 66, 311–330.

Bormann, F.H. and Likens, G.E. (1967) Nutrient cycling. *Science* 155, 424–429.

Bosch, J.M. and Hewlett, J.D. (1982) A review of catchment experiments to determine the effect of vegetation changes on water yield and evapotranspiration. *Journal of Hydrology* 55, 3–23.

Bradley, C.E. and Smith, D.G. (1986) Plains cottonwood recruitment and survival on a prairie meandering river floodplain, Milk River, southern Alberta and northern Montana. *Canadian Journal of Botany* 64, 1433–1442.

Briske, D.D., Fuhlendorf, S.D. and Smeins, F.E. (2003) Vegetation dynamics on rangelands: a critique of the current paradigms. *Journal of Applied Ecology* 40, 601–614.

Briske, D.D., Fuhlendorf, S.D. and Smeins, F.E. (2006) A unified framework for assessment and application of ecological thresholds. *Rangeland Ecology and Management* 59, 225–236.

Brooker, R.W. and Kikvidze, Z. (2008) Importance: an overlooked concept in plant interaction research. *Journal of Ecology* 96, 703–708.

Brooks, K.N., Ffolliott, P.F., Gregersen, H.M. and Thames, J.L. (1991) *Hydrology and Management of Watersheds*. Iowa State University Press, Ames, Iowa.

Brown, A.E., Zhang, L., McMahon, T.A., Western, A.W. and Vertessy, R.A. (2005) A review of paired catchment studies for determining changes in water yield resulting from alterations in vegetation. *Journal of Hydrology* 310, 28–61.

Brown, D.G. (1994) Beetle folivory increases

resource availability and alters plant invasion in monocultures of goldenrod. *Ecology* 75, 1673–1683.

Burke, I.C. (1989) Control of nitrogen mineralization in a sagebrush steppe landscape. *Ecology* 70, 1115–1126.

Caldwell, M. (1985) Cold desert. In: Chabot, B.F. and Mooney, H.A. (eds) *Physiological Ecology of North American Plant Communities.* Chapman and Hall, New York, pp. 198–212.

Caldwell, M.M., Dawson, T.E. and Richards, J.H. (1998) Hydraulic lift: consequences of water efflux from the roots of plants. *Oecologia* 113, 151–161.

Caldwell, T.G., Young, M.H., Zhu, J. and McDonald, E.V. (2008) Spatial structure of hydraulic properties from canopy to interspace in the Mojave Desert. *Geophysical Research Letters* 35, L19406.

Callaway, R.M. and Walker, L.R. (1997) Competition and facilitation: a synthetic approach to interactions in plant communities. *Ecology* 78, 1058 1066.

Campbell, G.S. (1977) *An Introduction to Environmental Biophysics.* Springer-Verlag, Berlin.

Chambers, J.C., Roundy, B.A., Blank, R.R., Meyer, S.E. and Whittaker, A. (2007) What makes great basin sagebrush ecosystems invasible by *Bromus tectorum*? *Ecological Monographs* 77, 117–145.

Chanasyk, D.S., Mapfumo, E., Willms, W.D. and Naeth, M.A. (2004) Quantification and simulation of soil water on grazed fescue watersheds. *Rangeland Ecology and Management* 57, 169–177.

Chapin III, F.S., Matson, P.A. and Mooney, H.A. (2002) *Principles of Terrestrial Ecosystem Ecology.* Springer, New York.

Chesson, P. (2000) Mechanisms of maintenance of species diversity. *Annual Review of Ecology and Systematics* 31, 343–366.

Cleary, M.B., Pendall, E. and Ewers, B.E. (2010) Aboveground and belowground carbon pools after fire in mountain big sagebrush steppe. *Rangeland Ecology and Management* 63, 187–196.

Clements, F.E. (1936) Nature and structure of the climax. *The Journal of Ecology* 24, 252–284.

Connell, J.H. and Slayter, R.O. (1977) Mechanisms of succession in natural communities and their role in community stability and organization. *The American Naturalist* 111, 1119–1144.

Corbin, J.D. and D'Antonio, C.M. (2004) Competition between native perennial and exotic annual grasses: implications for an historic invasion. *Ecology* 85, 1273–1283.

Crawley, M.J., Harvey, P.H. and Purvis, A. (1996) Comparative ecology of the native and alien floras of the British Isles. *Philosophical Transactions of the Royal Society of London, Series B* 351, 1251–1259.

Damgaard, C. and Fayolle, A. (2010) Measuring the importance of competition: a new formulation of the problem. *Journal of Ecology* 98, 1–6.

Davis, M.A. and Pelsor, M. (2001) Experimental support for a resource-based mechanistic model of invasibility. *Ecology Letters* 4, 421–428.

Davis, M.A., Grime, J.P. and Thompson, K. (2000) Fluctuating resources in plant communities: a general theory of invasibility. *Journal of Ecology* 88, 528–534.

Dawson, T.E. (1993) Hydraulic lift and water use by plants: implications for water balance, performance and plant-plant interactions. *Oecologia* 95, 565–574.

DeCant, J.P. (2008) Russian olive, *Elaeagnus angustifolia*, alters patterns in soil nitrogen pools along the Rio Grande river, New Mexico, USA. *Wetlands* 28, 896–904.

Dickson, T.L. and Foster, B.L. (2011) Fertilization decreases plant biodiversity even when light is not limiting. *Ecology Letters* 14, 380–388.

D'Odorico, P., Fuentes, J.D., Pockman, W.T., Collins, S.L., He, Y., Medeiros, J.S., DeWekker, S. and Litvak, M.E. (2010) Positive feedback between microclimate and shrub encroachment in the northern Chihuahuan desert. *Ecosphere* 1, Art. 17.

Donkor, N.T., Hudson, R.J., Bork, E.W., Chanasyk, D.S. and Naeth, M.A. (2006) Quantification and simulation of grazing impacts on soil water in boreal grasslands. *Journal of Agronomy and Crop Science* 192, 192–200.

Duff, G.A., Myers, B.A., Williams, R.J., Eamus, D., O'Grady, A. and Fordyce, I.R. (1997) Seasonal patterns in soil moisture, vapor pressure deficit, tree canopy cover and pre-dawn water potential in a northern Australia savanna. *Australian Journal of Botany* 45, 211–224.

Eamus, D., Hatton, T., Cook, P. and Colvin, C. (2006) *Ecohydrology: Vegetation Function, Water and Resource Management.* CSIRO Publishing, Collingwood, Victoria, Australia.

Ehleringer, J.R., Phillips, S.L., Schuster, W.S.F. and Sandquist, D.R. (1991) Differential utilization of summer rains by desert plants. *Oecologia* 88, 430–434.

Elmore, A.J., Mustard, J.F., Hamburg, S.P. and Manning, S.J. (2006) Agricultural legacies in the Great Basin alter vegetation cover, composition, and response to precipitation. *Ecosystems* 9, 1231–1241.

Elton, C.S. (1958) *The Ecology of Invasions by Plants and Animals*. Methuen & Co., London.

Eviner, V.T. (2004) Plant traits that influence ecosystem processes vary independently among species. *Ecology* 85, 2215–2229.

Eviner, V.T., Chapin III, F.S. and Vaughn, C.E. (2006) Seasonal variations in plant species effects on soil N and P dynamics. *Ecology* 87, 974–986.

Ewing, H.A., Weathers, K.C., Templer, P.H., Dawson, T.E., Firestone, M.K., Elliott, A.M. and Boukili, V.K.S. (2009) Fog water and ecosystem function: heterogeneity in a California redwood forest. *Ecosystems* 12, 417–433.

Facelli, J.M. and Temby, A.M. (2002) Multiple effects of shrubs on annual plant communities in arid lands of South Australia. *Austral Ecology* 27, 422–432.

Fahnestock, J.T. and Knapp, A.K. (1994) Plant responses to selective grazing by bison: interactions between light, herbivory and water stress. *Vegetatio* 115, 123–131.

Field, C.B., Chapin III, F.S., Matson, P.A. and Mooney, H.A. (1992) Response of terrestrial ecosystems to the changing atmosphere: a resource-based approach. *Annual Review of Ecology and Systematics* 23, 210–235.

Gause, G.F. (1934) *The Struggle for Existence*. Zoological Institute of the University of Moscow, Moscow.

Gleason, H.A. (1926) The individualistic concept of the plant association. *Bulletin of the Torrey Botanical Club* 53, 7–26.

Goldberg, D.E. and Barton, A.M. (1992) Patterns and consequences of interspecific competition in natural communities: a review of field experiments with plants. *The American Naturalist* 139, 771–801.

Goldberg, D.E., Turkington, R., Olsvig-Whittaker, L. and Dyer, A.R. (2001) Density dependence in an annual plant community: variation among life history stages. *Ecological Monographs* 71, 423–446.

Grace, J.B. (1991) A clarification of the debate between Grime and Tilman. *Functional Ecology* 5, 583–587.

Graham, R.C., Hirmas, D.R., Wood, Y.A. and Amrhein, C. (2008) Large near-surface nitrate pools in soils capped by desert pavement in the Mojave Desert, California. *Geology* 36, 259–262.

Griffith, A.B. and Loik, M.E. (2010) Effects of climate and snow depth on *Bromus tectorum* population dynamics at high elevation. *Oecologia* 164, 821–832.

Grime, J.P. (1977) Evidence for the existence of three primary strategies in plants and its relevance to ecological and evolutionary theory. *The American Naturalist* 111, 1169–1194.

Gross, K.L. and Werner, P.A. (1982) Colonizing abilities of 'biennial' plant species in relation to ground cover: implications for their distributions in a successional sere. *Ecology* 63, 921–931.

Gurevitch, J., Fox, G.A., Wardle, G.M., Inderjit and Taub, D. (2011) Emergent insights from the synthesis of conceptual frameworks for biological invasions. *Ecology Letters* 14, 407–418.

Hardin, G. (1960) The competitive exclusion principle. *Science* 131, 1292–1297.

Härdtle, W., Niemeyer, M., Niemeyer, T., Assmann, T. and Fottner, S. (2006) Can management compensate for atmospheric nutrient deposition in heathland ecosystems. *Journal of Applied Ecology* 43, 759–769.

Harris, G.A. (1967) Some competitive relationships between *Agropyron spicatum* and *Bromus tectorum*. *Ecological Monographs* 37, 89–111.

Haynes, R.J. (1986) The decomposition process: mineralization, immobilization, humus formation, and degradation. In: Haynes R.J. (ed.) *Mineral Nitrogen in the Plant-Soil System*. Academic Press, Orlando, Florida, pp. 52-126.

He, Y., D'Odorico, P., De Wekker, S.F.J., Fuentes, J.D. and Litvak, M. (2010) On the impact of shrub encroachment on microclimate conditions in the northern Chihuahuan desert. *Journal of Geophysical Research* 115, D21120.

Heger, T. and Trepl, L. (2003) Predicting biological invasions. *Biological Invasions* 5, 313–321.

Heil, M., Hilpert, A., Kaiser, W. and Linsenmair, K.E. (2000) Reduced growth and seed set following chemical induction of pathogen defence: does systemic acquired resistance (SAR) incur allocation costs. *Journal of Ecology* 88, 645–654.

Hill, J.P., Germino, M.J., Wraith, J.M., Olson, B.E. and Swan, M.B. (2006) Advantages in water relations contribute to greater photosynthesis in *Centaurea maculosa* compared with established grasses. *International Journal of Plant Science* 167, 269–277.

Hobbs, R.J. and Huenneke, L.F. (1992) Disturbance, diversity, and invasion: implications for conservation. *Conservation Biology* 6, 324–337.

Holdredge, C., Bertness, M.D., von Wettberg, E. and Silliman, B.R. (2010) Nutrient enrichment enhances hidden differences in phenotype to drive a cryptic plant invasion. *Oikos* 119, 1776–1784.

Holt, R.D. (2009) Bringing the Hutchinsonian niche into the 21st century: ecological and evolutionary perspectives. *Proceedings of the National Academy of Science* 106, 19659–19665.

Hooker, T., Stark, J., Norton, U., Leffler, A.J., Peek, M. and Ryel, R. (2008) Distribution of ecosystem C and N within contrasting vegetation types in a semiarid rangeland in the Great Basin, USA. *Biogeochemistry* 90, 291–308.

Hooper, D.U. and Vitousek, P.M. (1998) Effects of plant composition and diversity on nutrient cycling. *Ecological Monographs* 68, 121–149.

Howe, W.H. and Knopf, F.L. (1991) On the imminent decline of Rio Grande cottonwoods in central New Mexico. *The Southwestern Naturalist* 36, 218–224.

Hubbell, S.P. (2001) *The Unified Theory of Biodiversity and Biogeography*. Princeton University Press, Princeton, New Jersey.

Hubbell, S.P., Foster, R.B., O'Brien, S.T., Harms, K.E., Condit, R., Wechsler, B., Wright, S.J. and Loo de Lao, S. (1999) Light-gap disturbances, recruitment limitation, and tree diversity in a neotropical forest. *Science* 283, 554–557.

Hutchinson, G.E. (1957) Concluding Remarks. *Cold Spring Harbor Symposium on Quantitative Biology* 22, 415–427.

Inouye, R.S. (2006) Effects of shrub removal and nitrogen addition on soil moisture in sagebrush steppe. *Journal of Arid Environments* 65, 604–618.

James, J.J. (2008) Effect of soil nitrogen stress on the relative growth rate of annual and perennial grasses in the Intermountain West. *Plant and Soil* 310, 201–210.

James, J.J., Svejcar, A.J. and Rinella, M.J. (2011) Demographic processes limiting seedling recruitment in arid grassland restoration. *Journal of Applied Ecology* 48, 961–969.

Johansson, M.E. and Keddy, P.A. (1991) Intensity and asymmetry of competition between plant pairs of different degrees of similarity: an experimental study on two guilds of wetland plants. *Oikos* 60, 27–34.

Jones, C.G., Lawton, J.H. and Shachak, M. (1994) Organisms as ecosystem engineers. *Oikos* 69, 373–386.

Katz, G.L. and Shafroth, P.B. (2003) Biology, ecology and management of *Elaeagnus angustifolia* L. (Russian olive) in western North America. *Wetlands* 23, 763–777.

Keddy, P. (1989) *Competition*. Chapman & Hall, London.

Klironomos, J.N. (2002) Feedback with soil biota contributes to plant rarity and invasiveness in communities. *Nature* 417, 67–70.

Kobe, R.K. (1999) Light gradient partitioning among tropical tree species through differential seedling mortality and growth. *Ecology* 80, 187–201.

Kulmatiski, A., Beard, K.H. and Stark, J.M. (2006)

Exotic plant communities shift water-use timing in a shrub-steppe ecosystem. *Plant and Soil* 288, 271–284.

LaMalfa, E.M. and Ryel, R.J. (2008) Differential snowpack accumulation and water dynamics in aspen and conifer communities: Implication for water yield and ecosystem function. *Ecosystems* 11, 569–581.

Leffler, A.J. and Evans, A.S. (1999) Variation in carbon isotope composition among years in the riparian tree *Populus fremontii*. *Oecologia* 119, 311–319.

Leffler, A.J., England, L.E. and Naito, J. (2000) Vulnerability of Fremont cottonwood (*Populus fremontii* Wats.) individuals to xylem cavitation. *Western North American Naturalist* 60, 204–210.

Leffler, A.J., Ryel, R.J., Hipps, L., Ivans, S. and Caldwell, M.M. (2002) Carbon acquisition and water use in a northern Utah *Juniperus osteosperma* (Utah juniper) population. *Tree Physiology* 22, 1221–1230.

Leffler, A.J., Peek, M.S., Ryel, R.J., Ivans, C.Y. and Caldwell, M.M. (2005) Hydraulic redistribution through the root systems of senesced plants. *Ecology* 86, 633–642.

LeMaitre, D.C., Scott, D.F. and Colvin, C. (1999) A review of information on interactions between vegetation and groundwater. *Water SA* 25, 137–152.

Levine, J.M., Vilá, M., D'Antonio, C.M., Dukes, J.S., Grigulis, K. and Lavorel, S. (2003) Mechanisms underlying the impacts of exotic plant invasions. *Proceedings of the Royal Society B* 270, 775–781.

Li, Y.M. and Ghodrati, M. (1994) Preferential transport of nitrate through soil columns containing root channels. *Soil Science Society of America Journal* 58, 653–659.

Low, A.P., Stark, J.M. and Dudley, L.M. (1997) Effects of soil osmotic potential on nitrification, ammonification, N-assimilation, and nitrous oxide production. *Soil Science* 162, 16–27.

Ludwig, F., Dawson, T.E., Prins, H.H.T., Berendse, F. and de kroon, H. (2004) Below-ground competition between trees and grasses may overwhelm the facilitative effects of hydraulic lift. *Ecology Letters* 7, 623–631.

Ludwig, J.A., Wilcox, B.P., Breshears, D.D., Tongway, D.J. and Imeson, A.C. (2005) Vegetation patches and runoff-erosion as interacting ecohydrological processes in semiarid landscapes. *Ecology* 86, 288–297.

MacArthur, R.A. and Wilson, E.O. (1967) *The Theory of Island Biogeography*. Princeton University Press, Princeton.

MacDougall, A.S. and Turkington, R. (2005) Are

invasive species the drivers or passengers of change in degraded ecosystems? *Ecology* 86, 42–55.

Maestre, F.T., Bautista, S. and Cortina, J. (2003) Positive, negative, and net effects in grass–shrub interactions in Mediterranean semiarid grasslands. *Ecology* 84, 3186–3197.

Maron, J.L. and Connors, P.G. (1996) A native nitrogen-fixing shrub facilitates weed invasion. *Oecologia* 105, 302–312.

Marrs, R.H. (1993) Soil fertility and nature conservations in Europe – theoretical considerations and practical management solutions. *Advances in Ecological Research* 24, 241–300.

Matson, P., Lohse, K.A. and Hall, S.J. (2002) The globalization of nitrogen deposition: consequences for terrestrial ecosystems. *Ambio* 31, 113–119.

May, R.M. (1977) Thresholds and breakpoints in ecosystems with a multiplicity of stable states. *Nature* 269, 471–477.

Mazzola, M.B., Chambers, J.C., Blank, R.R., Pyke, D.A., Schupp, E.W., Allcock, K.G., Doescher, P.S. and Nowak, R.S. (2011) Effects of resource availability and propagule supply on native species recruitment in sagebrush ecosystems invaded by *Bromus tectorum*. *Biological Invasions* 13, 513–526.

McClaran, M.P., Moore-Kucera, J., Martens, D.A., van Haren, J. and Marsh, S.E. (2008) Soil carbon and nitrogen in relation to shrub size and death in a semi-arid grassland. *Geoderma* 145, 60–68.

McCulley, R.L., Burke, I.C. and Lauenroth, W.K. (2009) Conservation of nitrogen increases with precipitation across a major grassland gradient in the Central Great Plains of North America. *Oecologia* 159, 571–581.

McFarlane, D.J. and Williamson, D.R. (2002) An overview of water logging and salinity in southwestern Australia as related to the 'Ucarro' experimental catchment. *Agricultural Water Management* 53, 5–29.

McKane, R.B., Johnson, L.C., Shaver, G.R., Nadelhoffer, K.J., Rastetter, E.B., Fry, B., Giblin, A.E., Kielland, K., Kwiatkowski, B.L., Laundre, J.A. and Murray, G. (2002) Resource-based niches provide a basis for plant species diversity and dominance in arctic tundra. *Nature* 415, 68–71.

Messier, C., Parent, S. and Bergeron, Y. (1998) Effects of overstory and understory vegetation on the understory light environment in mixed boreal forests. *Journal of Vegetation Science* 9, 511–520.

Milchunas, D.G. and Lauenroth, W.K. (1993) Quantitative effects of grazing on vegetation and soils over a global range of environments. *Ecological Monographs* 63, 327–366.

Miller, A.J. and Cramer, M.D. (2004) Root nitrogen acquisition and assimilation. *Plant and Soil* 274, 1–36.

Mueggler, W.F. and Blaisdell, J.P. (1958) Effects on associated species of burning, rotobeating, spraying, and railing sagebrush. *Journal of Range Management* 11, 61–66.

Naeth, M.A. and Chanasyk, D.S. (1995) Grazing effects on soil water in Alberta foothills fescue grasslands. *Journal of Range Management* 48, 528–534.

Noy-Meir, I. (1973) Desert ecosystems: environments and producers. *Annual Review of Ecology and Systematics* 4, 25–51.

Nye, P.H. and Tinker, P.B. (1977) *Solute Movement in the Soil-Root System.* University of California Press, Berkeley, California.

Odum, E.P. (1969) The strategy of ecosystem development. *Science* 164, 262–270.

Odum, E.P. (1994) *Ecological and General Systems: An Introduction to Systems Ecology.* University Press of Colorado, Niwot, Colorado.

Ovington, J.D. (1962) Quantitative ecology and woodland ecosystem concept. *Advances in Ecological Research* 1, 103–192.

Peck, A.J. (1978) Salinisation of non-irrigated soils and associated streams: a review. *Australian Journal of Soil Research* 16, 157–168.

Peek, M.S. and Forseth, I.N. (2003) Enhancement of photosynthesis and growth of an aridland perennial in response to soil nitrogen pulses generated by mule deer. *Environmental and Experimental Botany* 49, 169–180.

Peek, M.S., Leffler, A.J., Ivans, C.Y., Ryel, R.J. and Caldwell, M.M. (2005) Fine root distribution and persistence under field conditions of three co-occurring Great Basin species of different life form. *New Phytologist* 165, 171–180.

Perry, L.G., Blumenthal, D.M., Monaco, T.A., Paschke, M.W. and Redente, E.F. (2010) Immobilizing nitrogen to control plant invasion. *Oecologia* 163, 13–24.

Pianka, E.R. (1987) The subtlety, complexity and importance of population interactions when more than two species are involved. *Revista Chilena de Historia Natural* 60, 351–361.

Pickett, S.T.A., Collins, S.L. and Armesto, J.J. (1987) Models, mechanisms and pathways of succession. *The Botanical Review* 53, 335–371.

Porazinska, D.L., Bardgett, R.D., Blaauw, M.B., Hunt, H.W., Parsons, A.N., Seastedt, T.R. and Wall, D.H. (2003) Relationships at the aboveground-belowground interface: plants, soil biota, and soil processes. *Ecological Monographs* 73, 377–395.

Prevéy, J.S., Germino, M.J. and Huntly, N.J. (2010a) Loss of foundation species increases population growth of exotic forbs in sagebrush steppe. *Ecological Applications* 20, 1890–1902.

Prevéy, J.S., Germino, M.J., Huntly, N.J. and Inouye, R.S. (2010b) Exotic plants increase and native plants decrease with loss of foundation species in sagebrush steppe. *Plant Ecology* 207, 39–51.

Pyšek, P. and Richardson, D.M. (2007) Traits associated with invasive alien plants: where do we stand. In: Nentwig, W. (ed.) *Biological Invasions.* Springer-Verlag, Berlin, pp. 97–125.

Rees, M., Grubb, P.J. and Kelly, D. (1996) Quantifying the impact of competition and spatial heterogeneity on the structure and dynamics of a four-species guild of winter annuals. *The American Naturalist* 147, 1–32.

Richards, J.H. and Caldwell, M.M. (1987) Hydraulic lift: substantial nocturnal water transport between soil layers by *Artemisia tridentata* roots. *Oecologia* 73, 486–489.

Roscher, C., Bessler, H., Oelmann, Y., Engels, C., Wilcke, W. and Schulze, E.D. (2009) Resources, recruitment limitation and invader species identity determine pattern of spontaneous invasion in experimental grasslands. *Journal of Ecology* 97, 32–47.

Rosenzweig, M.L. (1971) Paradox of enrichment: destabilization of exploitation ecosystems in ecological time. *Science* 171, 385–387.

Ryel, R.J., Ivans, C.Y., Peek, M.S. and Leffler, A.J. (2008) Functional differences in soil water pools: a new perspective on plant water use in water-limited ecosystems. *Progress in Botany* 69, 397–422.

Ryel, R.J., Leffler, A.J., Ivans, C., Peek, M.S. and Caldwell, M.M. (2010) Functional differences in water-use patterns of contrasting life forms in Great Basin steppelands. *Vadose Zone Journal* 9, 548–560.

Schlesinger, W.H., Reynolds, J.F., Cunningham, G.L., Huenneke, L.F., Jarrell, W.M., Virginia, R.A. and Whitford, W.G. (1990) Biological feedbacks in global desertification. *Science* 247, 1043–1048.

Schmitz, O.J. (2008) Effects of predator hunting mode on grassland ecosystem function. *Science* 319, 952–954.

Schuman, G.E., Reeder, J.D., Manley, J.T., Hart, R.H. and Manley, W.A. (1999) Impact of grazing management on the carbon and nitrogen balance of a mixed-grass rangeland. *Ecological Applications* 9, 65–71.

Scott, M.L., Auble, G.T. and Friedman, J.M. (1997) Flood dependency of cottonwood establishment along the Missouri River, Montana, USA. *Ecological Applications* 7, 677–690.

Seyfried, M.S., Schwinning, S., Walvoord, M.A., Pockman, W.T., Newman, B.D., Jackson, R.B. and Phillips, F.M. (2005) Ecohydrological control of deep drainage in arid and semiarid regions. *Ecology* 86, 277–287.

Shafroth, P.B., Friedman, J.M. and Ischinger, L.S. (1995) Effects of salinity on establishment of *Populus fremontii* (cottonwood) and *Tamarix ramosissima* (saltceder) in southwestern United States. *Great Basin Naturalist* 55, 58–65.

Shea, K. and Chesson, P. (2002) Community ecology theory as a framework for biological invasions. *Trends in Ecology and Evolution* 17, 170–176.

Sheley, R.L., Mangold, J.M. and Anderson, J.L. (2006) Potential for successional theory to guide restoration of invasive-plant-dominated rangeland. *Ecological Monographs* 76, 365–379.

Sher, A.A. and Hyatt, L.A. (1999) The disturbed resource-flux invasion matrix: a new framework for patterns of plant invasion. *Biological Invasions* 1, 107–114.

Sperry, J.S. and Hacke, U.G. (2002) Desert shrub water relations with respect to soil characteristics and plant functional type. *Functional Ecology* 16, 367–378.

Sperry, L.J., Belnap, J. and Evans, R.D. (2006) *Bromus tectorum* invasion alters nitrogen dynamics in an undisturbed arid grassland ecosystem. *Ecology* 87, 603–615.

Stein, C.M., Johnson, D.W., Miller, W.W., Powers, R.F., Young, D.A. and Glass, D.W. (2010) Snowbrush (*Ceanothus velutinus* Dougl) effects on nitrogen availability in soils and solutions from a Sierran ecosystem. *Ecohydrology* 3, 79–87.

Stella, J.C., Battles, J.J., McBride, J.R. and Orr, B.K. (2010) Riparian seedling mortality from simulated water table recession, and the design of sustainable flow regimes on regulated rivers. *Restoration Ecology* 18, 284–294.

Sturges, D.L. (1973) Soil moisture response to spraying big sagebrush the year of treatment. *Journal of Range Management* 26, 444–447.

Sturges, D.L. (1993) Soil-water and vegetation dynamics through 20 years after big sagebrush control. *Journal of Range Management* 46, 161–169.

Sutherland, J.P. (1974) Multiple stable points in natural communities. *The American Naturalist* 108, 859–873.

Tansley, A.G. (1935) The use and abuse of vegetational concepts and terms. *Ecology* 16, 284–307.

Tecco, P.A., Dìaz, S., Cabido, M. and Urcelay, C. (2010) Functional traits of alien plants across contrasting climatic and land-use regimes: do aliens join the locals or try harder than them? *Journal of Ecology* 98, 17–27.

Theoharides, K.A. and Dukes, J.S. (2007) Plant invasion across space and time: factors affecting non-indigenous species success during four stages of invasion. *New Phytologist* 176, 256–273.

Thompson Tew, D., Morris, L.A., Lee Allen, H. and Wells, C.G. (1986) Estimates of nutrient removal, displacement and loss resulting from harvest and site preparation of a *Pinus taeda* plantation in the piedmont of North Carolina. *Forest Ecology and Management* 15, 257–267.

Tiedemann, A.R. and Klemmedson, J.O. (1986) Long-term effects of mesquite removal on soil characteristics: I. Nutrients and bulk density. *Soil Science Society of America Journal* 60, 472–475.

Tilman, D. (1982) *Resource Competition and Community Structure.* Princeton University Press, Princeton, New Jersey.

Turner, C.L. and Knapp, A.K. (1996) Responses of a C_4 grass and three C_3 forbs to variation in nitrogen and light in tallgrass prairie. *Ecology* 77, 1738–1749.

Tyree, M.T., Kolb, K.J., Rood, S.B. and Patiño, S. (1994) Vulnerability to drought-induced cavitation of riparian cottonwoods in Alberta: a possible factor in the decline of the ecosystem. *Tree Physiology* 14, 455–466.

Vilá, M. and Pujadas, J. (2001) Land-use and socio-economic correlates of plant invasions in European and North African countries. *Biological Conservation* 100, 397–401.

Walter, H. (1971) *Natural Savannas: Ecology of Tropical and Subtropical Vegetation.* Oliver and Boyd, Edinburgh, UK.

Walvoord, M.A., Phillips, F.M., Stonestrom, D.A., Evans, R.D., Hartsough, P.C., Newman, B.D. and Striegl, R.G. (2003) A reservoir of nitrate beneath desert soils. *Science* 302, 1021–1024.

Wambolt, C.L. and Payne, G.F. (1986) An 18-year comparison of control methods for Wyoming big sagebrush in southwestern Montana. *Journal of Range Management* 39, 316–319.

Wan, S.Q., Hui, D.F. and Luo, Y.Q. (2001) Fire effects on nitrogen pools and dynamics in terrestrial ecosystems: a meta-analysis. *Ecological Applications* 11, 1349–1365.

Webb, R.H. and Leake, S.A. (2006) Ground-water surface-water interactions and long-term change in riverine riparian vegetation in the southwestern United States. *Journal of Hydrology* 320, 302–323.

Weigelt, A., Bol, R. and Bardgett, R.D. (2005) Preferential uptake of soil nitrogen forms by grassland plant species. *Oecologia* 142, 627–635.

Welden, C.W. and Slauson, W.L. (1986) The intensity of competition versus its importance: an overlooked distinction and some implications. *The Quarterly Review of Biology* 61, 23–44.

West, N.E. and Young, J.A. (2000) Intermountain valleys and lower mountain slopes. In: Barbour, M.G. and Billings, W.D. (eds) *North American Terrestrial Vegetation.* Cambridge University Press, Cambridge, UK, pp. 256–284.

Westoby, M., Walker, B. and Noy-Meir, I. (1989) Opportunistic management for rangelands not at equilibrium. *Journal of Range Management* 42, 266–274.

White, P.S. and Pickett, S.T.A. (1985) Natural disturbance and patch dynamics: an introduction. In: Pickett, S.T.A. and White, P.S. (eds) *The Ecology of Natural Disturbance and Patch Dynamics.* Academic Press, New York.

Williams, D.G. and Ehleringer, J.R. (2000) Intra- and interspecific variation for summer precipitation use in pinyon-juniper woodlands. *Ecological Monographs* 70, 517–537.

Williamson, M. and Fitter, A. (1996) The varying success of invaders. *Ecology* 77, 1661–1666.

Wright, I.J., Reich, P.B., Westoby, M., Ackerly, D.D., Baruch, Z., Bongers, F., Cavender-Bares, J., Chapin, T., Cornelissen, J.H.C., Diemer, M., Flexas, J., Garnier, E., Groom, P.K., Gulias, J., Hikosaka, K., Lamont, B.B., Lee, T., Lee, W., Lusk, C., Midgley, J.J., Navas, M.-L., Niinemets, U., Oleksyn, J., Osada, N., Poorter, H., Poot, P., Prior, L., Pyankov, V.I., Roumet, C., Thomas, S.C., Tjoelker, M.G., Veneklaas, E.J. and Villar, R. (2004) The worldwide leaf economics spectrum. *Nature* 428, 821–827.

Young, J.A., Evans, R.A. and Major, J. (1972) Alien plants in the Great Basin. *Journal of Range Management* 25, 194–201.

Zedler, P.H. and Scheid, G.A. (1988) Invasion of *Carpobrotus edulis* and *Salix lasiolepis* after fire in a coastal chaparral site in Santa Barbara County, California. *Madroño* 35, 196–201.

Zhang, D., Hui, D., Luo, Y. and Zhou, G. (2008) Rates of litter decomposition in terrestrial ecosystems: global patterns and controlling factors. *Journal of Plant Ecology* 1, 85–93.

Zhang, Y.-K. and Schilling, K.E. (2006) Effects of land cover on water table, soil moisture, evapotranspiration, and groundwater recharge: a field observation and analysis. *Journal of Hydrology* 319, 328–338.

Zink, T.A. and Allen, M.F. (1998) The effects of organic amendments on the restoration of a coastal sage scrub habitat. *Restoration Ecology* 6, 52–58.

5 Invasive Plant Impacts on Soil Properties, Nutrient Cycling, and Microbial Communities

Thomas A. Grant III and Mark W. Paschke

Department of Forest, Rangeland, and Watershed Stewardship, Colorado State University, USA

Introduction

Prior to the 19th century biological invasions were probably infrequent, natural events that contributed to a diverse and ever evolving patchwork of species and eco-systems. The dramatic increase in the human population and our influence over the earth's biotic and abiotic systems has created a world that is dominated by the effects of humans (Vitousek *et al.*, 1997). In general, humans have rapidly increased the movement of flora and fauna across what were once natural barriers to migration or survival and the consequences are now termed 'biological invasions' (Elton, 1958). All organisms influence and modify their environment, whether through competitive interactions that alter species composition, the addition or removal of resources, or the transformation of habitat and large-scale ecosystem processes. To successfully manage invasive species, it is critical that we understand how these species interact and modify their novel environment and use this knowledge to improve the efficiency and effectiveness of management practices. This chapter will specifically address the impacts of invasive plants on the soil, including the physical and chemical composition of soil, litter decomposition and biogeochemical cycling of nutrients, and soil microbial communities. The primary emphasis will focus on the impacts of invasive species on these systems and utilizing this knowledge to apply ecological principles to the management of ecosystem processes.

An ecosystem is a representation of the biotic elements and abiotic variables that interact at a given place and time. Any dramatic change in the species composition, either above- or belowground, will likely affect the composition and function of the whole. An understanding of the direct and indirect impacts of invasive species on a system is critical to delineate if we are to manage natural and utilitarian ecosystems effectively and with the goal of conserving ecosystem diversity and function. Invasion by exotic plants represents a fundamental change in a system and a unique management challenge for human society, both ethically and economically. From a Clementsian framework of successional dynamics, fol-lowing invasion, a late-seral community of native species may no longer be a possible outcome of long-term ecosystem develop-ment, because the whole system has changed and its trajectory is different (Clements, 1916). Within a Gleasonian individualistic perspective of plant dynamics, the key players have changed and they will influence the system in a unique and different manner (Gleason, 1926). History is no longer a guide to the interaction of species and systems, therefore how they will interact and the resulting community composition is unknown. As the plant community com-position changes due to the invaders, so will many of the ecological processes that link the above- and belowground elements of the ecosystem (Wardle *et al.*, 2004), including decomposition, nutrient cycling, rhizosphere exudation, microbial composition and function. It is likely that the change in plant community will influence the amount and type of animal herbivory, with cascading effects on soil properties, root growth responses from herbivory, and nutrient inputs from animal defecation. Lastly, a change in vegetation can alter the disturbance regime of a system, particularly fire cycles and intensity. Overall, any striking change in the plant community will have diverse effects throughout the whole realm of plant–soil–microbe interactions.

Invasive species tend to be highly successful through their rapid growth and reproduction, high fecundity, great dispersal abilities, and apparent predisposition to establish in disturbed areas. Many invasive species have been removed from their evolutionary constraints or have been released from enemies and are often found proliferating in disturbed areas that have available resources. These aggressive and opportunistic traits do not facilitate their dominance in the species' native range and this conundrum has been called the 'invasive plant paradox' (Rout and Callaway, 2009). It may be possible that invasive species dramatically modify the novel habitat to increase their fitness or negatively affect their neighbors and competitors. Most organisms attempt to do this in their struggle to reproduce and survive, but are usually constrained by resources or other species, especially in the long evolutionary timeframe of an ecosystem's development. These interactions, both direct and indirect, may be part of a plant–soil feedback that facilitates invasion through either positive or negative interactions. Feedbacks may facilitate fecundity of the invader, negatively influence competitors, or modify soil to the extent that the system is fundamentally different and less hospitable to certain organisms.

Much ecological research has focused on factors such as competition, dispersal, resource use efficiency, herbivory, and predation to understand plant community composition, succession, and invasion. In-depth research has provided quantitative information about how plants, animals, soil, and the atmosphere interact and ecosystems are modified, but the focus on competition and resource efficiency has not completely addressed many questions about vegetation dynamics (Klironomos, 2002), especially concerning the paradox of invasion. Competitive ability has been used to describe why one organism, species, or population displaces another or is more successful in the capture of a resource and the subsequent effect on its survival and fecundity. Obviously some species are better at acquiring a resource than another, but can the simplistic focus on one nutrient, growth habit or reproductive strategy explain the diversity of ecosystems or form a rationale for successful invasion by an exotic species? By focusing on a species' multi-scale direct and indirect interactions within an ecosystem and the subsequent impacts of a change in biodiversity, we may see the outcomes of competition are rooted in the interactions and feedbacks of a complex system. It may be possible to view these interactions as the mechanism that structures a community and facilitates invasion. This concept removes the idea that one species independently is a superior competitor and focuses on an organism or species' interactions with its environment as forming the mechanisms for the outcome of competitive scenarios. Theoretically, competition is removed from the attributes of an individual and placed in the realm of a complex and dynamic web of direct and indirect interactions of varying strengths that include both resource and non-resource connections (Hierro and Callaway, 2003). Given this systems-based approach or supposition, the focus of invasive species management should be on the system's interactions and the species' impacts and not just the inherent capabilities of the species, because the exotic entity cannot avoid being a constituent of a dynamic and inherently interconnected system.

Invasive species have dramatically impacted most ecosystems throughout the world and caused a rapid change to their diversity and function. Although plant species invasions are a natural part of ecosystem development and succession, the rate and scale of exotic invasions have increased due to anthropogenic modification to the environment and movement of propagules in an increasingly globalized world (Mack *et al.*, 2000). Restoration and control efforts on the most problematic invasive species are marginally successful at small scales, but not at landscape levels. The impacts of invasive species to agricultural production, recreation, and natural ecosystems are vast and costly, yet these are only the obvious outcomes of invasion and efficient management requires an understanding of how an exotic species successfully invades a system. To improve management or control of exotic species it is critical to understand the direct and indirect impacts of species on basic ecosystem processes and how these changes influence our ability to manage systems. Exotic species fundamentally modify nutrient cycling, litter decomposition, soil microbial communities, and physical properties of soil; simple removal of undesirable species or revegetating areas by seeding with native species may be ineffective in restoring systems to pre-invasion states or former ecosystem functioning. By addressing the novel conditions and processes that have developed due to dominance by an exotic species we can develop a clearer understanding of how to effectively manage systems, determine realistic restoration goals, and possibly prevent additional invasions. The purpose of this chapter is to review the processes and interactions by which exotic plant species alter the physical, chemical, and microbial characteristics of soil and provide a basis for applying these concepts to innovative soil management strategies that can reduce dominance of invasive plants and conserve diverse, functioning natural ecosystems.

Impacts of Invasive Species on Soil

The impacts of invasive species are pervasive. In many invasions, the whole plant community has changed from a diverse polyculture, represented by many different functional traits, to a simplified system that is dominated by functionally similar organisms. Broadly speaking, the energy flow of the system increases as the biomass and net primary productivity of the invaded system are modified. The consequences are directly and indirectly proliferated throughout the ecosystem. Aboveground the changes in the plant community are obvious, but it is at the plant–soil interface and belowground that scientific inquiry has begun to focus on questions concerning the effects of invasive plant litter on decomposition and nutrient cycling, and the subsequent changes in microbial community structure and function (Hawkes *et al.*, 2005). Many of these activities are strongly regulated by microorganisms, which will also change as their primary habitat, the rhizosphere, is modified by changes in species-specific root types, architecture, biomass, and physical and chemical characteristics of the soil. Dramatic changes in biochemistry of plant root exudates occur as functionally and biochemically diverse species are replaced by one dominant or functionally similar species. This cascade of interactions begins with the invasion by an exotic species and in all likelihood most plant–soil–microbe interactions are impacted, although the degree and direction is variable. Wolfe and Klironomos (2005) propose three linkages that are directly impacted by invasive species: (i) plant community composition and ecosystem processes; (ii) plant community composition and soil community composition; and (iii) soil community composition and ecosystem processes (Fig. 5.1). Exotic plant invasion directly affects these three primary linkages and potentially modifies the plant community, soil community, and ecosystem processes and function. In addition, numerous indirect interactions occur and dramatically increase the complexity of managing plant invasions based on ecological principles and successional theory (Krueger-Mangold *et al.*, 2006; Sheley *et al.*, 1996). Current research is attempting to understand the impacts of invasion on the

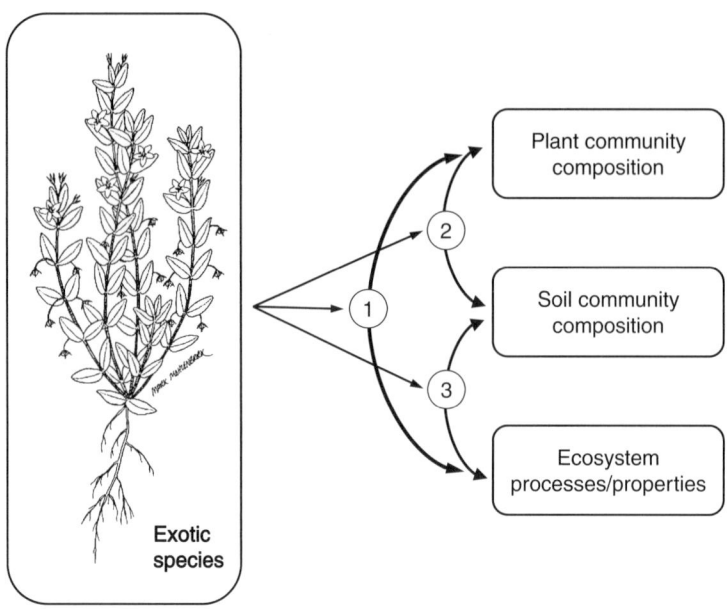

Fig. 5.1. Impacts of an invasive plant on the direct interactions and linkages between the plant community, soil community, and ecosystem processes. Adapted/redrawn from Wolfe and Klironomos (2005). *Anagallis arvensis* L. (scarlet pimpernel) illustration by Robert H. Mohlenbrock (Robert H. Mohlenbrock @ USDA-NRCS PLANTS Database / USDA SCS. 1991. *Southern wetland flora: Field office guide to plant species.* South National Technical Center, Fort Worth, Texas, USA).

soil and apply this information to the feedbacks between the complex systems that influence the invasion of exotic species (see Eviner and Hawkes, Chapter 7, this volume).

Direct and indirect interactions between plants and soil

The impacts of invasive species will influence both the potential vegetation and the ability to restore ecosystems to any semblance of their pre-invasion vegetation state and function. To comprehensively understand the impacts of an invasive species, it is critical to provide an overview of the direct and indirect interactions of an exotic species in its novel environment. Figure 5.2 attempts to illustrate a simplified scenario of the potential interactions between a plant and soil. Aboveground, the type and quantity of plant litter will likely change with invasion. Decomposition will be modified as

biochemically different material is added to the system, often in larger quantities than pre-invasion (Ehrenfeld, 2003). Within the litter:soil interface, the microbial community will adapt to the new conditions and energy sources, producing different decomposition rates and fluxes of nutrients. Indirect effects of increased litter layers and reduced solar radiation could include modifications to soil moisture, temperature, and microsites for plant establishment. Disturbance regimes, such as the influence of fire on litter, soil nutrients, or nascent plants, have been shown to be indirectly affected by invasion and could have long-term feedbacks that alter the successional trajectory and potential vegetation of the community. Change in the plant-community composition may also indirectly affect the amount or type of herbivory. Grazing can stimulate root growth and influence soil bulk density and nutrient inputs from fecal matter. Additionally, the feeding preferences of animals will likely change with a shift in

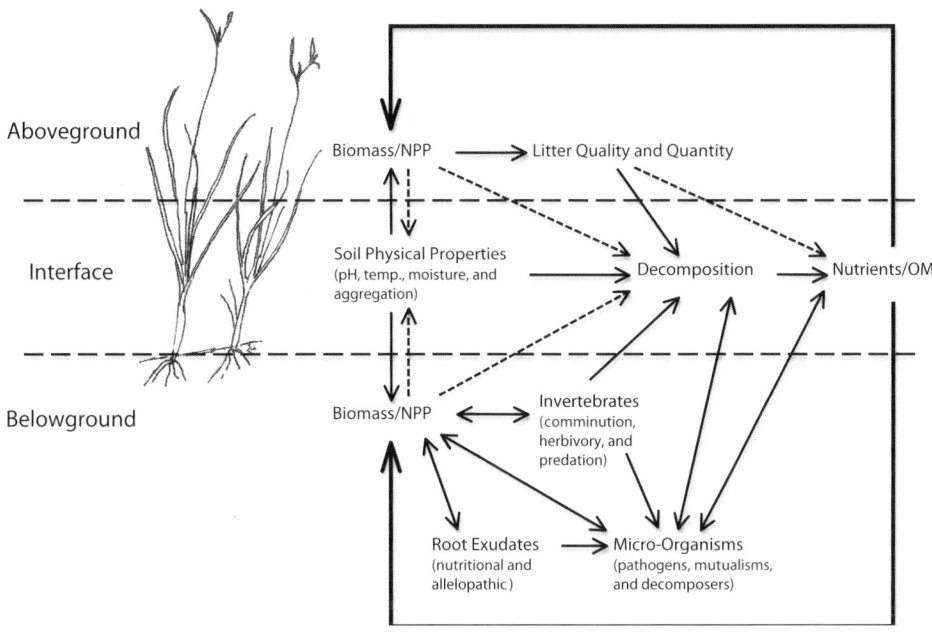

Fig. 5.2. Potential direct and indirect interactions in soils impacted by exotic plant invasion. Solid and dashed lines represent direct and indirect interactions, respectively. Illustration by Janet Wingate and reprinted with artist's permission.

vegetation. Belowground, nutrient inputs from the litter and microbial activity are indirectly modified by invasion, as are the habitats for microorganisms due to the change in the physical and biochemical composition of roots. The rhizosphere will change as the root architecture becomes more homogenized by physically and functionally similar roots of the invading species. Biochemically, root exudates and decomposing root masses will modify the rhizosphere and energy sources for microorganisms, therefore influencing the diversity and function of soil microbes.

Understanding direct and indirect interactions develops a framework for the myriad of connections that drive a system and provides a conceptual basis to apply ecological principles to the management of natural systems. The complexity of ecological systems has long made it difficult to understand how an invasive plant or anthropomorphic management action modifies an ecosystem, but as our knowledge of an ecosystem's above- and belowground interconnectedness increases so does our ability to develop management programs based on ecological principles. Historically, much of vegetation management consisted of simplistic, one-dimensional approaches, such as removing an unwanted species or augmenting desirable species. This approach ignored each species' influence on other aspects of the system (microsites, litter decomposition, soil nutrients, or microbial communities) and ultimately the system as a whole. Ecosystems are collections of numerous interacting organisms and abiotic conditions that cannot easily be compartmentalized or isolated. Progressive, process-based management can utilize ecological principles to manipulate ecosystem processes and direct a system to a desired state, possibly with significantly less disturbance than historical eradication and revegetation programs. The following sections of this chapter will highlight specific examples related to direct and indirect impacts from invasive plants on litter decomposition, nutrient cycling and biogeochemistry, microbial community diversity and function, invertebrates, physical properties of soil, and allelopathy. Throughout this discussion on the impacts and interactions of invasive plants, the concept of utilizing ecological principles to develop innovative management practices will be highlighted. Management of ecological processes can include the manipulation of nutrient inputs (litter) or cycling rates, pathogens or plant-growth promoting microbes, invertebrate herbivory, and chemical interference (allelopathy). The consequences of process-based management of systems will probably not be immediate or obvious, but through the subtle manipulation of ecological processes it may be possible to restore diversity and resilience to ecosystems.

Invasive plant litter and decomposition

A logical starting place to study the effects of an invasive species on the soil system is the plant litter being added to the environment. Every species has a unique biochemical composition (Ehrenfeld, 2006) and varying amounts of potential litterfall that will influence the system differently. The quantity and quality of plant litter will directly impact the potential nutrients, biogeochemical cycling, and microbial diversity of a site. Litter decomposes at different rates based on the quality or type of litter and the environment, which will directly affect the amount and type of nutrients flowing into the soil. Invasive plants have been shown to generally have greater biomass or net primary productivity than adjacent native species (Ehrenfeld, 2003), greater rates of decomposition (Ehrenfeld, 2003; Ashton et al., 2005), and higher amounts of nitrogen (N) in litter, especially if the species is capable of symbiotic N fixation with bacteria. A review of numerous studies by Ehrenfeld (2003) found that 9 of 13 papers reported increased litter decomposition with invasion.

Decomposition occurs through three primary pathways: leaching of soluble materials, the comminution (the physical or mechanical breakdown) of biomass, and the conversion of fixed carbon to CO_2, H_2O, or energy via oxidation or catabolism (Seastedt, 1984). These pathways reduce complex, organic structures into simpler compounds, which are used as an energy source for soil fauna and flora, affect the physical properties of the soil (i.e. soil organic matter), and ultimately provide nutrients for plants. Nitrogen is an essential nutrient for plants and is one of the most commonly limiting nutrients in the decomposition of plant matter, because it regulates both the growth and turnover of microbial communities (Heal et al., 1997). Nitrogen is often the focus of invasive plant studies, because it is a dominant driver of plant growth and is essential to both plants and microbes. The influence of soil carbon:nitrogen ratios on decomposition and microbial diversity began to be acknowledged in the 1920s (Heal et al., 1997). Currently, lignin:nitrogen and polymer:nitrogen ratios are also used to analyze decomposition pathways. In general, leaf litter with higher N content decomposes more rapidly due to preferential colonization by bacteria and fungi populations (Melillo et al., 1982), although C:N ratios are dynamic as decomposition progresses and microorganisms turnover. The microbial mineralization of N and subsequent death and decay of microbes is the primary source of N for most plants (Knops et al., 2002; Schmidt et al., 2007), although this traditional view of microbial control of N cycling has recently been challenged by the concept of plants being able to directly compete with microbes for organic N that has been depolymerized by the extracellular enzymes of microbes (Schimel and Bennett, 2004; Chapman et al., 2006). An understanding of the decomposition of plant matter is necessary for the development of ecosystem management practices that acknowledge the complex interactions between plants, soil, and microbes; and ultimately how these processes influence plant communities and successional dynamics.

Plants directly affect nutrient cycling, edaphic characteristics, soil fauna, and microorganism communities of an ecosystem by their litter (quantity and quality) (Wardle *et al.*, 2004; Georgieva *et al.*, 2005; Chapman *et al.*, 2006) and root exudates released to the rhizosphere (Bais *et al.*, 2006). Wardle *et al.* (2004) proposed the concept of plants as the integrator of above- and belowground feedbacks, but emphasized the difficulties in understanding the mechanisms due to the complexity of organisms and environments involved. Due to the increased net primary productivity (NPP) of many invasive species (Ehrenfeld, 2003), litter quality and amounts represent a starting point for studies of plant invasions and their impacts. Different species have markedly divergent litter qualities and these physical characteristics modify their nutrient composition, decomposition rates, and potentially the biogeochemistry and microbial diversity of a system. Additionally, leaf litter and root exudates of invasive plants may contribute to indirect chemical interference or allelopathy (Bonner, 1950; Muller, 1966; Bais *et al.*, 2003). In the context of invasive plants, changes in the quality and amount of litter within a system may represent a critical tipping point for an invasive species to modify many aspects of the system, including the rate of decomposition (*sensu* Ehrenfeld, 2003), the flux of nutrients (Evans *et al.*, 2001), and the microorganisms involved (Wardle *et al.*, 2004).

The decomposition environment can be dramatically different between invaded and non-invaded regions of the same ecosystem. Studies have shown that in the eastern hardwood forests of the USA, the litter of invasive species decomposes more rapidly than native plant litter and invaded ecosystems decompose litter faster regardless of the litter's origin (Ashton *et al.*, 2005). A separate study did not determine differences in decomposition between sites dominated by either native or invasive species, but found that the loss rates of phosphorus, lignin, and trace elements from litterbags was reduced in invaded sites (Pritekel *et al.*, 2006). The concept that an individual species

can alter N cycling and create a self-perpetuating positive feedback system due to its litter quality and quantity has been thoroughly evaluated, although more recent research integrates the mechanisms facilitating the feedback processes by incorporating the analysis of soil micro-organisms (Hawkes *et al.*, 2005, 2006) and fauna (De Deyn *et al.*, 2003), edaphic characteristics (Goslee *et al.*, 2003; Grant *et al.*, 2003), and the differences between organic and inorganic N use by plants (Schimel and Bennett, 2004; Chapman *et al.*, 2006) into the complex systems.

Process-based management of natural systems requires understanding ecological processes and developing innovative techniques that utilize scientific principles to achieve land management objectives. The manipulation of litter quantities or qualities may be a possible management technique to reduce the flow of energy through an invaded system (Fig. 5.2). The goal of manipulating litter inputs would be to stress the invasive species through limiting resource inputs and possibly increasing competition with species that are adapted to lower resource levels. This method is predicated on the assumption that reduced resources would add sufficient stress to the invasive species to influence vegetation dynamics. Management of litter inputs would require knowledge of the system and how invasive species have modified decomposition, microbial communities, and nutrient cycling; otherwise it may be ineffective or have unintended consequences. Two potential strategies for litter management should be evaluated further: (i) removal of litter; or (ii) addition of low quality litter (i.e. high C:N ratio). The removal of litter in an invaded system could reduce nutrient inputs and potentially stress the invasive species because they have high productivity and subsequent nutrient demands. Experiments incorporating carbon amendments into soils of invaded systems or to reduce invasion in areas undergoing revegetation or restoration have had limited success in reducing N availability and invasive plants (Perry *et al.*, 2010). Augmentation with low quality litter may achieve similar goals as soil carbon amendments, although the technique requires experimentation and outcomes could take years to be measurable. Another possible ecological consequence of litter manipulation would be the changes to microsites for seed germination or establishment. The addition of litter could modify the microsites (temperature, sunlight, moisture) and make them less hospitable for germination by invasive species, although this requires in-depth knowledge of an invasive species' autecology. The inherent connection between plant litter and plant-available nutrients emphasizes the importance of considering ecological principles related to decomposition as a potential method to manage ecosystem processes and direct a system towards a desirable vegetation state.

Nutrient cycling and biogeochemistry

In the mid-20th century a mechanistic understanding of decomposition led to the formation of many modern ecological theories and subsequently stressed the importance of nutrient transformations and cycling in the maintenance of ecosystem function (Heal et al., 1997). Plant productivity and diversity is inexplicably linked to nutrients, although understanding this interaction has proven difficult. Historically, a majority of research on invasive species and biogeochemical cycling of nutrients has focused on N, since it is a primary limitation of productivity in terrestrial ecosystems (LeBauer and Treseder, 2008). Many invasive species form symbiotic relationships with N-fixing bacteria and this interaction is capable of dramatically increasing the amount of N in a system (Ehrenfeld, 2003). In contrast, other invasive species can reduce the amount of N symbiotically fixed by native species (Wardle et al., 1994) and thus decrease the amount of plant-available N in a soil. Changes in plant species composition due to invasion will likely modify a system's biogeochemistry as changes occur in the quantity and quality of litter inputs, root architecture and exudates, and microbial communities. Whether the plant or microbes drive these changes in nutrients has long been debated, but will only briefly be discussed here as an introduction to the topic. At the center of this question are two competing yet likely co-occurring processes: (i) microbes control N cycling and plants can only access inorganic nutrients that remain after microbial turnover (Knops et al.'s (2002) microbial N loop); and (ii) plants are actively competing with microbes for organic N that is made available due to extracellular depolymerization by microbes (Schimel and Bennett, 2004; Chapman et al., 2006). Knops et al.'s (2002) theory infers that the type of litter and its quantities do not affect nutrient cycling, at least not as much as site or species-specific impacts on N inputs and losses that are based on factors such as fire, leaching, atmospheric deposition, and symbiotic N fixation. Additionally, because the microbial-N loop controls the flux of N, Knops et al.'s (2002) theory places less importance on issues of litter quality and quantity, therefore invasion by an exotic would not greatly affect N levels, assuming that there is no loss or gain of species capable of symbiotic N fixation or modifying major disturbance cycles. An alternate theory posits that plants compete with microbes for organic N and that the consequences of a change in species due to invasion will have great ramifications on the system's biogeochemistry, both as the litter type and quantity change, but also as the net primary productivity (NPP) and nutrient requirements of the invasive species are modified. Given the latter scenario, the subsequent impacts of an invasive plant species' dominance on an ecosystem will include all aspects of nutrient cycling, including effects on the richness, diversity, and functioning of the microbial community.

A review of the impacts of exotic plants on nutrient cycling by Ehrenfeld (2003) provides an excellent summary of our current knowledge. An overarching theme of this review is that general trends are difficult to identify and both positive and negative impacts from invasion are found for C, N, and water (Ehrenfeld, 2003). This lack of uniformity should not be a surprise given the incredible variability of natural systems and the context dependency of ecological theory. Regardless of the variability highlighted in Ehrenfeld's (2003) review, the following list includes many important generalizations concerning the impacts of exotic species on soils:

1. Exotic plants often have greater aboveground biomass, net primary productivity, higher shoot to root ratios, and faster growth rates than co-occurring native species.
2. Exotic plant litter usually decomposes faster than co-occurring native species.
3. Exotic plant soils generally have more extractable inorganic N than soil from co-occurring native species.
4. Soils of exotic species frequently have increased rates of N mineralization and nitrification.
5. Exotic plants capable of symbiotic N fixation can dramatically affect N cycling.
6. Exotic species can affect symbiotic and non-symbiotic N-fixing microorganisms that are associated with native plants.
7. Exotic species can influence the spatial distribution and temporal flux of nutrients, even if overall quantities of nutrients are not affected.
8. Both positive and negative changes in soil carbon, N, and water are associated with exotic species.

When considering the effect of an invasive plant on nutrient dynamics, it is important to assess all aspects of nutrient cycling. The following examples will focus on N due to the breadth of research on this essential nutrient. The plant–soil N cycle receives additions from atmospheric fixation by bacteria or lightning, transformations by mineralization, immobilization in living tissues, and losses by denitrification and leaching. Each of these components can be impacted by a change in the dominant vegetation and different species will have unique impacts on N cycling, including indirect effects related to a species changing disturbance regimes that can dramatically affect N (i.e. fire). In general, many invasive species increase the amount

of plant-available, inorganic N and the rates of mineralization, nitrification, and atmospheric fixation (*sensu* Ehrenfeld, 2003), although exceptions exist. The increased NPP of invaded areas may also require higher amounts of N to support larger standing crops, but also deposit greater amounts of litter. The cyclical nature of plant productivity, decomposition, nutrient availability, and plant nutrient requirements are important to consider when assessing how invasive species affect a system.

A study of wiener-leaf or saltlover (*Halogeton glomeratus*) invasion in the cold deserts of Utah, USA documents distinct increases in nitrate and phosphorus concentrations in the invaded area compared to the adjacent ecotone and uninvaded areas (Duda *et al.*, 2003). Interestingly, the ecotone had significantly less ammonium than the adjacent invaded or uninvaded areas, which could signify higher mineralization and nitrification rates. Duda *et al.* (2003) also detected much higher salts (Na, Ca, K), organic matter, and bacterial functional diversity (BIOLOG substrate analysis) in the *H. glomeratus* infestations. Due to a limited experimental design, the authors could not rule out the possibility that wiener-leaf preferentially invaded soils with specific nutrient and microbial characteristics, although they state that 'the possibility of pre-existing gradients fails to explain the patterns in our data.' Cheatgrass (*Bromus tectorum*) is one of the most problematic and widespread noxious weeds in the western USA. A study in Canyonlands National Park of Utah, USA found that cheatgrass invasion was associated with a decrease in the amount of plant-available N and N mineralization, and increased N immobilization (Evans *et al.*, 2001). The researchers found that the changes to N dynamics were linked to increased litter quantity, changes in the quality of litter and in the amount of soil organic matter directly modifying soil carbon to nitrogen ratios. Surprisingly, the impacts of cheatgrass on the N cycle occurred within 2 years of invasion. Additional research has documented higher nutrient levels in invaded habitats of Europe for several invasive species and prescribed the effect to increased NPP of the invasive species compared to native vegetation (Vanderhoeven et al., 2005; Dassonville et al., 2007). In a comparison of five exotic species (Fallopia japonica, Heracleum mantegazzianum, Prunus serotina, Rosa rugosa, and Solidago gigantea) in Belgium, Vanderhoeven et al. (2005) found significant increases in potassium and manganese in the invaded sites compared to the adjacent uninvaded areas. Another study in Belgium by Dassonville et al. (2007) concluded that F. japonica increased nutrient cycling and topsoil fertility. These few examples illustrate the variability of invasive species' impacts to the many different aspects of nutrient cycling and the capacity for invasive species to modify ecosystem function.

Novel nutrient acquisition strategies may aid invasion by an exotic species, especially if the system being invaded is limited in a specific nutrient or lacks a species capable of a unique strategy (i.e. actinorrhizal nitrogen fixation). Due to the organism's inherent ability to obtain nutrients, an empty niche may be available for the exotic species to fill. Invasive species that are capable of dinitrogen-fixing symbioses with bacteria may have an advantage in the novel environment and the diazotrophic relationship may have large-scale and long-term impacts on nutrient availability, vegetation composition, and disturbance dynamics of the system. The invasion of the Hawaiian Islands by Myrica faya (firetree) represents a pivotal point in our scientific understanding of how an invasive species can impact an ecosystem, modify nutrient dynamics, and alter successional processes. Myrica faya is a small tree from the Canary and Azores islands that invaded relatively young volcanic substrates in Hawaii beginning in the late 1800s. The invasive tree is capable of forming symbiotic relationships with N_2-fixing actinomycetes (Frankia spp.). No other species in this ecosystem develops actinorrhizal symbioses and this relationship dramatically changed the inputs and amount of biologically available N in these nitrogen limited systems (Vitousek et al., 1987;

Vitousek and Walker, 1989). N_2 fixation by M. faya increased the amount of N in the system and dramatically altered ecosystem development by increasing the amount of exotic species following the decline of M. faya (Vitousek and Walker, 1989; Adler et al., 1998) and increasing the potential for fire (Adler et al., 1998). Only by a thorough study of the impacts of invasion and the discovery of a novel source for N input to the system was it possible to understand the long-term impacts of M. faya on vegetation succession, ecosystem function, and large-scale disturbances (fire).

Russian olive (Elaeagnus angustifolia) is an intentionally introduced tree species that is currently getting a great amount of attention in the western USA. It invades riparian systems and modifies large scale biogeochemical cycling. The species also forms symbiotic associations with Frankia spp. and has leaf and litter N levels that are nearly double that of the native cottonwoods (Populus spp.) (Katz and Shafroth, 2003). Control or restoration of Russian olive-invaded areas has proven difficult and has primarily focused on chemical and mechanical methods, although managing rivers to simulate historic flood regimes and promote recruitment of native cottonwoods is being tested (Lesica and Miles, 2001). The significant amount of N added to invaded areas may have long-term effects on revegetation success or promote invasion by exotic forbs or grasses. Similar issues have been noted with invasive N_2-fixing black locust (Robinia pseudoacacia) in Europe and parts of Asia (Weber, 2003).

Strategic management of nutrients to reduce invasive plant populations is context specific and requires knowledge of how a target species or plant community will react to changes in nutrient cycling (Perry et al., 2010). Managing weeds by manipulating ecological processes attempts to utilize the inherent transformations and interactions of an ecosystem to achieve a management goal. Nutrient cycling is inexplicitly linked to litter inputs, decomposition, and microbial communities. Some species will probably not be affected by anthropogenic attempts to reduce nutrients and stress invasive species, while others may become less competitive and make it easier to establish desirable species. Overall, nutrients are only one component of the complex interactions that determine vegetation dynamics. Generally, when managing early seral, r-selected weedy species, reducing nutrient availability and increasing competition from aggressive native species may be a successful approach. Conversely, some invasive species do not fit into the early-seral concept and may not be affected by nutrient reductions (i.e. long-lived rhizomatous species such as *Acroptilon repens*). Hypothetically, adding nutrients and revegetating with aggressive early-seral, native species could assist in establishing species that can compete with the more K-selected invasive species or at least create an opening for establishment of competitive species. Applying Davis *et al*.'s (2000) theory of fluctuating resources and the subsequent invasibility of an ecosystem to the manipulation of nutrients and revegetation (pseudo-invasion) with native species may provide effective methods to restore dominance of native plants in certain situations. Manipulation of a system's biogeochemistry to direct plant community dynamics must be based on a strong understanding of ecological principles that drive processes, otherwise unintended consequences may occur.

Microbial communities

The soil is often represented as a black box in ecological experiments and studies of plant invasion (Kardol *et al.*, 2006; Kulmatiski and Beard, 2011). The diversity of prokaryotic species (bacteria and archaea) is potentially in the millions, while only approximately 4500 species have been identified (Torsvik *et al.*, 2002). Approximately 170,000 soil organisms have been identified with the largest group being fungi (Wall and Virginia, 1997; Wall and Moore, 1999). Our limited understanding of soil organisms and their interactions with plants and the environment has made it difficult to incorporate these complex systems into ecological

theory, much less management of plant invasions. Considering the role of microorganisms in the mineralization of nutrients, decomposition, N fixation, soil aggregation and aeration, and their positive or negative growth effects on plants, it is essential to assess the impacts of invasive species on microbial communities. Field and greenhouse experiments have documented different soil microbial communities in the soils of different plant species (Kourtev *et al.*, 2002) and the influence of unique plant species on the differentiation of the soil microbial community (Westover *et al.*, 1997). Recent research has documented how the soil microbial community differs between native and invasive plant species (Hawkes *et al.*, 2005, 2006; Klein *et al.*, 2006) and that these changes in composition and function can affect nutrient cycling and availability. It is broadly acknowledged that soil microorganisms mediate or regulate nutrient cycling in the soil and our recent understanding of the indirect effects of plant invasion on microbial composition highlights the importance of understanding the cascade of effects invasion will cause on soil microbe composition, nutrient cycling, and potentially pathogen accumulation (Fig. 5.2). These effects can influence the dominance of invasive species through feedback cycles, but will also affect our ability to target specific ecosystem processes to achieve management goals in an efficient and timely manner. If the impacts of invasion fundamentally alter the composition of microorganisms and ultimately the biogeochemical functioning of ecosystems, it becomes critical to recognize these changes and adapt management practices to the novel conditions created by the invasive species. The following examples highlight how plant invasion can change the soil microbial community and indirectly affect ecosystem processes.

The impacts of invasive species on soil microorganisms are often determined by microbial mediated changes to the system's biogeochemistry and research has frequently focused on N due to plants' heavy reliance on this essential macronutrient. A recent study by Hawkes *et al.* (2005) documented increased amounts of nitrifying bacteria and unique DNA signatures (restriction length patterns based on polymerase chain reaction (PCR) methods) in experimentally grown monocultures of exotic grasses compared to monocultures of a native grass, forb, or polycultural mixtures of exotic and native species. The authors related these changes in the microbial community to different plant compositions and linked the impacts of the exotic grass(es) and modified soil microbial community to functional changes in the system's nutrient cycling. A field-based study in the forests of the northeastern USA documented different microbial community composition and function in the soils of two invasive and one native understory species (Kourtev et al., 2002). The modification of soil communities was strongest in the rhizosphere soils, but surprisingly the impact was also documented in nearby bulk soil. Using canonical correlation analysis the study found that changes in the function of the soils were correlated to changes in the microbial composition and structure. A greenhouse experiment by the same researchers and with the same plant species replicated the field results and also identified increased nitrification rates and pH in the soil of one exotic species (Kourtev et al., 2003).

The belowground diversity of arbuscular mycorrhizal fungi (AMF) has been directly related to the functioning and stability of plant communities. At low AMF diversity, plant community composition has been shown to fluctuate greatly (lack of stability), while high AMF diversity promoted greater nutrient capture and productivity (van der Heijden et al., 1998). We expect exotic, invasive species to have different AMF communities than neighboring native species due to inherent physiological and phenological differences between plant species, but invasive species have also been shown to cause changes in the fungal diversity of co-occurring native species following invasion (Hawkes et al., 2006). The consequences of change to the AMF community of an invaded area are unknown, but are potentially an important mechanism in successful plant invasion (Callaway et al.,

2004) depending on the role of fungi in plant nutrient uptake, nutrient immobilization and turnover of fungal hyphae, and plant root responses to fungal infection (carbon exudation). Busby et al. (in press) have suggested that the recovery of AMF communities after invasion by exotic cheatgrass may be dependent on the species identity of native plant used for restoration. Wolfe and Klironomos (2005) provide an excellent overview of specific invasive species and their documented effects on the structure and function of native soil communities. Based on the examples provided, invasive species generally decreased AMF, fungi abundance, or diversity, although the results varied between species (Wolfe and Klironomos, 2005).

An interesting feedback system was recently documented in India in which an exotic, invasive plant (Jack in the bush, Chromolaena odorata) promotes the growth of a native, generalist soil pathogen (Fusarium sp.) and subsequently creates a negative effect on native plant species (Mangla et al., 2008). The root exudates of C. odorata were shown to promote Fusarium growth in non-invaded soils and activated carbon (AC) reduced the promotion of the fungi by the root exudates. This unique feedback pathway illustrates the complexity and variability in the reaction between plants and microorganisms, regardless of their origin (i.e. home versus foreign). Although not a direct impact of an invasive species on the soil, several experiments have documented exotic species experiencing less negative impact from microbial pathogens (Mitchell and Power, 2003) or accumulating pathogens at a slower rate than native species (Klironomos, 2002; Eppinga et al., 2006) and have hypothesized that this release from enemies contributes to the invaders' success.

Microorganisms represent an incredible breadth of diversity, although our understanding of species and functional groups is limited. The regulation of decomposition and nutrient cycling by microbes requires in-depth research and elucidation if we intend to manage ecosystems through the manipulation of ecological processes.

Molecular techniques (PCR and DNA community profiling techniques and sequencing), fatty acid analysis (phospholipid fatty acids, fatty acid methyl esters), and carbon substrate utilization (i.e. Biolog plates and substrate induced respiratory responses (SIR)) methods are beginning to classify and describe the functional traits of microorganisms. As this knowledge base expands and the methods become more consistent and less expensive, the classification of a plant community's microbially regulated nutrient cycling will make it possible to incorporate microbial communities in land management practices. When we understand how an invasive species modifies the microbial community, and subsequently decomposition and biogeochemistry, it will be realistic to attempt to modify microorganisms in ways that enhance desirable plant communities and suppress unwanted species. This will probably focus on managing N fluxes and availability or utilizing species' specific pathogens. Ecological principles guided by system and species-specific knowledge can lead innovative management practices that manipulate processes to achieve land management goals, although, because microbes are the smallest and most numerous, their interactions may be the most complex to grasp for management purposes. This novel type of management will require a massive increase in our understanding of plant–microbe interactions and the potential feedbacks that could ensue when we begin to tinker with complex systems.

Soil invertebrates

Soil fauna play an important role in many ecological processes including decomposition of biomass via comminution, root herbivory, movement of nutrients, and modification of bacterial and fungal communities through feeding activities. These activities can influence plant succession directly and indirectly, although the impacts of soil fauna and how to incorporate these effects into ecological management remain controversial or unknown. Early studies utilized litterbags

with differing mesh sizes to exclude certain soil biota (Seastedt, 1984; Huhta, 2006), while others have applied chemicals (i.e. naphthalene) or X-rays to eliminate arthropods (Newell et al., 1987). Both biocide methods have been shown to have non-target effects on soil fungi (Newell et al., 1987), although the results of litterbag studies have overwhelmingly shown that soil fauna increase the decomposition of plant litter (Huhta, 2006) and can have variable effects on nutrient cycling. A well-replicated field study in Ohio used electroshocking of soil to reduce earthworm populations without non-target effects on microarthropods, nematodes, or microorganisms (Bohlen et al., 1995). In the context of invasive plant research and management, the role of soil invertebrates is relevant due to their influence on decomposition rates and nutrient availability. Comminution affects the size and surface area of litter, which consequently modifies many factors that regulate invertebrates and microorganisms, including: microsites, predator–prey relationships due to size limitations, and access to water or nutrients. Although bacteria and fungi are the primary decomposers of organic matter, soil fauna are intricately involved in the physical processing and movement of plant biomass (fecal excretion) (Davidson and Grieve, 2006), and the direct or indirect regulation of soil microorganism communities by their feeding habitats (Seastedt, 1984). Due to difficulties identifying and describing soil fauna, the organisms are frequently grouped into the following functional groups based on Brussaard et al. (1997). Macrofauna consist of root herbivore insects, termites, ants, and earthworms. Mesofauna include mites, collembola, and enchrytraeids. Microfauna are protozoas, ciliates, and nematodes. The macro- and mesofauna generally decrease particle size of plant litter and indirectly increase surface area (habitat for microorganisms) and mobilization of nutrients. Microfauna and some mesofauna (mites and collembola) graze on fungal spores and bacteria. Soil fauna can directly impact plants by root herbivory, spreading of pathogens, and modification of soil microorganism populations through predation. Invasive species will directly change the inputs to the soil (litter, roots, and exudates) and therefore the habitat and food sources for soil fauna are likely to be modified with invasion. In order to adequately understand the impacts of invasive plants, the interactions of soil invertebrates with microorganisms, the rhizosphere, and the physical properties of soil must be incorporated into ecological studies. Previous studies and several review papers (Seastedt, 1984; Huhta, 2006) provide a general framework for incorporating the effects of soil fauna into plant–soil interactions and plant successional dynamics. Many studies are based upon litter quality (i.e. labile or resistant) and initial soil N levels as important factors in the interactions of microflora and -fauna. Huhta (2006) summarized the role of soil fauna in N-limited systems as 'generally enhance(ing) decomposition and mineralisation, whereas in the presence of excess nitrogen they have little effect.'

Soil fauna may alter succession due to selective predatory effects upon the dominant species, therefore releasing the sub-dominant plant species from competition and facilitating the succession and development of a more diverse and heterogeneous vegetation community (i.e. less dominance by any species based upon Simpson's evenness index) (De Deyn et al., 2003). In the context of invasive species, Mayer et al. (2005) documented increases in decomposition when macro-detritivores were allowed access to litterbags and the amount of decomposition correlated positively with increasing cover of an invasive grass (*Festuca arundinacea*). In general, our understanding of the interactions between soil invertebrates and invasive species is too limited to base management practices upon. The addition or removal of soil fauna are possible, although the practice must be based on ecological principles and a sound understanding of how the management will affect the system and target species. The primary manners in which soil fauna can influence vegetation are through root herbivory, comminution of litter (increased

decomposition and possibly nutrients), and predation of microorganisms (pathogens, N fixers, decomposers) (Fig. 5.2). The utilization of soil fauna to influence any of these ecological processes will require thorough knowledge of the specific plant–soil system to successfully achieve management goals.

Soil physical properties

The physical properties of soil are the results of long-term interaction between a region's geology, climate, and biota. Hans Jenny (1941) first described the formation of soil with a simple function:

soil = *f* (climate, organisms, topography, parent material, and time)

Each of these factors will affect the composition of soil and subsequent plant–soil feedbacks. The incredible variability in soil development and interactions with plants will make it difficult to identify consistent impacts on the soil due to plant invasion, but an understanding of the soils, plants, and microbial species involved can identify novel ecological interactions and improve our management practices minimizing invasive species or restoring native flora. As the medium for plants, microbes, and nutrients to interact, the soil is often represented as a black box due to its complexity and poorly understood systems. Yet it is also the arena in which many of the impacts from a change in biodiversity or dominance by an invasive species will be manifested. Impacts of invasion on the soil's physical properties are highly variable and probably species and site specific, but can include the following characteristics: soil moisture content, salinity, pH, organic matter content, soil aggregation, and microclimate effects.

Invasive species frequently have greater biomass than surrounding native vegetation and therefore probably require more nutrients and water, but also increase litter inputs to the system. Consequently, these dramatic changes to the productivity, water usage, nutrient cycling, and inputs to the system will affect many aspects of the soil, including pH. In New Zealand, studies have shown that mouse-eared hawkweed (*Hieracium pilosella*) decreased soil pH by approximately 0.5 units (McIntosh *et al.*, 1995). Conversely, Kourtev *et al.* (2003) documented increased soil pH in greenhouse incubations of the exotic Japanese stiltgrass (*Microstegium vimineum*) compared to a native blueberry. Ehrenfeld's (2003) review highlights the variability of soil pH following occupation by an exotic species. A continental-scale study of soil microbial composition in North and South America found that differences in plant species richness and diversity were largely explained by soil pH and plant community (Fierer and Jackson, 2006), although the paper did not directly address invasive species.

Water use by invasive species may alter evapotranspiration and overall water usage rates, which can cause changes in soil moisture content, water table levels, and salinity. The impacts of saltcedar (*Tamarix* spp.) in the southwestern USA are hotly debated, because the major river systems it infests supply water to millions of people and have numerous contractual and international obligations. The prolific invasive species was thought to use more water than native species and was targeted for eradication by states and municipalities in order to salvage water for anthropocentric uses. Early measurements of saltcedar water usage (>200 gallons per tree per day) may have been inaccurate and overestimated the economic benefits of control methods (Owens and Moore, 2007). Water salvage experiments that focused on saltcedar control have not been as successful as expected (Shafroth *et al.*, 2005). The impacts of saltcedar on riverine and groundwater are difficult to measure consistently due to issues of scale. Plant water use measurements are conducted at the leaf, stem, plant, or ecosystem scale and comparisons across scales can be inconsistent, although a general trend for greater water use by invasive species in drier, hotter climates and at larger scales has been documented (Cavaleri and Sack, 2010). The impacts of

saltcedar on water resources remain unclear, but the species has significant effects on soil salinity (Nagler *et al.*, 2008) and riparian forest structure.

Soil aggregation can be used as a segway to generically describe soil quality, because more stable aggregates are less prone to erosion and hold greater amounts of water, nutrients, and carbon (Batten *et al.*, 2005). Soil structure or aggregation is frequently studied through the quantification of glomalin, a glycoprotein that is produced by arbuscular mycorrhizal fungi (AMF) and is positively correlated with the stability of soil aggregates (Lutgen and Rillig, 2004). Because many invasive plants affect soil microbial communities, it is of interest to determine if the impacts cascade to soil aggregate stability or soil quality. A study of chemically and mechanically controlled spotted knapweed (*Centaurea stoebe*) infestations found that total glomalin levels and AMF hyphal lengths were negatively correlated with percent cover of the invasive plant, but did not detect a reduction in aggregate stability (Lutgen and Rillig, 2004). The authors stated that soil aggregate water stability was initially high at the study sites and that 'spotted knapweed may exert a deleterious effect on soil structure' in areas with lower initial stability. Preliminary evidence of the invasive Jack in the bush (*Chromolaena ordorata*) improving soil structure through the promotion of earthworm activity was documented in eucalyptus plantations in the Congo (Mboukou-Kimbatsa *et al.*, 2007). The variability of responses in soil aggregate stability following invasion exemplifies the species specificity of impacts and the importance of considering initial soil conditions and the wide variety of ecosystems when assessing an exotic species impact.

Allelopathy and invasive plants

Although the concept of allelopathy or chemical interference between plants has long been postulated, many difficulties have been encountered in the detection and quantification of this elusive interaction.

The success of many invasive plant species has been attributed to allelopathy, primarily through the soil matrix, and therefore it is important to consider the potential impacts of allelopathic invasive species on soil and potential management or restoration. Allelopathy specifically describes the release of a chemical into the environment by a plant or microorganism, via exudation, volatization, or transformation of biomass, which has a direct or indirect positive or negative effect on another species (Rice, 1984). Currently, the discussion of allelopathy focuses on negative or inhibitory effects on a plant due to the release of a chemical by another species and usually ignores potential stimulatory effects. If chemical interference can be determined to affect plant growth and the availability of water or nutrients within natural plant communities, the consequences of allelopathic induced changes in vegetation composition or succession within ecosystems should be considered in addition to the traditional view of ecological theory that is based primarily upon the competition for resources (Bonner, 1950; Muller, 1966). A major point of contention concerning allelopathy has assumed that the effects are direct and therefore measurable. We emphasize the importance of understanding that the impacts of potentially allelopathic root exudates are most likely weak, indirect and will occur over long time frames and involve multiple scales, including interactions with microbes and subsequent changes to soil chemistry. For these reasons, we promote the concept of soil chemical ecology over allelopathy, because it highlights the complexity of interactions in a plant–soil system and removes much of the historical controversy surrounding allelopathy (Inderjit and Weiner, 2001).

Potential allelopathic interactions in chick pea (*Cicer arietinum*) was first noted by Theophrastus around 300 BC (Rice, 1984). Although many years have passed, there is still much confusion about both the definition of allelopathy and its detection in natural ecological systems. Agricultural problems related to 'soil sickness' brought the issue of chemical interference between plants into the scientific realm in the early 1800s. In 1832, a system of crop rotation was developed by the botanist A.P. DeCandolle based upon his research into the interspecific inhibitory effects of certain agricultural species upon others (Bonner, 1950). The early theories on chemical interference or allelopathy were dismissed by many researchers as information concerning the depletion of soil nutrients and competition for these minerals and water formed the prevailing theory in plant interactions (Bonner, 1950). Additional research in the early to mid-1900s detected toxic substances in or around many plant species, including: the leaves of a desert shrub (*Encelia farinosa*), the leaves and roots of black walnut (*Juglans nigra*), the roots of smooth brome (*Bromus inermis*), and the soils of peach and rubber tree plantations (Bonner, 1950). The major problem in determining that chemical inhibition was influencing the vegetation composition and plant community succession related to the lack of evidence connecting the known phytotoxins to their release from the plant, accumulation in the soil, and the mechanism that negatively affects the surrounding vegetation. Many of the early experiments could detect phytotoxic chemicals, but they could only be correlated with the inhibition of neighboring plants. Direct evidence regarding the release of the chemical and how it interacts with the soil and neighboring plants is still proving difficult to document. The fact that most allelopathic effects are weak relative to other factors suggests that any impacts on neighboring plants would play out on longer time scales than are typically considered in ecological studies. Slow and chronic antagonistic effects would be difficult to document against a backdrop of other competitive processes. The field of allelopathy was thus heavily criticized in the mid-1900s due to the correlative nature of the studies and the vast amount of research supporting competition for resources as the primary driver in plant interactions and successional dynamics.

The subject of allelopathy came into the forefront of science again in the 1960s, primarily due to the work of C.H. Muller on the bare zones surrounding the aromatic shrubs of coastal California. Muller's (1966) well-known paper indicated that certain shrubs produce phytotoxic substances that could inhibit the establishment of seedlings (intra- and interspecific) and therefore are important factors in the diversity and successional changes in a plant community. Although Muller had considered the effects of herbivory and granivory on the bare zones, his work and allelopathy as a whole came under intense scrutiny with the publication of Bartholomew's (1970) paper concerning the role of animals in the bare zones between the shrub and grassland communities.

To this day, the separation of chemical interference and resource competition is still the largest methodological hurdle in the drive to detect and understand allelopathy (Muller, 1966; Weidenhamer, 1996; Wardle et al., 1998; Romeo, 2000; Ridenour and Callaway, 2001; Inderjit and Callaway, 2003). The importance of conducting appropriately controlled laboratory experiments (Inderjit and Dakshini, 1995) (i.e. realistic toxin concentrations) and similar field-based experiments (Inderjit et al., 2001) is essential to isolating chemical interference from resource competition. Experiments on allelopathy should include density-dependent factors (competitors and/or chemicals) and methodologies that attempt to separate resource competition from chemical interference, such as activated carbon or resource addition and removal (Romeo, 2000; Inderjit and Callaway, 2003). Methodologically, allelopathy research needs dramatic improvement in field-based studies and more in-depth knowledge of the chemical agents and their potential multi-functionality over the ecologically meaning-ful time spans that weak interactions are likely to play out.

Select Methods to Manage Soil Processes for Invasive Species Control

Traditional management of invasive species has focused on eradication with chemical, mechanical, or biological control methods and occasional follow up with revegetation with native plants or restoration of a specific ecosystem process (i.e. hydrology). Although these control techniques have proven effective in some situations, they are also cost prohibitive and frequently have non-target or unintended effects. Novel eco-system management practices attempt to utilize our understanding of ecosystem interactions and processes to manipulate plant communities towards a desired outcome. The manipulation of soil nutrients, microorganisms, invertebrates, and soil chemistry may provide low impact methods to achieve resource management objectives, although many contingencies exist when working with specific species and eco-systems. The immediate results of these novel practices may not be as dramatic as eradication methods, but by attempting to work within an existing system we reduce the amount of disturbance and may promote more resilient and stable ecosystems in the long term. Here we offer a few illustrative examples.

Carbon addition and fertilization

High N availability has been shown to facilitate invasion by exotics (sensu Perry et al., 2010), primarily because N is a major limiting resource in most communities (LeBauer and Treseder, 2008) and promotes the rapid growth of early-seral species, often to the detriment of late-seral native species. Nitrogen levels have been dramatically increased worldwide due to anthropogenic fixation and increased agricultural growth of N-fixing legumes (soybeans and lucerne) (Vitousek et al., 1997). Long-term research in the shortgrass steppe of Colorado, USA has shown that fertilization with N increased the abundance of annual grasses and forbs compared to perennial plant growth in control and carbon-addition treatments (Paschke et al., 2000). Control methods that extend the focus beyond eradication and directly address the causes of invasion (i.e. high N availability) are more likely to prevent reinvasion and achieve long-term management objectives (Perry *et al.*, 2010). Many methods exist to reduce N in an ecosystem including carbon addition, burning, grazing, biomass and top soil removal. Carbon addition causes N to be temporarily immobilized in microorganisms, while the other methods remove N from the system. A reasonably large body of evidence has shown that carbon addition reduces or prevents plant invasion, primarily by reducing the growth of invasive species that have high nutrient demands and therefore increasing the competitiveness of native species, or disrupting feedbacks between exotic plants and nutrient cycling (Perry *et al.*, 2010). Carbon is usually added in the form of sugar, sawdust, or wood chips and sequesters N by increasing microbial growth and activity, thus causing more N to be immobilized in microbes. Alpert (2010) provides an excellent overview of the results of carbon addition in different ecosystems, amounts of carbon required, and a timeline of impacts. The monetary cost and application of carbon to a system can be expensive and disruptive, especially on a large scale, although the method may be practical for the management of important, high value sites.

Conceptually similar to carbon addition, removal of litter or biomass may be a potential tool to reduce invasive plant populations, especially if litter and nutrient cycling in the invaded habitat are enhanced or accelerated by the exotic species' presence. The removal of plant material will reduce nutrient inputs to the system and reduce the competitiveness or growth of species that have high nutrient demands. Another potential technique may be the addition of low quality litter to a system. The decomposition of low nutrient biomass will affect nutrient cycling, microbial communities, and could have beneficial management outcomes. If applied in conjunction, these two methods may be a low impact, cost-effective method to slowly reduce populations of invasive plants. The challenge however is to concomitantly establish desired native species in the site with lowered nutrient availability.

Activated carbon as a tool for management

In recent years, activated carbon (AC) has had resurgence as an experimental treatment to study allelopathic interactions between plants and potentially as a tool to minimize the impacts of some invasive plants. Activated carbon is known to sorb large, organic molecules indiscriminately and therefore is a 'blunt' tool for studying the highly complex chemical interactions in the plant–soil interface. It is produced from charcoal, wood, or nutshells and has an incredibly high surface to volume ratio. The large surface area and pore volume gives the compound the ability to sequester organic molecules, including phytotoxic root exudates and organic nutrients. Many suspected allelopathic compounds are secondary metabolites (Muller, 1966), such as sesquiterpene lactone or polyacetylenes, and presumably would be bound by AC. Experimentally, AC has been shown to be effective in modifying competitive outcomes between native and invasive plants (Mahall and Callaway, 1992; Callaway and Aschehoug, 2000; Abhilasha *et al.*, 2008), although many methodological hurdles exist in our understanding and utilization of AC. In addition to AC's effects on chemical interference, recent research on the impacts of AC to microbial communities support the compound's role in aiding the restoration of native plant communities beyond a treatment consisting of only seeding with native species (Kulmatiski, 2011).

Due to the physical and chemical properties of AC, additional 'non-target' effects are known to occur and can confound experimental results (Lau *et al.*, 2008). A study by Ridenour and Callaway (2001) found that AC decreased the rate of water loss in soil. It has been hypothesized that AC could decrease microbial activity due to the sequestering of organic compounds and a subsequent reduction in bacterial transformations of N (Kulmatiski and Beard, 2006). Kulmatiski and Beard (2006) found that organic N and C were decreased in the presence of AC, but inorganic nitrate increased, possibly due to a decrease in microbial activity. A separate study found an alteration of the microbial community composition with the addition of AC and an increase in carbon sequestration compared to other organic adsorbing compounds included in the experiment (Pietikainen *et al.*, 2000).

A recent review of AC by Lau *et al.* (2008) documented experimental artifacts in the use of AC and variable species-specific responses to the amendment. The study found an increase of plant biomass with AC addition in most of the species studied, increased potting soil pH, and positive or negative changes in the amounts of specific nutrients. Plant-available N was significantly increased in the presence of AC even though the amount of N in the AC was relatively small (0.549% to 0.637% depending upon the source of AC). Methodologically, the confounding effects of AC may be minimized or 'controlled' by an understanding of the treatment's effect on individual species and in interspecific competitive scenarios with and without the addition of AC and fertilizer (Lau *et al.*, 2008). Due to the indiscriminate nature of AC, its value as an experimental or more importantly a management tool depends heavily on the experimental design utilized and an evaluation of the interactions with the species being studied. An important factor limiting the use of AC in management settings is the need to incorporate it into soils to be effective. Regardless of the problems and contingencies associated with AC use, it is one of few tools available to potentially minimize the impacts of invasive plants in the soil or alter the microbial community (Kulmatiski, 2011) and additional research may yield practical applications for the compound.

Manipulation of soil microbial and invertebrate populations

The impacts of invasive species may be managed by progressive methods that directly or indirectly modify soil microbial or invertebrate communities. Although research in this area is limited and many contingencies exist, the manipulation of soil microbial or invertebrate communities is a potential method to minimize or modify the negative consequences of invasion or assist in the restoration of invaded communities (Boyetchko, 1996). Inoculation with N-fixing bacteria has long been used to promote plant growth or higher N levels in the soil and it may be possible to add soil microfauna or flora that will promote native species or have a detrimental effect on invasive plants, potentially via pathogen accumulation. Similar to biological control with insects, the addition of pathogens from an invasive species' native habitat may help to suppress the species. Another method is to reduce or remove microbes or invertebrates with X-rays or chemicals, possibly to reduce pathogen accumulation on native or rare species, which have been shown to accumulate pathogens more rapidly than exotic plants (Klironomos, 2002; Mitchell and Power, 2003). This action must be specifically tailored to a known ecological interaction with the invasive species. De Deyn *et al.* (2003) showed how invertebrates enhanced secondary succession of European grassland vegetation by suppressing dominant early seral species. Increasing an invaded community's succession towards later-seral vegetation may promote diversity and slowly diminish the dominance of invasive species without physically disturbing the system. It is possible to manage earthworm populations using electroshocking without non-target effects (Bohlen *et al.*, 1995) and indirectly affect microbial activity (Binet *et al.*, 1998) and potentially nutrient cycling. The impacts of earthworms, native and invasive, have been overlooked in many ecosystems and may provide novel management strategies. Naturally any application of these practices must have a well-defined goal and understanding of the plant–soil–microbial interactions.

A novel concept promoted by Harris (2009) utilizes fungal:bacterial ratios across dominant vegetation groups as a measure of a system's successional development and suggests that the manipulation of microbial communities may be used to enhance the restoration of degraded systems towards a specific late-seral plant community. The use of microbes as a 'restoration shortcut' probably depends upon whether microorganisms facilitate vegetation dynamics or are followers of the change (Harris, 2009). In all likelihood, this will be species and system dependent, but the concept of microbial manipulation is worthy of additional research and management applications especially if known pathogens or beneficial symbioses can be appropriately applied. Few studies have researched the impacts of microbial inoculation on invasive plants or restoration projects and the outcomes have been mixed (Stacy et al., 2005; Rowe et al., 2007, 2009; Abhilasha et al., 2008). Another approach may be to manage the species composition of plant communities either through plant species additions or removals in order to exact desired changes in soil microbial communities (Boyetchko, 1996; Busby et al., in press). Most of these novel management practices are based on the Natural Enemies theory (Elton, 1958), Enemy Release Hypothesis (Mitchell and Power, 2003; Levine et al., 2006; Mitchell et al., 2006), or an understanding of how a species accumulates pathogens (Eppinga et al., 2006), and will require extensive knowledge of the system and proper experimentation prior to use as a management tool.

An important premise of this type of resource management is a relatively clear understanding of the organisms and interactions involved, otherwise unintended outcomes will occur. Clear examples of non-target effects by management have occurred in biological control with insects and belowground manipulations must strive to avoid these mistakes (Louda et al., 1997; Pemberton, 2000; Pemberton and Cordo, 2001). It is critical to avoid introducing potentially harmful or invasive microorganisms into systems in which our comprehension is limited (van der Putten et al., 2007).

Conclusion

Invasive plant species impact all aspects of soil, including litter decomposition, nutrient cycling, soil fauna and flora, and the physical characteristics of soil. The complex interactions between these organisms and entities require in-depth research to clarify our understanding of plant–soil feedbacks, yet recent research is beginning to provide information that can be used to manage basic ecological processes for the control of invasive species. Historically, invasive plant species were managed by a top-down, command and control approach based on removing the species or propagules, often repeatedly. Although this approach can be effective in the short term, it creates disturbance and frequently promotes reinvasion. Additionally, chemical and mechanical control methods are expensive, energy intensive, and may have undesirable effects on the environment and perpetuate disturbance cycles.

As our understanding of dynamic biological systems increases, we have the knowledge and skills to manage invasive species through the manipulation of basic ecosystem function or processes. The modification of nutrient availability due to microbial function or inputs from decomposition can influence plant community dynamics, although it will not be immediate and will rarely cause the complete removal of an unwanted species. Similarly, a change in microbial pathogens or levels of root herbivory will be unlikely to create a dramatic difference in vegetation, at least in the short term. Soil amendments or fertilization can alter microorganisms and plant dynamics leading to different successional trajectories and plant communities. Vegetation, soils, and microbial systems are interconnected and their linkages are beginning to be understood. Potential management methods are based on tinkering in the incredibly complex feedbacks between plants, soil, and microbes in order to find a balance or stability wherein higher species diversity is reached, while preventing a non-native species from becoming dominant. Few invasive species form monocultures in their native ranges. Managing ecological processes and successional trends may require a paradigm shift away from the simplistic terms of native and exotic, as it promotes a style of management based on supporting resilient and diverse ecosystems through an understanding of the ecological interactions that drive systems. It does not mean that chemical and mechanical control of invasive species have no place in management, only that we need to approach ecosystem management with long-term goals and understand how our actions will impact the entire system as we attempt to support and restore natural and resilient biotic complexes.

References

Abhilasha, D., Quintana, N., Vivanco, J. and Joshi, J. (2008) Do allelopathic compounds in invasive *Solidago canadensis* s.l. restrain the native European flora? *Journal of Ecology* 96, 993–1001.

Adler, P.B., D'Antonio, C.M. and Tunison, J.T. (1998) Understory succession following a Dieback of *Myrica faya* in Hawai'i Volcanoes National Park. *Pacific Science* 52, 69–78.

Alpert, P. (2010) Amending invasion with carbon: after fifteen years, a partial success. *Rangelands* 32, 12–15.

Ashton, I.W., Hyatt, L.A., Howe, K.M., Gurevitch, J. and Lerdau, M.T. (2005) Invasive species accelerate decomposition and litter nitrogen loss in a mixed deciduous forest. *Ecological Applications* 15, 1263–1272.

Bais, H.P., Vepachedu, R., Gilroy, S., Callaway, R.M. and Vivanco, J.M. (2003) Allelopathy and exotic plant invasion: from molecules and genes to species interactions. *Science* 301, 1377–1380.

Bais, H.P., Weir, T.L., Perry, L.G., Gilroy, S. and Vivanco, J.M. (2006) The role of root exudates in rhizosphere interations with plants and other organisms. *Annual Review of Plant Biology* 57, 233–266.

Bartholomew, B. (1970) Bare zone between California shrub and grassland communities – role of animals. *Science* 170, 1210–1212.

Batten, K.M., Six, J., Scow, K.M. and Rillig, M.C. (2005) Plant invasion of native grassland on serpentine soils has no major effects upon selected physical and biological properties. *Soil Biology and Biochemistry* 37, 2277–2282.

Binet, F., Fayolle, L. and Pussard, M. (1998) Significance of earthworms in stimulating soil microbial activity. *Biology and Fertility of Soils* 27, 79–84.

Bohlen, P.J., Parmelee, R.W., Blair, J.M., Edwards, C.A. and Stinner, B.R. (1995) Efficacy of methods for manipulating earthworm populations in large-scale field experiments in agroecosystems. *Soil Biology and Biochemistry* 27, 993–999.

Bonner, J. (1950) The role of toxic substances in the interactions of higher plants. *Botanical Review* 16, 51–65.

Boyetchko, S.M. (1996) Impact of soil microorganisms on weed biology and ecology. *Phytoprotection* 77, 41–56.

Brussaard, L., Behan-Pelletier, V.M., Bignell, D.E., Brown, V.K., Didden, W., Folgarait, P., Fragoso, C., Freckman, D.W., Gupta, V., Hattori, T., Hawksworth, D.L., Klopatek, C., Lavelle, P., Malloch, D.W., Rusek, J., Soderstrom, B., Tiedje, J.M. and Virginia, R.A. (1997) Biodiversity and ecosystem functioning in soil. *Ambio* 26, 563–570.

Busby, R.R., Gebhart, D.L., Stromberger, M.E., Meiman, P.J. and Paschke, M.W. (2011) Early seral plant species interactions with an arbuscular mycorrhizal fungi commuity are highly variable. *Applied Soil Ecology* (In press)

Callaway, R.M. and Ascheoug, E.T. (2000) Invasive plants versus their new and old neighbors: a mechanism for exotic invasion. *Science* 290, 2075–2075.

Callaway, R.M., Thelen, G.C., Rodriguez, A. and Holben, W.E. (2004) Soil biota and exotic plant invasion. *Nature* 427, 731–733.

Cavaleri, M.A. and Sack, L. (2010) Comparative water use of native and invasive plants at multiple scales: a global meta-analysis. *Ecology* 91, 2705–2715.

Chapman, S.K., Langley, J.A., Hart, S.C. and Koch, G.W. (2006) Plants actively control nitrogen cycling: uncorking the microbial bottleneck. *New Phytologist* 169, 27–34.

Clements, F.E. (1916) *Plant Succession: an analysis of the development of vegetation.* Carnegie Institution of Washington (no. 242), Washington, DC.

Dassonville, N., Vanderhoeven, S., Gruber, W. and Meerts, P. (2007) Invasion by *Fallopia japonica* increases topsoil mineral nutrient concentrations. *Ecoscience* 14, 230–240.

Davidson, D.A. and Grieve, I.C. (2006) Relationships between biodiversity and soil structure and function: evidence from laboratory and field experiments. *Applied Soil Ecology* 33, 176–185.

Davis, M.A., Grime, J.P. and Thompson, K. (2000) Fluctuating resources in plant communities: a general theory of invasibility. *Journal of Ecology* 88, 528–534.

De Deyn, G.B., Raaijmakers, C.E., Zoomer, H.R., Berg, M.P., de Ruiter, P.C., Verhoef, H.A., Bezemer, T.M. and van der Putten, W.H. (2003) Soil invertebrate fauna enhances grassland succession and diversity. *Nature* 422, 711–713.

Duda, J.J., Freeman, D.C., Emlen, J.M., Belnap, J., Kitchen, S.G., Zak, J.C., Sobek, E., Tracy, M. and Montante, J. (2003) Differences in native soil ecology associated with invasion of the exotic annual chenopod, *Halogeton glomeratus*. *Biology and Fertility of Soils* 38, 72–77.

Ehrenfeld, J.G. (2003) Effects of exotic plant invasions on soil nutrient cycling processes. *Ecosystems* 6, 503–523.

Ehrenfeld, J.G. (2006) A potential novel source of information for screening and monitoring the impact of exotic plants on ecosystems. *Biological Invasions* 8, 1511–1521.

Elton, C.S. (1958) *The Ecology of Invasions by Animals and Plants*. The University of Chicago, Chicago, Illinois.

Eppinga, M.B., Rietkerk, M., Dekker, S.C., De Ruiter, P.C. and Van der Putten, W.H. (2006) Accumulation of local pathogens: a new hypothesis to explain exotic plant invasions. *Oikos* 114, 168–176.

Evans, R.D., Rimer, R., Sperry, L. and Belnap, J. (2001) Exotic plant invasion alters nitrogen dynamics in an arid grassland. *Ecological Applications* 11, 1301–1310.

Fierer, N. and Jackson, R.B. (2006) The diversity and biogeography of soil bacterial communities. *Proceedings of the National Academy of Sciences of the United States of America* 103, 626–631.

Georgieva, S., Christensen, S., Petersen, H., Gjelstrup, P. and Thorup-Kristensen, K. (2005) Early decomposer assemblages of soil organisms in litterbags with vetch and rye roots. *Soil Biology and Biochemistry* 37, 1145–1155.

Gleason, H.A. (1926) The individualistic concept of the plant association. *Bulletin of the Torrey Botanical Club* 53, 7–26.

Goslee, S.C., Beck, K.G. and Peters, D.P.C. (2003) Distribution of Russian knapweed in Colorado: climate and environmental factors. *Journal of Range Management* 56, 206–212.

Grant, D.W., Peters, D.P.C., Beck, G.K. and Fraleigh, H.D. (2003) Influence of an exotic species, *Acroptilon repens* (L.) DC. on seedling

emergence and growth of native grasses. *Plant Ecology* 166, 157–166.

Harris, J. (2009) Soil microbial communities and restoration ecology: facilitators or followers? *Science* 325, 573–574.

Hawkes, C.V., Wren, I.F., Herman, D.J. and Firestone, M.K. (2005) Plant invasion alters nitrogen cycling by modifying the soil nitrifying community. *Ecology Letters* 8, 976–985.

Hawkes, C.V., Belnap, J., D'Antonio, C. and Firestone, M.K. (2006) Arbuscular mycorrhizal assemblages in native plant roots change in the presence of invasive exotic grasses. *Plant and Soil* 281, 369–380.

Heal, O.W., Anderson, J.M. and Swift, M.J. (1997) Plant litter quality and decomposition: an historical overview. In: Cadisch, G. and Giller, K.E. (eds) *Driven by Nature: Plant Litter Quality and Decomposition*. CAB International, Wallingford, UK, pp. 3–32.

Hierro, J.L. and Callaway, R.M. (2003) Allelopathy and exotic plant invasion. *Plant and Soil* 256, 29–39.

Huhta, V. (2006) The role of soil fauna in ecosystems: a historical review. *Pedobiologia* 50, 489–495.

Inderjit and Callaway, R.M. (2003) Experimental designs for the study of allelopathy. *Plant and Soil* 256, 1–11.

Inderjit and Dakshini, K.M.M. (1995) On laboratory bioassays in allelopathy. *Botanical Review* 61, 28–44.

Inderjit and Weiner, J. (2001) Plant allelochemical interference or soil chemical ecology? *Perspectives in Plant Ecology Evolution and Systematics* 4, 3–12.

Inderjit, Kaur, M. and Foy, C.L. (2001) On the significance of field studies in allelopathy. *Weed Technology* 15, 792–797.

Jenny, H. (1941) *Factors of Soil Formation; a system of quantitative pedology*. McGraw-Hill, New York.

Kardol, P., Bezemer, T.M. and van der Putten, W.H. (2006) Temporal variation in plant-soil feedback controls succession. *Ecology Letters* 9, 1080–1088.

Katz, G.L. and Shafroth, P.B. (2003) Biology, ecology and management of *Elaeagnus angustifolia* L. (Russian olive) in western North America. *Wetlands* 23, 763–777.

Klein, D.A., Paschke, M.W. and Heskett, T.L. (2006) Comparative fungal responses in managed plant communities infested by spotted (*Centaurea maculosa* Lam.) and diffuse (*C-diffusa* Lam.) knapweed. *Applied Soil Ecology* 32, 89–97.

Klironomos, J.N. (2002) Feedback with soil biota

contributes to plant rarity and invasiveness in communities. *Nature* 417, 67–70.

Knops, J.M.H., Bradley, K.L. and Wedin, D.A. (2002) Mechanisms of plant species impacts on ecosystem nitrogen cycling. *Ecology Letters* 5, 454–466.

Kourtev, P.S., Ehrenfeld, J.G. and Haggblom, M. (2002) Exotic plant species alter the microbial community structure and function in the soil. *Ecology* 83, 3152–3166.

Kourtev, P.S., Ehrenfeld, J.G. and Haggblom, M. (2003) Experimental analysis of the effect of exotic and native plant species on the structure and function of soil microbial communities. *Soil Biology and Biochemistry* 35, 895–905.

Krueger-Mangold, J.M., Sheley, R.L. and Svejcar, T.J. (2006) Toward ecologically-based invasive plant management on rangeland. *Weed Science* 54, 597–605.

Kulmatiski, A. (2011) Changing soils to manage plant communities: activated carbon as a restoration tool in ex-arable fields. *Restoration Ecology* 19, 102–110.

Kulmatiski, A. and Beard, K.H. (2006) Activated carbon as a restoration tool: potential for control of invasive plants in abandoned agricultural fields. *Restoration Ecology* 14, 251–257.

Kulmatiski, A. and Beard, K.H. (2011) Long-term plant growth legacies overwhelm short-term plant growth effects on soil microbial community structure. *Soil Biology and Biochemistry* 43, 823–830.

Lau, J.A., Puliafico, K.P., Kopshever, J.A., Steltzer, H., Jarvis, E.P., Schwarzlander, M., Strauss, S.Y. and Hufbauer, R.A. (2008) Inference of allelopathy is complicated by effects of activated carbon on plant growth. *New Phytologist* 178, 412–423.

LeBauer, D.S. and Treseder, K.K. (2008) Nitrogen limitation of net primary productivity in terrestrial ecosystems is globally distributed. *Ecology* 89, 371–379.

Lesica, P. and Miles, S. (2001) Natural history and invasion of Russian olive along eastern Montana rivers. *Western North American Naturalist* 61, 1–10.

Levine, J.M., Pachepsky, E., Kendall, B.E., Yelenik, S.G. and HilleRisLambers, J. (2006) Plant-soil feedbacks and invasive spread. *Ecology Letters* 9, 1005–1014.

Louda, S.M., Kendall, D., Connor, J. and Simberloff, D. (1997) Ecological effects of an insect introduced for the biological control of weeds. *Science* 277, 1088–1090.

Lutgen, E.R. and Rillig, M.C. (2004) Influence of spotted knapweed (*Centaurea maculosa*) management treatments on arbuscular mycorrhizae and soil aggregation. *Weed Science* 52, 172–177.

Mack, R.N., Simberloff, D., Lonsdale, W.M., Evans, H., Clout, M. and Bazzaz, F.A. (2000) Biotic invasions: causes, epidemiology, global consequences, and control. *Ecological Applications* 10, 689–710.

Mahall, B.E. and Callaway, R.M. (1992) Root communication mechanisms and intracommunity distributions of two Mojave Desert shrubs. *Ecology* 73, 2145–2151.

Mangla, S., Inderjit and Callaway, R.M. (2008) Exotic invasive plant accumulates native soil pathogens which inhibit native plants. *Journal of Ecology* 96, 58–67.

Mayer, P.M., Tunnell, S.J., Engle, D.M., Jorgensen, E.E. and Nunn, P. (2005) Invasive grass alters litter decomposition by influencing macrodetritivores. *Ecosystems* 8, 200–209.

Mboukou-Kimbatsa, I., Bernhard-Reversat, F., Loumeto, J.J., Ngao, J. and Lavelle, P. (2007) Understory vegetation, soil structure and soil invertebrates in Congolese eucalypt plantations, with special reference to the invasive plant *Chromolaena odorata* and earthworm populations. *European Journal of Soil Biology* 43, 48–56.

McIntosh, P.D., Loeseke, M. and Bechler, K. (1995) Soil changes under mouse-ear hawkweed (*Hieracium pilosella*). *New Zealand Journal of Ecology* 19, 29–34.

Melillo, J.M., Aber, J.D. and Muratore, J.F. (1982) Nitrogen and lignin control of hardwood leaf litter decomposition dynamics. *Ecology* 63, 621–626.

Mitchell, C.E. and Power, A.G. (2003) Release of invasive plants from fungal and viral pathogens. *Nature* 421, 625–627.

Mitchell, C.E., Agrawal, A.A., Bever, J.D., Gilbert, G.S., Hufbauer, R.A., Klironomos, J.N., Maron, J.L., Morris, W.F., Parker, I.M., Power, A.G., Seabloom, E.W., Torchin, M.E. and Vazquez, D.P. (2006) Biotic interactions and plant invasions. *Ecology Letters* 9, 726–740.

Muller, C.H. (1966) The role of chemical inhibition (allelopathy) in vegetational composition. *Bulletin of the Torrey Botanical Club* 93, 332–351.

Nagler, P.L., Glenn, E.P., Didan, K., Osterberg, J., Jordan, F. and Cunningham, J. (2008) Wide-area estimates of stand structure and water use of *Tamarix* spp. on the Lower Colorado River: implications for restoration and water management projects. *Restoration Ecology* 16, 136–145.

Newell, K., Frankland, J.C. and Whittaker, J.B. (1987) Effects on microflora of using naphthalene or x-rays to reduce arthropod populations in the field. *Biology and Fertility of Soils* 3, 11–13.

Owens, M.K. and Moore, G.W. (2007) Saltcedar water use: realistic and unrealistic expectations. *Rangeland Ecology and Management* 60, 553–557.

Paschke, M.W., McLendon, T. and Redente, E.F. (2000) Nitrogen availability and old-field succession in a shortgrass steppe. *Ecosystems* 3, 144–158.

Pemberton, R.W. (2000) Predictable risk to native plants in weed biological control. *Oecologia* 125, 489–494.

Pemberton, R.W. and Cordo, H.A. (2001) Potential and risks of biological control of *Cactoblastis cactorum* (Lepidoptera: Pyralidae) in North America. *Florida Entomologist* 84, 513–526.

Perry, L.G., Blumenthal, D.M., Monaco, T.A., Paschke, M.W. and Redente, E.F. (2010) Immobilizing nitrogen to control plant invasion. *Oecologia* 163, 13–24.

Pietikainen, J., Kiikkila, O. and Fritze, H. (2000) Charcoal as a habitat for microbes and its effect on the microbial community of the underlying humus. *Oikos* 89, 231–242.

Pritekel, C., Whittemore-Olson, A., Snow, N. and Moore, J.C. (2006) Impacts from invasive plant species and their control on the plant community and belowground ecosystem at Rocky Mountain National Park, USA. *Applied Soil Ecology* 32, 132–141.

Rice, E.L. (1984) *Allelopathy*. Academic Press, Orlando, Florida.

Ridenour, W.M. and Callaway, R.M. (2001) The relative importance of allelopathy in interference: the effects of an invasive weed on a native bunchgrass. *Oecologia* 126, 444–450.

Romeo, J.T. (2000) Raising the beam: moving beyond phytotoxicity. *Journal of Chemical Ecology* 26, 2011–2014.

Rout, M.E. and Callaway, R.M. (2009) An invasive plant paradox. *Science* 324, 734–735.

Rowe, H.I., Brown, C.S. and Claassen, V.P. (2007) Comparisons of mycorrhizal responsiveness with field soil and commercial inoculum for six native montane species and *Bromus tectorum*. *Restoration Ecology* 15, 44–52.

Rowe, H.I., Brown, C.S. and Paschke, M.W. (2009) The influence of soil inoculum and nitrogen availability on restoration of high-elevation steppe communities invaded by *Bromus tectorum*. *Restoration Ecology* 17, 686–694.

Schimel, J.P. and Bennett, J. (2004) Nitrogen mineralization: challenges of a changing paradigm. *Ecology* 85, 591–602.

Schmidt, S.K., Costello, E.K., Nemergut, D.R.,

Cleveland, C.C., Reed, S.C., Weintraub, M.N., Meyer, A.F. and Martin, A.M. (2007) Biogeochemical consequences of rapid microbial turnover and seasonal succession in soil. *Ecology* 88, 1379–1385.

Seastedt, T.R. (1984) The role of microarthropods in decomposition and mineralization processes. *Annual Review of Entomology* 29, 25–46.

Shafroth, P.B., Cleverly, J.R., Dudley, T.L., Taylor, J.P., Van Riper, C., Weeks, E.P. and Stuart, J.N. (2005) Control of Tamarix in the Western United States: implications for water salvage, wildlife use, and riparian restoration. *Environmental Management* 35, 231–246.

Sheley, R.L., Svejcar, T.J. and Maxwell, B.D. (1996) A theoretical framework for developing successional weed management strategies on rangeland. *Weed Technology* 10, 766–773.

Stacy, M.D., Perryman, B.L., Stahl, P.D. and Smith, M.A. (2005) Brome control and microbial inoculation effects in reclaimed cool-season grasslands. *Rangeland Ecology and Management* 58, 161–166.

Torsvik, V., Ovreas, L. and Thingstad, T.F. (2002) Prokaryotic diversity – magnitude, dynamics, and controlling factors. *Science* 296, 1064–1066.

van der Heijden, M.G.A., Klironomos, J.N., Ursic, M., Moutoglis, P., Streitwolf-Engel, R., Boller, T., Wiemken, A. and Sanders, I.R. (1998) Mycorrhizal fungal diversity determines plant biodiversity, ecosystem variability and productivity. *Nature* 396, 69–72.

van der Putten, W.H., Klironomos, J.N. and Wardle, D.A. (2007) Microbial ecology of biological invasions. *Isme Journal* 1, 28–37.

Vanderhoeven, S., Dassonville, N. and Meerts, P. (2005) Increased topsoil mineral nutrient concentrations under exotic invasive plants in belgium. *Plant and Soil* 275, 169–179.

Vitousek, P.M. and Walker, L.R. (1989) Biological invasion by *Myrica faya* in Hawaii – plant demography, nitrogen-fixation, ecosystem effects. *Ecological Monographs* 59, 247–265.

Vitousek, P.M., Walker, L.R., Whiteaker, L.D., Muellerdombois, D. and Matson, P.A. (1987) Biological invasion by Myrica-Faya alters ecosystem development in Hawaii. *Science* 238, 802–804.

Vitousek, P.M., Mooney, H.A., Lubchenco, J. and Melillo, J.M. (1997) Human domination of Earth's ecosystems. *Science* 277, 494–499.

Wall, D.H. and Moore, J.C. (1999) Interactions underground – soil biodiversity, mutualism, and ecosystem processes. *Bioscience* 49, 109–117.

Wall, D.H. and Virginia, R.A. (1997) The World Beneath Our Feet: Soil Biodiversity and

Ecosystem Functioning. In: Raven, P.H. (ed.) *Nature and Human Society: The Quest for a Sustainable World*. National Academy of Sciences Press, Washington, DC, pp. 225–241.

Wardle, D.A., Nicholson, K.S., Ahmed, M. and Rahman, A. (1994) Interference effects of the invasive plant *Carduus nutans* L. against the nitrogen fixation ability of *Trifolium repens* L. *Plant and Soil* 163, 287–297.

Wardle, D.A., Nilsson, M.C., Gallet, C. and Zackrisson, O. (1998) An ecosystem-level perspective of allelopathy. *Biological Reviews of the Cambridge Philosophical Society* 73, 305–319.

Wardle, D.A., Bardgett, R.D., Klironomos, J.N., Setala, H., van der Putten, W.H. and Wall, D.H. (2004) Ecological linkages between above-ground and belowground biota. *Science* 304, 1629–1633.

Weber, E. (2003) *Invasive Plant Species of the World: a reference guide to environmental weeds*. CAB International, Wallingford, UK.

Weidenhamer, J.D. (1996) Distinguishing resource competition and chemical interference: overcoming the methodological impasse. *Agronomy Journal* 88, 866–875.

Westover, K.M., Kennedy, A.C. and Kelley, S.E. (1997) Patterns of rhizosphere microbial community structure associated with co-occurring plant species. *Journal of Ecology* 85, 863–873.

Wolfe, B.E. and Klironomos, J.N. (2005) Breaking new ground: soil communities and exotic plant invasion. *Bioscience* 55, 477–487.

Part II

Principles and Practices to Influence
Ecosystem Change

6

Weather Variability, Ecological Processes, and Optimization of Soil Micro-environment for Rangeland Restoration

Stuart P. Hardegree, Jaepil Cho, and Jeanne M. Schneider

US Department of Agriculture, Agricultural Research Service, USA

Introduction

Precipitation, solar radiation, wind speed, air temperature, and humidity are principal drivers controlling energy and water flux in plant communities. Climate is defined as the long-term average representation of these variables, and their seasonal pattern. Rangelands are generally characterized by an arid or semi-arid climate with plant communities dominated by grassland, shrub-steppe, and savanna vegetation. Gross climatic variability generally determines the suitability of both native and introduced plant materials for rangeland restoration and rehabilitation (Shown *et al.*, 1969; Shiflet, 1994; Barbour and Billings, 2000; Vogel *et al.*, 2005; USDA, 2006). Unfortunately, the micro-environmental requirements for germination, emergence, and seedling establishment are much more restrictive than the longer term climatic requirements for maintenance of mature plant communities (Call and Roundy, 1991; Peters, 2000; Hardegree *et al.*, 2003).

Weather is also the combined expression of climatic variables, but over a relatively short time scale. Rangelands are relatively dry in a climatological sense, but exhibit high spatial and temporal variability in weather (Rajagopalan and Lall, 1998). Figure 6.1 demonstrates annual and seasonal variability in precipitation for a rangeland site in southern Idaho, USA that receives 295 mm of average annual precipitation. Weather is even more variable at the level of seasonal distribution (Fig. 6.1; Table 6.1), which is the relevant scale for critical phases of germination, emergence, and seedling establishment.

Individual plants respond to their local micro-environment in which weather inputs are moderated by interactions with other plants, the soil surface, and lower soil layers (Campbell and Norman, 1998). Soil development is itself affected by weather and climate, but also by the resident vegetation, underlying parent material, topography, soil age, and other factors, all of which are highly variable over space (Jenny, 1980). The soil micro-environment is most variable in near-surface layers that respond quickly to changes in atmospheric variables (Flerchinger and Hardegree, 2004). Deeper soil layers can store precipitation and buffer establishing plants from unfavorable weather conditions, but the majority of mortality events among seeded species occur relatively near the soil surface at, or before, the relatively vulnerable stage of seedling emergence (James *et al.*, 2011). Existing plant cover shades the soil surface and provides protection from wind-driven evaporation, but competition for moisture from invasive weeds can be a dominant limiting factor for water availability in the seedbed. Seedbed preparation and planting

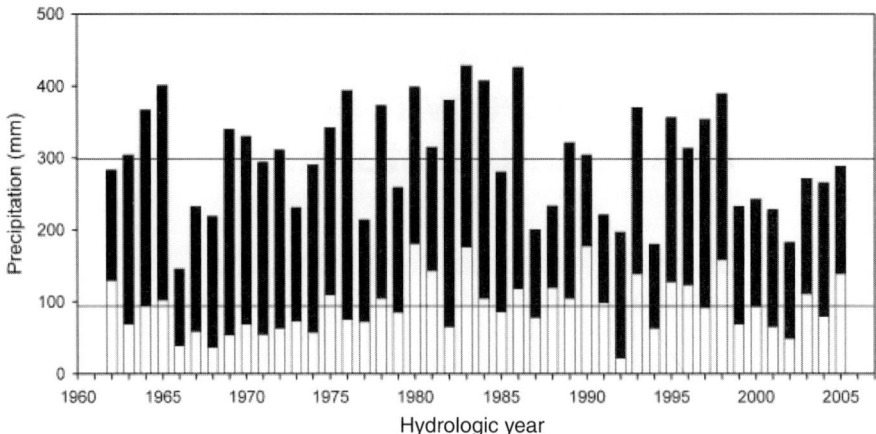

Fig. 6.1. Annual (black bars) and March–May precipitation (white bars) for Boise, Idaho, USA. Mean annual precipitation and standard error of the mean (SE) for this location is 295 ± 11 mm but the standard deviation of the mean (SD) is 68 mm with a coefficient of variation (CV) of 24%. March–May precipitation is relatively more variable with a mean and SE of 93 ± 6 mm, an SD of 38 mm, and CV of 41%.

Table 6.1. Mean monthly precipitation (mm) for rangelands with similar annual precipitation. Numbers in parentheses represent the standard deviation of the mean.

	Boise (USA)	Urumqi (China)	Windhoek (Namibia)	Neuquen (Argentina)	Ivanhoe (Australia)
January	35 (20)	8 (6)	64 (60)	13 (20)	33 (43)
February	27 (17)	9 (7)	77 (61)	17 (25)	30 (42)
March	31 (18)	19 (13)	70 (65)	26 (36)	29 (36)
April	29 (18)	31 (20)	24 (22)	22 (29)	20 (28)
May	33 (26)	33 (23)	4 (7)	19 (18)	29 (25)
June	22 (18)	33 (23)	2 (7)	26 (26)	23 (19)
July	7 (8)	28 (21)	1 (2)	18 (19)	24 (20)
August	7 (12)	21 (21)	0 (1)	15 (17)	24 (20)
September	14 (16)	24 (17)	2 (3)	19 (23)	23 (19)
October	20 (15)	21 (14)	10 (18)	28 (25)	31 (31)
November	34 (16)	18 (10)	18 (21)	19 (22)	24 (23)
December	36 (22)	13 (8)	27 (27)	19 (29)	24 (27)
Annual	295 (68)	258 (74)	299 (141)	241 (144)	314 (131)

methods are designed to optimize micro-environmental conditions for planted species, to increase the number of favorable microsites for germination and establishment, and to mitigate or control competition for water and other resources from undesirable species (Roundy and Call, 1988; Call and Roundy, 1991; Sheley *et al.*, 1996, 2006; Krueger-Mangold *et al.*, 2006).

Published research on rangeland restor-ation seldom addresses the issue of weather variability *per se*. Hardegree *et al.* (2011) surveyed the rangeland planting literature for the western USA and observed that less than 60% of studies reported weather conditions, and less than half were replicated for year effects that would have included weather as a variable factor. In studies that did report weather conditions, successful establishment was almost always associated with what would be considered average or above average precipitation during the study period. This implies that climatic thresholds exist, below which, management actions have little effect on establishment success. In practice, most rangeland restoration activities make limited use of weather and climate information. Climate and gross soil differences are generally considered only in the selection of appropriate plant materials based on the potential distribution of mature-plant communities. Seedbed preparation and planting techniques are designed to optimize seedbed micro-environment but these methodologies are used prescriptively regardless of historical or potential weather conditions. Weather information is mostly used retrospectively to explain seeding failure.

Rangeland restoration practices are generally viewed in the context of a successional model that recognizes the causes of succession and ecological processes that are amenable to management (Pickett *et al.*, 1987; Westoby *et al.*, 1989; Whisenant, 1999; Bestelmeyer *et al.*, 2003; Roundy, 2005; Krueger-Mangold *et al.*, 2006; Sheley *et al.*, 2006). Sheley *et al.* (2010) and James *et al.* (2010) outline these processes, and the ecological principles that define management alternatives for invasive plant management and rangeland restoration. All of the ecological processes underlying succession are directly or indirectly affected by weather and climate. Sheley *et al.* (2010) also describe a generalized planning cycle that includes initial site assessment, monitoring, adaptive management, and reassessment of management effects. Site assessment and adaptive management are two steps in the planning cycle that can be significantly improved by incorporation of weather and climate information. In the following sections, we use the ecologically based model described by Sheley *et al.* (2010) to structure a discussion of weather and climate impacts on some key processes affecting succession, the micro-environmental underpinnings of some important restoration management tools, and strategies to incorporate weather information into the restoration-planning cycle.

Weather, Climate, and Successional Processes

Site availability and optimization of seedbed micro-environment

Disturbance is the principal ecological process affecting site availability for rangeland plant establishment (Sheley *et al.*, 2010). Soil disturbance can favor invasive over desirable species, but some type of mechanical disturbance is generally necessary to create safe sites for establishment of desirable, late-successional species (James *et al.*, 2010). Seedbed preparation and planting methods are designed to reduce water loss and mitigate adverse thermal conditions in the seed zone. This is generally accomplished through mechanical disturbance, soil firming and surface modification, control of seeding depth, and less frequently, the application of soil surface amendments (Roundy and Call, 1988; Sheley *et al.*, 1996).

Soil surface modification is often used to increase water availability to the seed. Alternative seedbed preparation strategies can alter seedbed micro-environment by improving seed–soil contact, reducing the amount of surface area subject to evaporation, improving infiltration and water holding capacity, and creating microsites that capture and retain water (McGinnies, 1959; Roundy *et al.*, 1992). Animal trampling, land imprinting, pitting, furrowing, and rolling treatments have all been used to create microsites that capture or preserve moisture, and to improve hydraulic conductivity by pressing surface-applied seeds into the soil (Hyder and Sneva, 1956; McGinnies, 1962; Haferkamp *et al.*, 1987; Winkel and Roundy, 1991; Roundy *et al.*, 1992). Post-disturbance soil firming can improve hydraulic conductivity to drilled seeds by reducing soil surface area and soil macroporosity (Hyder and Sneva, 1956; McGinnies, 1962). Various studies have also reported differential establishment success relative to the position of soil surface features that affect local micro-environment (Bragg and Stephens, 1979; Eckert *et al.*, 1986; Roundy *et al.*, 1992). Surface modification, however, can also cause seed burial beyond establishment depth (Kincaid and Williams, 1966; Slayback and Renney, 1972; Winkel *et al.*, 1991a).

Mulch application can reduce water loss and moderate soil surface temperatures but is generally considered too expensive for extensive rangeland use (Lavin *et al.*, 1981; McGinnies, 1987; Ethridge *et al.*, 1997). Hardegree *et al.* (2011) found that mulch application improved seedling establishment in only 62% of 21 rangeland seeding studies surveyed. Mulch application has a more consistent record, however, for erosion control and soil stabilization, which can have positive secondary effects on soil microsite availability (Bautista *et al.*, 1996; Brockway *et al.*, 2002; Benik *et al.*, 2003).

Depth of planting is a critical factor in successful plant establishment as microsite favorability is depth dependent (Young *et al.*, 1990; Winkel and Roundy, 1991; Chambers and MacMahon, 1994; Ott *et al.*, 2003). The physical rationale for depth recommendations is based on a tradeoff between increased water availability and increased energy requirements for emergence as a function of depth (Roundy and Call, 1988; Call and Roundy, 1991). Evidence for depth effects is generally limited to relatively small but detailed studies conducted in the laboratory, and less frequently, in the field (Kinsinger, 1962; Vogel, 1963; Hull, 1964). There are many rangeland seeding studies that compare establishment success of broadcast and planted seeds (e.g. Nelson *et al.*, 1970; Wood *et al.*, 1982; Haferkamp *et al.*, 1987; Ott *et al.*, 2003), but very few have actually characterized post-planting seed depth (Winkel and Roundy, 1991; Winkel *et al.*, 1991a, b). These studies generally support the hypothesis that very small seeds establish more frequently from near-surface seed placement, larger seeds require soil cover for maximal performance, and seed performance drops dramatically below some threshold of depth (Hull, 1948; Stewart, 1950; Douglas *et al.*, 1960). Seeding depth recommendations can be fairly specific, but are based on rules of thumb regarding seeding depth as a function of seed size (Plummer *et al.*, 1968; Jensen *et al.*, 2001;

Monsen and Stevens, 2004; Lambert, 2005; Ogle *et al.*, 2008a, b). The decision to broadcast or plant seeds, however, is often mandated by topographic complexity and economic considerations.

Seeding rate and optimization of species availability

Rangeland restoration is often conducted in areas that have lost their source of native plant materials, and availability of desirable species can only be addressed by seeding. Seeding rate recommendations are linked to microclimatic considerations as increased seed numbers increase the probability of seeds reaching safe microsites, irrespective of active depth management (Harper *et al.*, 1965; Roundy *et al.*, 1992; Chambers, 1995). Most individual studies reporting effects of seeding rate on establishment success are not replicated sufficiently to evaluate interactions with annual and seasonal variability in weather conditions. The combined literature, however, supports the concept that higher seeding rates may enhance the likelihood of successful initial establishment (Vogel, 1987; Sheley *et al.*, 1999; Wiedemann and Cross, 2000; Williams *et al.*, 2002; Eisworth and Shonkwiler, 2006; Hardegree *et al.*, 2011). Broadcast seeding rates are generally recommended at 2–3 times the rates for planted seeds to increase the likelihood that sufficient seeds find safe sites for establishment (e.g. Nelson *et al.*, 1970; Wood *et al.*, 1982; Haferkamp *et al.*, 1987; Ott *et al.*, 2003).

Species performance and competition for resources

Climate is the primary criterion for selection of acceptable plant materials for rangeland restoration. This is generally acknowledged in most seeding guides in the form of tables that list species and cultivar suitability as a function of mean annual precipitation (e.g. Jensen *et al.*, 2001; Lambert, 2005; Ogle *et al.*, 2008b). Seeding guides may also cite climatic thresholds below which active seeding practices are not recommended due to the low probability of success (Anderson et al., 1957; Jordan, 1981). Vegetation distribution as a function of climate is the basis of most recommendations for native plant materials (e.g. Barbour and Billings, 2000; USDA, 2006), but cultivar-specific recommendations can also be based on plant materials evaluation and development by local and regional government, and academic organizations (Schwendiman, 1956; Harlan, 1960; Alderson and Sharp, 1994; Asay et al., 2003). Plant materials are often selected or bred for superior establishment, growth, and production within a targeted region or climatic regime (Schwendiman, 1958; Johnson and Asay, 1995; Asay et al., 2003). Current plant materials development and evaluation programs are increasingly focused on para-meters related to species performance and establishment under alternative conditions of weather and climate (Aguirre and Johnson, 1991; Johnson and Asay, 1995; Arredondo et al., 1998; Jensen et al., 2005).

Asay et al. (2001) have argued that the relatively harsh climatic conditions on rangelands may preclude the effective use of many native plant materials, and that it may be prudent to plant more easily established non-native species. Indeed, biodiversity and restoration objectives may require multiple-year strategies for replacement of non-native species only after initial site stabilization and suppression of annual weed competition (Bakker et al., 2003; Cox and Anderson, 2004). Biodiversity and restoration object-ives may need to be addressed only in years when climatic conditions are amenable (Holmgren and Scheffer, 2001; Hardegree et al., 2003; Cox and Anderson, 2004; Hardegree and Van Vactor, 2004).

Species performance can be optimized by selecting the most appropriate season for planting. Optimal planting season is determined by the historical pattern of precipitation and temperature, and the anticipated phenological development of planted species, existing desirable vegetation, and resident weeds. The general objective is to get seeds in the ground before the most favorable season for plant establishment (Plummer et al., 1968; Roundy and Call, 1988; Monsen and Stevens, 2004). The following serve as examples from rangelands in the western USA. Spring is generally the most favorable establishment period in Mediterranean-coastal and semi-arid interior rangelands that are subject to significant summer drought (Douglas et al., 1960; Nord et al., 1971; Harris and Dobrowolski, 1986). The summer monsoon is a critical establishment period in many arid desert environments (Jordan, 1981; Abbot and Roundy, 2003; Hereford et al., 2006). Plant establishment generally occurs in late-spring through early summer in relatively more moderate precipitation zones in the Great Plains (Robertson and Box, 1969; Hart and Dean, 1986; Ries and Hofmann, 1996) and late-spring through early fall in some higher elevation mountain sites (Hull, 1966; Lavin et al., 1973). Post-planting microclimate must be favorable for initial germination and emergence, but also needs to remain favorable during the vulnerable period of seedling establishment (Hyder et al., 1971; McGinnies, 1973; Frasier et al., 1987; Abbot and Roundy, 2003; James et al., 2011).

Dormant-fall seeding is a commonly recommended practice in the Intermountain West, USA. This practice places seeds in the ground well in advance of the optimal growing season to take advantage of all opportunities for germination, emergence, and growth during favorable periods in the winter and spring (Plummer et al., 1968; Nelson et al., 1970; Hart and Dean, 1986; Monsen and Stevens, 2004). Dormant-fall seeding is also recommended when wet spring weather precludes the use of mechanical seeding equipment and to mitigate effects of unpredictable spring weather (McGinnies, 1973; Hart and Dean, 1986). The timing of seeding may also be dependent on seasonal patterns of weed establishment and timing requirements of essential weed control measures (Robocker et al., 1965; Klomp and Hull, 1972). Eiswerth and Shonkwiler (2006) conducted meta-analysis of a large number of rangeland seedings and confirmed the relative benefits of fall/winter-dormant seeding on interior rangeland locations in

Nevada, USA. Very few experimental studies of seeding-season effects, however, are replicated in more than 1 or 2 years (Hull, 1948, 1974; Douglas *et al.*, 1960; Robocker *et al.*, 1965; Ries and Hofmann, 1996). Fall-dormant planting was found to be superior to spring planting in 73% of the individual studies reviewed by Hardegree *et al.* (2011) for rangeland seeding in the Great Basin, USA, although the majority of these studies were conducted in years of relatively favorable precipitation. Early emerging seedlings can take advantage of favorable conditions, but are also more vulnerable to periods of drought, extreme cold, and other mortality factors (James *et al.*, 2011).

Seedbed preparation and planting methods often include strategies to reduce competition for water by undesirable plants (Gonzalez and Dodd, 1979; Ott *et al.*, 2003; Mangold *et al.*, 2007). Cheatgrass (*Bromus tectorum* L.) is a dominant annual weed that has invaded millions of hectares of rangeland in the Intermountain West, USA (Knapp, 1996). Water availability in the seedbed is greatly affected by competition from cheatgrass (Fig. 6.2), which is relatively more efficient than native perennial grasses for both initial establishment and subsequent

resource utilization under conditions of water stress, and when soil temperatures are low in the fall, winter, and early spring (Harris and Wilson, 1970; Harris, 1977; Melgoza *et al.*, 1990; Roundy *et al.*, 2007; Hardegree *et al.*, 2010). Chemical or mechanical weed control is frequently necessary for successful establishment of desirable plant species in areas that are affected by invasive weeds (e.g. Evans and Young, 1978; Humphrey and Schupp, 2002; Mangold *et al.*, 2007). Hardegree *et al.* (2011) observed that out of 52 rangeland seeding studies surveyed that included an evaluation of weed control treatments, all but two concluded that weed control was either necessary, or at least beneficial to successful establishment.

Assessment, Monitoring, and Adaptive Management Tools

State-and-transition probabilities and assessment of weather variability

The need for rangeland restoration begins with a perception that an existing vegetation state is undesirable relative to some

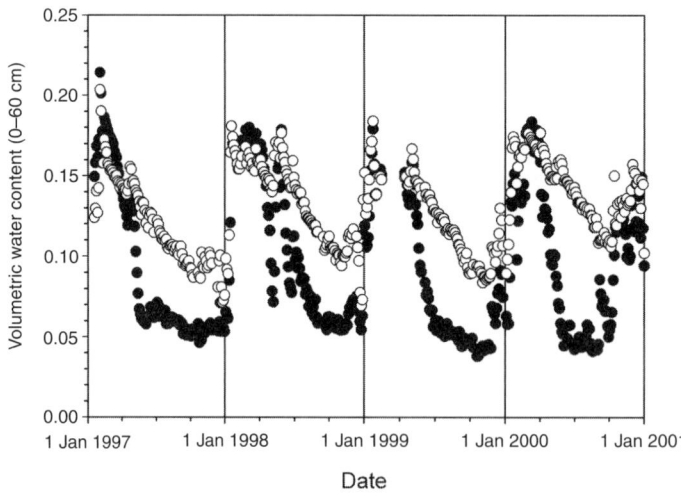

Fig. 6.2. Volume-averaged percent water content over the depth range of 0–60 cm on a loamy-sand (sandy, mixed mesic Xeric Haplargid) soil type in southwestern Idaho, USA. Soil water content was measured with time-domain-reflectometry sensors under replicated (n=3) and interspersed bare soil (open circles) and cheatgrass (closed circles) cover plots.

alternative state. Ecological site descriptions utilize state-and-transition models to define the range of vegetation states that currently exist, that may develop after additional site disturbance, or that may be achievable through management (see Brown and Bestelmeyer, Chapter 1, this volume). Current state-and-transition models acknowledge that the potential trajectories between undesirable and desirable states are limited by weather variability and that some transition pathways may require a specific and perhaps infrequent series of climatic events before a successful change in state can occur (Westoby et al., 1989; Batabyal and Godfrey, 2002; Bestelmeyer et al., 2003; Briske et al., 2008).

Figure 6.3 shows a simplified schematic of alternative states and transition pathways

for the example of sagebrush-steppe vegetation in the Great Basin region, USA. This vegetation type historically maintained a disturbance cycle with a period of 60–100 years that included wildfire, and a post-fire successional sequence that moved through a native bunchgrass phase, and increasing shrub component (Wyoming big sagebrush; *Artemisia tridentata* Nutt. ssp. *wyomingensis* Beetle and Young). Introduced annual grasses, such as cheatgrass, have disrupted this system over millions of hectares resulting in a 5–10 year cycle of wildfire followed by immediate and persistent annual weed cover. Cheatgrass can be expected to establish adequately on most sites and most years on sagebrush-steppe rangeland (Roundy et al., 2007), therefore, the fire/cheatgrass cycle represented in Fig. 6.3 is relatively robust under almost any weather scenario.

Sheley et al. (2011) discuss the initial evaluation of undesirable site conditions, and the use of rangeland health assessment to identify the ecological processes in need of repair (Pyke et al., 2002). We recommend that an additional tool for initial assessment be the evaluation of historical weather variability. This variability may provide valuable perspective on the probability that proposed management actions will have the desired effect on a given ecological process in any given year.

A minimal evaluation of weather should include an assessment of mean annual and monthly patterns of precipitation and temperature. Information on climatological means will be necessary for selection of appropriate plant materials, and will define the seasonality of conditions favorable for plant establishment and growth. Climatological means also will provide baseline conditions for interpreting deviation from average conditions in any given establishment year. Annual variability can be assessed by organizing and ranking historical data of the type displayed in Fig. 6.1 for the Boise precipitation record. Ranking can be used to assess where a given year falls in the spectrum of potential weather conditions and is especially useful in interpreting the relative success of previous management actions.

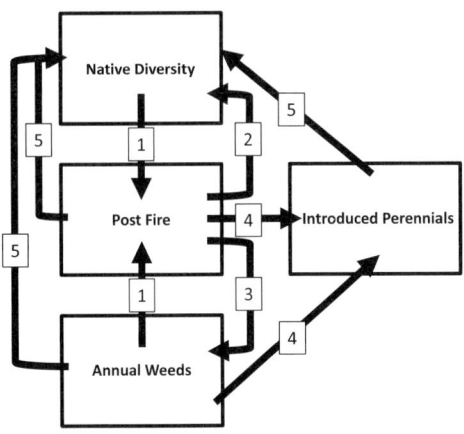

Fig. 6.3. State-and-transition model for sagebrush-bunchgrass rangeland in southern Idaho, USA that has been disturbed by introduced annual weeds. Transition 1 represents fire events, which can happen in any given year. Transition 2 represents natural recovery of native plant diversity in the absence of introduced annual weeds. Transition 3 represents type conversion if introduced weed seeds are present. Transition 4 represents a transition to an introduced perennial grass community, which may be possible only in moderate to favorable precipitation years. Transition 5 represents a transition to a diverse native community that may be possible only in favorable precipitation years with annual weed control (if necessary).

Monitoring and adaptive management

Monitoring is a critical part of ecologically based restoration planning as most management actions involve a relatively high degree of uncertainty (Sheley *et al.*, 2010). This uncertainty is exacerbated by weather variability, and a general lack of information on how weather impacts the relative success of alternative management strategies. Most individual studies in the range planting literature are insufficiently replicated to extract valid inferences about weather effects (Hardegree *et al.*, 2011). The majority of range planting studies, until fairly recently, have not measured critical environmental factors affecting success, such as soil temperature and water relations, but only report relative treatment effects (Call and Roundy, 1991; Vargas *et al.*, 2001). Range planting studies also tend to extrapolate results obtained from atypical sites and weather conditions over larger areas (Cox and Martin, 1984), and are seldom replicated in multiple seeding years to account for inter-annual weather variability (Casler, 1999). It is logistically difficult to obtain field data that replicates inter-annual variability. Unfortunately, previous studies do not include enough commonality in experimental design features to be subject to any detailed meta-analysis of general weather effects (Durlak and Lipsay, 1991; Michener, 1997; Osenberg *et al.*, 1999; Gurevitch *et al.*, 2001). It may be possible to develop guidelines, however, for establishing some common experimental design features for future studies that may be amenable to more sophisticated meta-analysis.

The stochastic nature of weather variability may require adoption of new paradigms for monitoring and evaluating alternative management strategies. Specifically, both restoration failure and success must be evaluated in the context of weather conditions during the period of establishment. Adaptive management alternatives should be viewed in the context of weather ranking during the establishment season being evaluated. If the seasonal conditions were significantly below average, it may not be necessary to abandon strategies that did not seem to work in that particular year. Lessons learned from successful management actions should also be weighed in the context of relative favorability in weather in that year. Multi-year evaluation should be considered when comparing alternative management treatments, and multi-year treatments may be necessary to achieve acceptable levels of establishment success at a given site.

Long-term weather forecasting opportunities

The most useful potential technology for enhancing establishment success lies in development and utilization of relatively long-range weather forecast technology specific to rangeland planting applications (e.g. Barnston *et al.*, 2000, 2005; Garbrecht and Schneider, 2007; Lim *et al.*, 2011). Similar technology is in relatively common use for more traditional agricultural applications (Doblas-Reyes *et al.*, 2006; Baigorria *et al.*, 2008; O'Lenic *et al.*, 2008). Long-term weather forecasts in many rangeland areas are often merely synoptic descriptions of historical weather patterns and not based on physical or empirical prediction of future weather conditions. Even low-resolution weather forecasts, however, would increase the probability of successful native plant establishment if management decisions at the time of seeding could be based on the anticipation of favorable conditions for seed germination, emergence, and seedling establishment (Hardegree *et al.*, 2003; Hardegree and Van Vactor, 2004). Weather forecasts could be used to initiate contingency plans in areas that have been previously identified for restoration, and for which pre-management logistics of equipment, personnel, and plant materials are in place (Westoby *et al.*, 1989; Bakker *et al.*, 2003).

Currently available long-term forecast information is probabilistic in format. Probability of Exceedance (PoE) distributions define the potential deviation of future weather conditions (typically over seasons) from the long-term mean (Barnston *et al.*,

2000; Schneider and Garbrecht, 2006; Garbrecht and Schneider, 2007; Lim *et al.*, 2011). In simpler terms, the forecasts predict the odds for whether the weather will shift toward wetter or drier, warmer or cooler, when compared to the reference period of record. These predictions apply to relatively large spatial domains and are currently subject to relatively high predictive errors for many parts of the globe. Long-term forecast predictions work better in some areas than others and are generally more accurate for temperature than for precipitation (Schneider and Garbrecht, 2006; Livezey and Timoveyeva, 2008; Lim *et al.*, 2011). In the USA, predictive accuracy is relatively higher where extreme weather events and sustained deviations from average are associated with the El Niño Southern Oscillation (ENSO) phenomenon. Long-term weather predictions in the USA are relatively more accurate in the southeast, southern Texas, desert southwest, California, Pacific Northwest, and northern Rocky Mountains, but are relatively less accurate in the Intermountain Great Basin. Other areas of the world, particularly tropical regions in or adjacent to the Pacific Ocean, enjoy a stronger and more reliable predictive signal. The Australian Government Bureau of Meteorology has developed a successful forecast system for ENSO-related impacts on agriculture and water resources in eastern and southern Australia (Lim *et al.*, 2011).

Weather and climate data resources

It can be challenging to locate accurate, continuous, long term, site-specific weather data. Much of the data available over the internet has not been quality assured, or may be a derivative product representing data averaged or interpolated over large areas.

The global database for historical weather data can be accessed through the US Department of Commerce, National Oceanic and Atmospheric Administration (NOAA), National Climate Data Center website (www.ncdc.noaa.gov/oa/ncdc.html). This site is the

repository for US weather information, but also has links to global weather resources, including the World Meteorological Organization that maintains a list of meteorological data sources by country (www.wmo.int/pages/members/members_en.html). NOAA has also been developing Regional Climate Centers that consolidate state weather and climate data (www.ncdc.noaa.gov/oa/climate/regionalclimatecenters.html), and can frequently provide advice on data access and quality assurance questions.

Many field sites will not have sufficient local information for detailed characterization of site variability. Some areas will have additional regional resources for interpolating weather records in areas that do not have local monitoring. Interpolated data will not be as accurate as actual observations but may be sufficient for many practical applications. Tools and databases for interpolation of historical weather data are available in much of the USA through sites such as the DAYMET US Data Center, which is provided by the Montana State University, Numerical Terradynamic Simulation Group (www.daymet.org/default.jsp) and the USDA Natural Resources Conservation Service National Water and Climate Center (www.wcc.nrcs.usda.gov/climate/prism.html).

Long-term (covering weeks to multiple months) weather forecast information is increasingly available in many countries. These forecasts are created through ensemble analysis of multiple global climate models, and analysis of statistical relationships between weather and oceanic state (e.g. ENSO) in different parts of the globe (Barnston *et al.*, 2000, 2005; Lim *et al.*, 2011). Many nations and nation groups have developed, or are developing, regional predictive capabilities such as the European Centre for Medium-Range Weather Forecasts (www.ecmwf.int). Long-term forecasts for precipitation and temperature for Australia are available through the Australian Bureau of Meteorology (www.bom.gov.au/climate/ahead), and for much of the globe through the International Research Institute for Climate and Society (http://portal.iri.columbia.edu/portal/ server.pt/). The principal repository of long-term forecast data and information in the USA is the NOAA National Weather Service's Climate Prediction Center (www.cpc.ncep. noaa.gov).

Incorporating weather variability and long-term forecasts in rangeland restoration efforts

As a first step, assessment of historical weather variability should be made for the rangeland site of interest. Monthly total precipitation and average daily air temperature are relatively easy to obtain and are sufficient to provide a baseline for assessing rangeland restoration options, and for considering the possible use of long-term forecasts. The World Meteorological Organization produces a product representing the most recent 30-year period, revised every decade (currently 1981–2010). It is highly desirable to find the most local monthly data possible, or barring that, a good interpolation for the area of interest, for these 30 years. Once obtained, monthly, seasonal, and annual data can be totaled (precipitation) and averaged (air temperature) to establish baseline data for ranking of current and historical time periods relative to average conditions for a given time period.

Forecasting applications are generally applied to 3-month intervals rather than individual monthly predictions (e.g. January–February–March, February–March–April, etc.). It is critical to understand that forecasts are probabilistic predictions and subject to relatively large errors. The authors recommend considering only those precipitation or air temperature forecasts predicting a shift in odds of at least 8% (Schneider and Garbrecht, 2006), with more serious consideration for larger forecast shifts, especially during moderate to strong ENSO conditions. Such strong forecasts are rarely offered for precipitation but may represent a reasonable guide for the efficient expenditure of rangeland restoration resources when they do occur.

References

Abbott, L.B. and Roundy, B.A. (2003) Available water influences field germination and recruitment of seeded grasses. *Journal of Range Management* 56, 56–64.

Aguirre, L. and Johnson, D.A. (1991) Root morphological development in relation to shoot growth in seedlings of four range grasses. *Journal of Range Management* 44, 341–346.

Alderson, J. and Sharp, W.C. (1994) *Grass Varieties in the United States.* Agricultural Handbook No 170. US Department of Agriculture, Washington, DC.

Anderson, D., Hamilton, L.P., Reynolds, H.G. and Humphrey, R.R. (1957) *Reseeding Desert Grassland Ranges in Southern Arizona.* Arizona Agricultural Experiment Station Bulletin 249, Tucson, Arizona.

Arredondo, J.T., Jones, T.A. and Johnson, D.A. (1998) Seedling growth of Intermountain perennial and weedy annual grasses. *Journal of Range Management* 51, 584–589.

Asay, K.H., Horton, W.H., Jensen, K.B. and Palazzo, A.J. (2001) Merits of native and introduced Triticeae grasses on semiarid rangelands. *Canadian Journal of Plant Science* 81, 45–52.

Asay, K.H., Chatterton, N.J., Jensen, K.B., Jones, T.A., Waldron, B.L. and Horton, W.H. (2003) Breeding improved grasses for semiarid rangelands. *Arid Land Research and Management* 17, 469–478.

Baigorria, G.A., Jones, J.W. and O'Brien, J.J. (2008) Potential predictability of crop yield using an ensemble climate forecast by a regional circulation model. *Agricultural and Forest Meteorology* 148, 1353–1361.

Bakker, J.D., Wilson, S.D., Christian, J.M., Li, X.D., Ambrose, L.G. and Waddington, J. (2003) Contingency of grassland restoration on year, site, and competition from introduced grasses. *Ecological Applications* 13, 137–153.

Barbour, M.G. and Billings, W.D. (2000) *North American Terrestrial Vegetation.* Cambridge University Press, New York.

Barnston, A.G., He, Y. and Unger, D.A. (2000) A forecast product that maximizes utility for state-of-the-art seasonal climate prediction. *Bulletin of the American Meteorological Society* 81, 1271–1279.

Barnston, A.G., Kumar, A., Goddard, L. and Hoerling, M.P. (2005) Improving seasonal prediction practices through attribution of climate variability. *Bulletin of the American Meteorological Society* 86, 59–72.

Batabyal, A.A. and Godfrey, E.B. (2002) Rangeland management under uncertainty: a conceptual approach. *Journal of Range Management* 55, 12–15.

Bautista, S., Bellot, J. and Vallejo, V.R. (1996) Mulching treatment for postfire soil conservation in a semiarid ecosystem. *Arid Soil Research and Rehabilitation* 10, 235–242.

Benik, S.R., Wilson, B.N., Biesboer, D.D., Hansen, B. and Stenlund, D. (2003) Evaluation of erosion control products using natural rainfall events. *Journal of Soil and Water Conservation* 58, 98–105.

Bestelmeyer, B.T., Brown, J.R., Havstad, K.M., Alexander, R., Chavez, G. and Herrick, J. (2003) Development and use of state-and-transition models for rangelands. *Journal of Range Management* 56, 114–126.

Bragg, T.B. and Stephens, L.J. (1979) Effects of agricultural terraces on the re-establishment of bluestem grasslands. *Journal of Range Management* 32, 437–441.

Briske, D.D., Bestelmeyer, B.T., Stringham, T.K. and Shaver, P.L. (2008) Recommendations for development of resilience-based state-and-transition models. *Rangeland Ecology and Management* 61, 359–367.

Brockway, D.G., Gatewood, R.G. and Paris, R.B. (2002) Restoring grassland savannas from degraded pinyon-juniper woodlands: effects of mechanical overstory reduction and slash treatment alternatives. *Journal of Environmental Management* 64, 179–197.

Call, C.A. and Roundy, B.A. (1991) Perspectives and processes in revegetation of arid and semiarid rangelands. *Journal of Range Management* 44, 543–549.

Campbell, G.S. and Norman, J.M. (1998) *An Introduction to Environmental Biophysics*, 2nd edn. Springer-Verlag, New York.

Casler, M.D. (1999) Repeated measures vs. repeated plantings in perennial forage grass trials: an empirical analysis of precision and accuracy. *Euphytica* 105, 33–42.

Chambers, J.C. (1995) Relationships between seed fates and seedling establishment in an alpine ecosystem. *Ecology* 76, 2124–2133.

Chambers, J.C. and MacMahon, J.A. (1994) A day in the life of a seed: movements and fates of seeds and their implications for natural and managed systems. *Annual Review of Ecology and Systematics* 25, 263–292.

Cox, J.R. and Martin, M.H. (1984) Effects of planting depth and soil texture on the emergence of four lovegrasses. *Journal of Range Management* 37, 204–205.

Cox, R.D. and Anderson, V.J. (2004) Increasing native diversity of cheatgrass-dominated rangeland through assisted succession. *Journal of Range Management* 57, 203–210.

Doblas-Reyes, F.J., Hagedorn, R. and Palmer, T.N. (2006) Developments in dynamical seasonal forecasting relevant to agricultural management. *Climate Research* 33, 19–26.

Douglas, D.S., Hafenrichter, A.L. and Klages, K.H. (1960) Cultural methods and their relation to establishment of native and exotic grasses in range seedings. *Journal of Range Management* 13, 53–57.

Durlak, J.A. and Lipsay, M.W. (1991) A practitioner's guide to metaanalysis. *American Journal of Community Psychology* 19, 291–332.

Eckert, R.E., Peterson, F.F., Meurisse, M.S. and Stephens, J.L. (1986) Effects of soil-surface morphology on emergence and survival of seedlings in big sagebrush communities. *Journal of Range Management* 39, 414–420.

Eiswerth, M.E. and Shonkwiler, J.S. (2006) Examining post-wildfire reseeding on arid rangeland: a multivariate tobit modelling approach. *Ecological Modelling* 192, 286–298.

Ethridge, D.E., Sherwood, R.D., Sosebee, R.E. and Herbel, C.H. (1997) Economic feasibility of rangeland seeding in the arid southwest. *Journal of Range Management* 50, 185–190.

Evans, R.A. and Young, J.A. (1978) Effectiveness of rehabilitation practices following wildfire in a degraded big sagebrush-downy brome community. *Journal of Range Management* 31, 185–188.

Flerchinger, G.N. and Hardegree, S.P. (2004) Modelling near-surface soil temperature and moisture for germination response predictions in post-wildfire seedbeds. *Journal of Arid Environments* 59, 369–385.

Frasier, G.W., Cox, J.R. and Woolhiser, D.A. (1987) Wet-dry cycle effects on warm-season grass seedling establishment. *Journal of Range Management* 40, 2–6.

Garbrecht, J.D. and Schneider, J.M. (2007) Climate forecast and prediction product dissemination for agriculture in the United States. *Australian Journal of Agricultural Research* 58, 966–974.

Gonzalez, C.L. and Dodd, J.D. (1979) Production response of native and introduced grasses to mechanical brush manipulation, seeding, and fertilization. *Journal of Range Management* 32, 305–309.

Gurevitch, J., Curtis, P.S. and Jones, M.H. (2001) Meta-analysis in ecology. *Advances in Ecological Research* 32, 199–247.

Haferkamp, M.R., Ganskopp, D.C., Miller, R.F. and Sneva, F.A. (1987) Drilling versus imprinting for establishing crested wheatgrass in the sagebrush-bunchgrass steppe. *Journal of Range Management* 40, 524–530.

Hardegree, S.P. and Van Vactor, S.S. (2004)

Microclimatic constraints and revegetation planning in a variable environment. *Weed Technology* 18, 1213–1215.

Hardegree, S.P., Flerchinger, G.N. and Van Vactor, S.S. (2003) Hydrothermal germination response and the development of probabilistic germination profiles. *Ecological Modelling* 167, 305–322.

Hardegree, S.P., Moffet, C.A., Roundy, B.A., Jones, T.A., Novak, S.J., Clark, P.E., Pierson, F.B. and Flerchinger, G.N. (2010) A comparison of cumulative-germination response of cheatgrass (*Bromus tectorum* L.) and five perennial bunchgrass species to simulated field-temperature regimes. *Environmental and Experimental Botany* 69, 320–327.

Hardegree, S.P., Jones, T.A., Roundy, B.A., Shaw, N.L. and Monaco, T.A. (2011) Assessment of range planting as a conservation practice. In: Briske, D.D. (ed.) *Conservation Benefits of Rangeland Practices: Assessment, Recommendations, and Knowledge Gaps.* Allen Press, Lawrence, Kansas. pp. 171–212

Harlan, J.R. (1960) Breeding superior forage plants for the Great Plains. *Journal of Range Management* 13, 86–89.

Harper, J.L., Williams, J.T. and Sagar, G.R. (1965) The behavior of seeds in soil I. The heterogeneity of soil surfaces and its role in determining the establishment of plants from seed. *Journal of Ecology* 53, 273–286.

Harris, G.A. (1977) Root phenology as a factor of competition among grass seedlings. *Journal of Range Management* 30, 172–177.

Harris, G.A. and Dobrowolski, J.P. (1986) Population dynamics of seeded species on northeast Washington semiarid sites 1948-1983. *Journal of Range Management* 39, 46–51.

Harris, G.A. and Wilson, A.M. (1970) Competition for moisture among seedlings of annual and perennial grasses as influenced by root elongation at low temperature. *Ecology* 51, 530–534.

Hart, R.H. and Dean, J.G. (1986) Forage establishment: weather effects on stubble vs. fallow and fall vs. spring seeding. *Journal of Range Management* 39, 228–230.

Hereford, R., Webb, R.H. and Longpre, C.I. (2006) Precipitation history and ecosystem response to multidecadal precipitation variability in the Mojave Desert region, 1893-2001. *Journal of Arid Environments* 67, 13–34.

Holmgren, M. and Scheffer, M. (2001) El Nino as a window of opportunity for the restoration of degraded arid ecosystems. *Ecosystems* 4, 151–159.

Hull, A.C. (1948) Depth, season and row spacing for planting grasses on southern Idaho range lands. *Journal of the American Society of Agronomy* 40, 960–969.

Hull, A.C. (1964) Emergence of cheatgrass and three wheatgrasses from four seeding depths. *Journal of Range Management* 17, 32–35.

Hull, A.C. (1966) Emergence and survival of intermediate wheatgrass and smooth brome seeded on a mountain range. *Journal of Range Management* 19, 279–283.

Hull, A.C. (1974) Seedling emergence and survival from different seasons and rates of seeding mountain rangelands. *Journal of Range Management* 27, 302–304.

Humphrey, L.D. and Schupp, E.W. (2002) Seedling survival from locally and commercially obtained seeds on two semiarid sites. *Restoration Ecology* 10, 88–95.

Hyder, D.N. and Sneva, F.A. (1956) Seed and plant-soil relations as affected by seedbed firmness on a sandy loam rangeland soil. *Soil Science Society of America Proceedings* 20, 416–419.

Hyder, D.N., Everson, A.C. and Bement, R.E. (1971) Seedling morphology and seeding failures with blue grama. *Journal of Range Management* 24, 287–292.

James, J.J., Smith, B.S., Vasquez, E.A. and Sheley, R.L. (2010) Principles for ecologically based invasive plant management. *Invasive Plant Science and Management* 3, 229–239.

James, J.J., Svejcar, T.J. and Rinella, M.J. (2011) Demographic processes limiting seedling recruitment in arid grassland restoration. *Journal of Applied Ecology* 48, 961–969.

Jenny, H. (1980) *The Soil Resource: Origin and Behavior.* Ecological Studies No. 37. Springer, New York.

Jensen, K.B., Horton, W.H., Reed, R. and Whitesides, R.E. (2001) *Intermountain Planting Guide.* Extension Publication AG510. Utah State University, Logan, Utah.

Jensen, K.B., Peel, M.D., Waldron, B.L., Horton, W.H. and Asay, K.H. (2005) Persistence after three cycles of selection in NewHy RS-Wheatgrass (*Elymus hoffmannii* K.B. Jensen & Asay) at increased salinity levels. *Crop Science* 45, 1717–1720.

Johnson, D.A. and Asay, K.H. (1995) Breeding and selection of grasses for improved drought response: a review. *Annals of Arid Zone* 34, 163–178.

Jordan, G.L. (1981) *Range Seeding and Brush Management on Arizona Rangelands.* Bulletin T81121. Agricultural Experiment Station, University of Arizona College of Agriculture, Tucson, Arizona.

Kincaid, D.R. and Williams, G. (1966) Rainfall

effects on soil surface characteristics following range improvement treatments. *Journal of Range Management* 19, 346–351.

Kinsinger, F.E. (1962) The relationship between depth of planting and maximum foliage height of seedlings of Indian ricegrass. *Journal of Range Management* 15, 10–13.

Klomp, G.J. and Hull, A.C. (1972) Methods for seeding three perennial wheatgrasses on cheatgrass ranges in southern Idaho. *Journal of Range Management* 25, 266–268.

Knapp, P.A. (1996) Cheatgrass (*Bromus tectorum* L.) dominance in the Great Basin desert – history, persistence, and influences to human activities. *Global Environmental Change – Human and Policy Dimensions* 6, 37–52.

Krueger-Mangold, J.M., Sheley, R.L. and Svejcar, T.J. (2006) Toward ecologically-based invasive plant management on rangeland. *Weed Science* 54, 597–605.

Lambert, S. (2005) *Guidebook to the Seeds of Native and Non-native Grasses, Forbs and Shrubs of the Great Basin.* Technical Bulletin 2005 04, Bureau of Land Management, Boise, Idaho.

Lavin, F., Gomm, F.B. and Johnson, T.N. (1973) Cultural, seasonal, and site effects on pinyon-juniper rangeland plantings. *Journal of Range Management* 26, 279–285.

Lavin, F., Johnsen, T.N. and Gomm, F.B. (1981) Mulching, furrowing, and fallowing of forage plantings on Arizona pinyon-juniper ranges. *Journal of Range Management* 34, 171–177.

Lim, E., Hendon, H.H., Anderson, D.L.T., Charles, A. and Alves, O. (2011) Dynamical, statistical-dynamical, and multimodel ensemble forecasts of Australian spring season rainfall. *Monthly Weather Review* 139, 958–975.

Livezey, R.A. and Timofeyeva, M.M. (2008) The first decade of long-lead US seasonal forecasts. *Bulletin of the American Meteorological Society* 89, 843–854.

Mangold, J.M., Poulsen, C.L. and Carpinelli, M.F. (2007) Revegetating Russian knapweed (*Acroptilon repens*) infestations using morphologically diverse species and seedbed preparation. *Rangeland Ecology and Management* 60, 378–385.

McGinnies, W.J. (1959) The relationship of furrow depth to moisture content of soil and to seedling establishment on range soil. *Agronomy Journal* 51, 13–14.

McGinnies, W.J. (1962) Effect of seedbed firming on the establishment of crested wheatgrass seedlings. *Journal of Range Management* 15, 230–234.

McGinnies, W.J. (1973) Effects of date and depth of planting on the establishment of three range grasses. *Agronomy Journal* 65, 120–123.

McGinnies, W.J. (1987) Effects of hay and straw mulches on the establishment of seeded grasses and legumes on rangeland and a coal strip mine. *Journal of Range Management* 40, 119–121.

Melgoza, G., Nowak, R.S. and Tausch, R.J. (1990) Soil water exploitation after fire: competition between *Bromus tectorum* (cheatgrass) and two native species. *Oecologia* 83, 7–13.

Michener, W.K. (1997) Quantitatively evaluating restoration experiments: research design, statistical analysis, and data management considerations. *Restoration Ecology* 5, 324–337.

Monsen, S.B. and Stevens, R. (2004) Seedbed preparation and seeding practices. In: Monson, S.B., Stevens, R. and Shaw, N.L. (eds) *Restoring Western Ranges and Wildlands*, Vol. 1. General Technical Report RMRS-GTR-136. US Department of Agriculture, Forest Service, Fort Collins, Colorado, pp. 121–154.

Nelson, J.R., Wilson, A.M. and Goebel, C.J. (1970) Factors influencing broadcast seeding in bunchgrass range. *Journal of Range Management* 23, 163–170.

Nord, E.C., Hartless, P.F. and Nettleton, W.D. (1971) Effects of several factors on saltbush establishment in California. *Journal of Range Management* 24, 216–223.

Ogle, D., St John, L., Cornwell, J., Stannard, M. and Holzworth, L. (2008a) *Pasture and Range Seedings: Planning-Installation-Evaluation-Management.* Technical Note, PM-TN-10, US Department of Agriculture, Natural Resources Conservation Service, Boise, Idaho.

Ogle, D., St John, L., Stannard, M. and Holzworth, L. (2008b) *Grass, Grass-like, Forb, Legume, and Woody Species for the Intermountain West.* Technical Note, TN Plant Materials No. 24, US Department of Agriculture, Natural Resources Conservation Service, Boise, Idaho.

O'Lenic, E.A., Unger, D.A., Halpert, M.S. and Pelman, K.S. (2008) Developments in operational long-range climate prediction at CPC. *Weather and Forecasting* 23, 496–515.

Osenberg, C.W., Sarnelle, O. and Goldberg, D.E. (1999) Meta-analysis in ecology: concepts, statistics, and applications. *Ecology* 80, 1103–1104.

Ott, J.E., McArthur, E.D. and Roundy, B.A. (2003) Vegetation of chained and non-chained seedings after wildfire in Utah. *Journal of Range Management* 56, 81–91.

Peters, D.P.C. (2000) Climatic variation and simulated patterns in seedling establishment of two dominant grasses at a semi-arid-arid grassland ecotone. *Journal of Vegetation Science* 11, 493–504.

Pickett, S.T.A., Collins, S.L. and Armesto, J.J. (1987) A hierarchical consideration of causes and mechanisms of succession. *Vegetatio* 69, 109–114.

Plummer, P.A., Christensen, D.R. and Monsen, S.B. (1968) *Restoring Big-game Range in Utah*. Publication No. 68-3, Utah Division of Fish and Game, Ephraim, Utah.

Pyke, D.A., Pellant, M., Shaver, P. and Herrick, J.E. (2002) Rangeland health attributes and indicators for qualitative assessment. *Journal of Range Management* 55, 584–597.

Rajagopalan, B. and Lall, U. (1998) Interannual variability in western US precipitation. *Journal of Hydrology* 210, 51–67.

Ries, R.E. and Hofmann, L. (1996) Perennial grass establishment in relationship to seeding dates in the northern Great Plains. *Journal of Range Management* 49, 504–508.

Robertson, T.E. and Box, T.W. (1969) Interseeding sideoats grama on the Texas high plains. *Journal of Range Management* 22, 243–245.

Robocker, W.C., Gates, D.H. and Kerr, H.D. (1965) Effects of herbicides, burning, and seeding date in reseeding an arid range. *Journal of Range Management* 18, 114–118.

Roundy, B.A. (2005) Plant succession and approaches to community restoration. In: Shaw, N.L., Pellant, M. and Monsen, S.B. (eds) *Proceedings of the Sage-Grouse Habitat Restoration Symposium*. US Department of Agriculture, Forest Service, Rocky Mountain Research Station, Fort Collins, Colorado, pp. 43–48.

Roundy, B.A. and Call, C.A. (1988) Revegetation of arid and semiarid rangelands. In: Tueller, P.T. (ed.) *Vegetation Science Applications for Rangeland Analysis and Management*. Kluwer Academic Publishers, Dordrecht, the Netherlands, pp. 607–635.

Roundy, B.A., Winkel, V.K., Khalifa, H. and Matthias, A.D. (1992) Soil water availability and temperature dynamics after one-time heavy cattle trampling and land imprinting. *Arid Soil Research and Rehabilitation* 6, 53–69.

Roundy, B.A., Hardegree, S.P., Chambers, J.C. and Whittaker, A. (2007) Prediction of cheatgrass field germination potential using wet thermal accumulation. *Rangeland Ecology and Management* 60, 613–623.

Schneider, J.M. and Garbrecht, J.D. (2006) Dependability and effectiveness of seasonal forecasts for agricultural applications. *Transactions ASABE* 49, 1737–1753.

Schwendiman, J.L. (1956) Improvement of native range through new grass introduction. *Journal of Range Management* 9, 91–95.

Schwendiman, J.L. (1958) Testing new range forage plants. *Journal of Range Management* 11, 71–76.

Sheley, R.L., Svejcar, T.J. and Maxwell, B.D. (1996) A theoretical framework for developing successional weed management strategies on rangeland. *Weed Technology* 10, 766–773.

Sheley, R.L., Jacobs, J.S. and Velagala, R.P. (1999) Enhancing intermediate wheatgrass establishment in spotted knapweed infested rangeland. *Journal of Range Management* 52, 68–74.

Sheley, R.L., Mangold, J.M. and Anderson, J.L. (2006) Potential for successional theory to guide restoration of invasive-plant-dominated rangeland. *Ecological Monographs* 76, 365–379.

Sheley, R., James, J., Smith, B. and Vasquez, E. (2010) Applying ecologically based invasive-plant management. *Rangeland Ecology and Management* 63, 605–613.

Sheley, R.L., James, J.J., Vasquez, E.A. and Svejcar, T.J. (2011) Using rangeland health assessment to inform successional management. *Invasive Plant Science and Management* 4, 356–366.

Shiflet, T.N. (1994) *Rangeland Cover Types of the United States*. Society for Range Management, Denver, Colorado.

Shown, L.M., Miller, R.F. and Branson, F.A. (1969) Sagebrush conversion to grassland as affected by precipitation, soil, and cultural practices. *Journal of Range Management* 22, 303–311.

Slayback, R.D. and Renney, C.W. (1972) Intermediate pits reduce gamble in range seeding in the southwest. *Journal of Range Management* 25, 224–227.

Stewart, G. (1950) Reseeding research in the intermountain region. *Journal of Range Management* 3, 52–59.

United States Department of Agriculture (USDA) (2006) *Land Resource Regions and Major Land Resource Areas of the United States, the Caribbean, and the Pacific Basin*. Agricultural Handbook, 296. US Department of Agriculture, Natural Resources Conservation Service, Washington, DC.

Vargas, M., Crossa, J., Van Eeuwijk, F., Sayre, K.D. and Reynolds, M.P. (2001) Interpreting treatment × environment interaction in agronomy trials. *Agronomy Journal* 93, 949–960.

Vogel, K.P. (1987) Seeding rates for establishing big bluestem and switchgrass with preemergence atrizine applications. *Agronomy Journal* 79, 509–512.

Vogel, K.P., Schmer, M.R. and Mitchell, R.B. (2005)

Plant adaptation regions: ecological and climatic classification of plant materials. *Rangeland Ecology and Management* 58, 315–319.

Vogel, W.G. (1963) Planting depth and seed size influence emergence of beardless wheatgrass seedlings. *Journal of Range Management* 16, 273–275.

Westoby, M., Walker, B. and Noy-Meir, I. (1989) Opportunistic management for rangelands not at equilibrium. *Journal of Range Management* 42, 266–274.

Whisenant, S.G. (1999) *Repairing Damaged Wildlands: A Process-Oriented, Landscape-Scale Approach*. Cambridge University Press, Cambridge, UK.

Wiedemann, H.T. and Cross, B.T. (2000) Disk chain effects on seeded grass establishment. *Journal of Range Management* 53, 62–67.

Williams, M.I., Schuman, G.E., Hild, A.L. and Vicklund, L.E. (2002) Wyoming big sagebrush density: effects of seeding rates and grass competition. *Restoration Ecology* 10, 385–391.

Winkel, V.K. and Roundy, B.A. (1991) Effects of cattle trampling and mechanical seedbed preparation on grass seedling emergence. *Journal of Range Management* 44, 176–180.

Winkel, V.K., Roundy, B.A. and Blough, D.K. (1991a) Effects of seedbed preparation and cattle trampling on burial of grass seeds. *Journal of Range Management* 44, 171–175.

Winkel, V.K., Roundy, B.A. and Cox, J.R. (1991b) Influence of seedbed microsite characteristics on grass seedling emergence. *Journal of Range Management* 44, 210–214.

Wood, M.K., Eckert, R.E., Blackburn, W.H. and Peterson, F.F. (1982) Influence of crusting soil surfaces on emergence and establishment of crested wheatgrass, squirreltail, thurber needlegrass and fourwing saltbush. *Journal of Range Management* 35, 282–287.

Young, J.A., Evans, R.A. and Palmquist, D. (1990) Soil surface characteristics and emergence of big sagebrush seedlings. *Journal of Range Management* 43, 358–367.

7

The Effects of Plant–Soil Feedbacks on Invasive Plants: Mechanisms and Potential Management Options

Valerie T. Eviner[1] and Christine V. Hawkes[2]

[1] *Department of Plant Sciences, University of California, USA*
[2] *Section of Integrative Biology, The University of Texas, USA*

Introduction

There are countless examples of management projects that have attempted to decrease or eradicate invasive species at a site, only to have them rapidly recolonize within a few years. While this is often attributed to reinvasion through propagules remaining at the site, or high propagule pressure from the surrounding landscape (Leung *et al.*, 2004; Lockwood *et al.*, 2005), this also may be due to invasive species changing site conditions to favor conspecifics over native species. Many studies have documented that invasive plants can impact numerous soil properties and processes (Leffler and Ryel, Chapter 4, this volume; Ehrenfeld, 2010), and that invader impacts on soil can influence competitive dynamics between plant species, often favoring the invaders (Callaway and Aschehoug, 2000; Reinhardt and Callaway, 2006; Batten *et al.*, 2008; Kulmatiski *et al.*, 2008; reviewed in Eviner *et al.*, 2010). Some of the effects of invasive species on soils can persist after the invader has been removed, making the system more susceptible to reinvasion (reviewed in Eviner and Hawkes, 2008; Kulmatiski and Beard, 2011). In these cases, restoration efforts must be focused not only on removing invasive species, but also counteracting their effects on soil characteristics and processes (Heneghan *et*

al., 2008; Harris, 2009; Eviner *et al.*, 2010). In this chapter, we explore the mechanisms driving plant–soil feedbacks in invaded systems, and potential management tools to alter these feedbacks to be more beneficial to natives over invasive species.

The role of plant–soil feedbacks in shaping plant communities and plant invasions

Plant–soil feedbacks occur when a shift in plant community composition changes soil conditions, and these altered soil conditions further alter the plant community. Positive feedbacks occur when a given plant species alters the soil in a way that promotes its own persistence and growth (either directly by enhancing its own growth and that of conspecifics, or indirectly via greater inhibition of the growth of other species compared to conspecifics). Conversely, a plant species can alter a soil to its own detriment, or in a way that promotes other species more than itself, resulting in a negative feedback. Recent research has shown that plant–soil feedbacks can play an important role in shaping succession, species coexistence, species dominance, range expansion, and the success of invasive species (reviewed in Bardgett *et al.*, 2005;

Kardol *et al.*, 2006; Manning *et al.*, 2008; van der Putten *et al.*, 2009).

Plant–soil feedbacks are of particular relevance in understanding and managing species invasions, because positive feedbacks are more common in invaded communities, while negative feedbacks are more prevalent in native communities (Klironomos, 2002; Kulmatiski and Kardol, 2008; Kulmatiski *et al.*, 2008; van der Putten *et al.*, 2009). Of particular concern are cases of 'invasional meltdown' (Simberloff and Von Holle, 1999), when one invasive species changes the soil to enhance not only itself, but also the invasion of other non-native species. For example the invasion of *Bromus tectorum* enhances invasion of *Taeniatherum caput-medusae*, and invasion of *Taeniatherum* increases invasion of exotic forbs (reviewed in Eviner *et al.*, 2010). Similarly, the invasive *Bromus inermis* alters the soil microbial community to enhance the growth of the invader *Euphoria esula* (Jordan *et al.*, 2008).

While, on average, invasive species are more likely than native species to create positive (or less negative) feedbacks, there are many exceptions to this general trend. Many plant–soil feedbacks are highly species-specific, so that a given invasive species may negatively impact a subset of native species, but not all of them, and different invaders are likely to impact different native species (Casper and Castelli, 2007; Manning *et al.*, 2008). In contrast to the general trends, some invasive species create soil conditions that generate negative feedbacks to conspecifics, while some native species create positive soil feedbacks to conspecifics and negative feedbacks to invaders (Kulmatiski *et al.*, 2004; van der Putten *et al.*, 2007). For example, in a shrub-steppe ecosystem in Washington State, USA, the native perennial grass, *Pseudoroegneria spicata*, alters soil in a way that decreases its own growth, but has even stronger negative effects on the invasive species *Centaurea diffusa*, reducing invader cover from 18% to 5% (Kulmatiski *et al.*, 2004). Promoting the specific native species that decrease the abundance of invasive plants can be a promising first step in restoration of native plant communities.

Invader plant–soil feedbacks enhance resilience of invaded state

Invasive species that generate positive feedbacks are of particular concern for conservation and restoration, because they often create a barrier to the reintroduction of native species. Regardless of what factors precipitated the initial success of an invader, established invasive species can alter the soil and create a 'novel ecosystem,' an alternative stable state that is difficult, if not impossible to revert back to the native state (Suding *et al.*, 2004; Seastedt *et al.*, 2008; Farrer and Goldberg, 2009; Hobbs *et al.*, 2009; Hardegree *et al.*, Chapter 6, this volume).

The degree of persistence versus reversibility of invader impacts on soils and associated ecosystem processes is a critical component of restoration potential. Some of the changes caused by invasive species may be rapidly reversible upon removal of the invader and do not require additional management. For example, decreased soil water availability caused by high plant transpiration rates should reverse quickly once the invasive plant species is removed. In contrast, alterations to soil properties such as soil structure, water infiltration, water holding capacity, carbon storage, and nitrogen cycling rates may persist for months to decades, even with active management (van der Putten *et al.*, 2009). In these cases, reinvasion is likely to take place before soil conditions can be restored, particularly if the altered state favors the invasive plant species relative to native species. For example, extensive erosion as a result of invasion of *Centaurea maculosa* (Lacey *et al.*, 1989) can take decades to centuries to reverse via soil formation processes and the gradual buildup of organic matter by the restored plant community. Such cases highlight the importance of disrupting invader–soil feedbacks early in the invasion process.

While invader-induced feedbacks may create a stable invaded state in an invader's new range, these feedbacks do not always operate in the invader's home range. For example, the negative effects of *C. diffusa* on its neighbors are much stronger in its invaded range than its home range (Callaway and Aschehoug, 2000). In their home ranges, the invasive species are usually subject to the same negative plant–soil feedbacks common to native plants in general (Reinhart *et al.*, 2003; reviewed in Reinhart and Callaway, 2006). The existence of controls over invaders in their home ranges (e.g. through soil feedbacks, natural enemies, competitive effects, or of the evolution of neighbor resistance to allelochemicals), suggests that there may be long-term potential to control invaders in their new ranges through approaches such as bio-control agents or selection for native plant species that are resistant to the invader effects. Both of these will likely happen over the long term, even without active management. With increasing time since invasion, invaders tend to lose their initial advantage due to escape from negative interactions in the new range (Hawkes, 2007), and the invaders' impacts on the native community decrease (Strayer *et al.*, 2006; reviewed in Diez *et al.*, 2010). Alternatively, the invasive species may evolve to have increased competitive ability, which can strengthen both its negative impacts on native species and positive feedbacks to conspecifics. For example, an invader that benefits from its own litter buildup may evolve to have more recalcitrant litter, strengthening the positive feedback (Eppinga *et al.*, 2011). Because few studies have documented long-term impacts of invasive species on communities and ecosystems (reviewed in Strayer *et al.*, 2006), we are still unable to predict whether long-term presence of a specific invader will control versus enhance invasion through changes in the strength and direction of feedbacks.

Mechanisms of Feedbacks, and Potential Management Tools

Plant–soil feedbacks can be mediated through many mechanisms, including plant-induced changes to soil structure, chemistry, and biota (Ehrenfeld *et al.*, 2005; reviewed in Casper *et al.*, 2008), as well as the litter layer (Farrer and Goldberg, 2009). A number of

these mechanisms can be important in any given invasion, and little is known about their relative importance or the extent to which they strengthen or counteract one another to create overall positive versus negative feedbacks. While some mechanisms have similar management approaches for counteracting their associated feedbacks (Table 7.1), effective management will require knowledge of how feedbacks are generated by a given invader. Identifying the mechanisms driving feedbacks for an invasive species is often not straightforward, and even when they can be identified, management of these feedbacks is still largely in the experimental stage. This chapter highlights promising approaches to managing plant–soil feedbacks, but we recognize that continued research on management strategies is required, both across invasive species and sites, to improve these management tools and our ability to predict which approaches will be most effective for a given invader.

Litter

Plant litter dynamics are an important driver of plant community structure and ecosystem processes (reviewed in Ehrenfeld *et al.*, 2005). In general, increased litter accumulation often decreases plant diversity in herbaceous communities (Grime, 1979; Foster and Gross, 1998). Litter alters surface and soil microclimate, directly inhibiting the establishment of select species (Facelli and Pickett, 1991) or enhancing key plant herbivores or pathogens (Lenz *et al.*, 2003; reviewed in Flory and Clay, 2010). These physical effects of litter are often initially more important than associated nutrient feedbacks, which can take longer to develop (Amatangelo *et al.*, 2008). While invasive plant species can aggressively compete for resource uptake, in some cases, the litter, rather than the live plant, is directly responsible for the invader's impacts on the plant community and soil conditions (Farrer and Goldberg, 2009; Holdredge and Bertness, 2011). There are many examples where litter accumulation drives both

Table 7.1. Mechanisms that drive plant–soil feedbacks and potential management tools.

Mechanism driving feedbacks	Potential management tools	Potential limitations
Litter	Litter removal through mowing, burning, grazing/trampling	Timing is critical – may enhance or control invasion Not always possible at all sites or across broad scales
Allelochemicals	Activated charcoal	Also can impact nutrient availability, or inhibit the activity of other compounds
	Removal of invaders + time	Reinvasion can reintroduce the allelochemicals quickly
	Plant non-sensitive 'transition' species	Limited information available on which species may not be sensitive, and which can transition to the ultimate desired community
Soil microbial community	Activated charcoal	Also can impact nutrient availability, or inhibit the activity of microbes
	Plant 'transition' species which can tolerate invader soil or promote pathogens of invader	Interactions of plants and microbes are highly species-specific, so there is limited information on which species are impacted by, or resistant to given changes in the microbial community, and limited information on which species can transition to the ultimate desired community
	Soil inoculation	Local sources that specifically enhance native growth are usually not readily available, not always effective in establishing desired microbes
Nitrogen	Removal of N – burning, mowing, grazing	Disturbance may also enhance invaders and decrease natives in the short term
	Carbon additions to sequester soil N	Mixed effectiveness in sequestering soil N; have been shown to enhance, as well as inhibit invasion by some species
	Topsoil removal	Can disrupt native microbial community and seed bank
	Plant native species that take up high quantities of N	Often invaders are more aggressive than natives in taking up N, but has been effective in conjunction with burning and carbon additions
Salinity	Promote leaching of salts out of the upper layers of soil; method depends on soil, but can include increased water additions or promotion of soil drainage through increasing soil pores (through roots or soil organisms)	Can take a long time to reverse salinization, irrigation water may also add salts
	Plant salt-tolerant natives	Will not necessarily reverse salinity, or aid in reestablishment of natives that were on the site before it became saline
	Remove topsoil	Also removes nutrients, soil microbes, and seeds

invasive plant species' impacts and feed-backs, including: *Typha × glauca* invasion into wetlands with associated increases in nitrogen availability and decreases in light and native species diversity and abundance (Farrer and Goldberg, 2009); *Taeniatherum caput-medusae* invasion into western US rangelands where its recalcitrant litter inhibits the germination of other species, leading to monotypic stands (Young *et al.*, 1971); and *Microstegium vimineum* invasion of northeastern US forests where the physical litter barrier inhibits native tree seedling establishment and reduces seedling survival through enhanced vole activity (Flory and Clay, 2010). Litter buildup can also promote fires, further leading to eco-system alterations that may benefit invasive over native species (reviewed in Davies and Svejcar, 2008). Plant litter inputs are also one of the main mechanisms driving species' impacts on soil chemistry, structure, and biology (reviewed in Eviner and Chapin, 2003a).

Litter accumulation does not always benefit invasive plants. In some cases, the accumulation of litter from invasive plants may also benefit native species. In California, USA, coastal sage scrub, accumulation of invasive grass litter benefits a suite of invasive grasses, but also enhances growth of native shrubs by enhancing soil moisture availability (Wolkovich *et al.*, 2009). In other cases, native species may negatively affect invasive plant species through native litter accumulation. The invasive *M. vimineum*, for example, which benefits from its own litter, has lower seedling survivorship in patches where the litter of native species builds up (Schramm and Ehrenfeld, 2010). Litter can play a key role in shaping the community, but the relative feedbacks to invasive and native species may need to be considered in litter management strategies.

Management

Grazing, mowing, and burning are effective for litter removal and often increase native species in invaded stands (Sheley *et al.*, 2007; reviewed in Holdredge and Bertness, 2011). The seasonality of litter removal has strong impacts on which species benefit, because timing of these disturbances can also greatly impact seed production (Pollak and Kan, 1996; DiTomaso *et al.*, 2006; Holdredge and Bertness, 2011).

Allelochemicals

A number of studies have suggested that some invasive species decrease the per-formance of native plant species through the release of allelochemicals: organic compounds that are either directly phyto-toxic, or inhibit the activity of microbes that are symbiotic with plants (Wardle *et al.*, 1998; Ridenour and Callaway, 2001; reviewed in Bais *et al.*, 2006). For example, *Alliaria petiolata* can decrease the growth of native plant species by releasing compounds that decrease arbuscular mycorrhizal fungi (Stinson *et al.*, 2006). Similarly, *Carduus nutans* releases compounds that inhibit nodulation and nitrogen fixation in legumes, which is likely the cause for this invasive plant species decreasing the growth of a neighboring legume species (Wardle *et al.*, 1993, 1994). Other invaders that negatively impact native communities by releasing allelochemicals include: *C. maculosa* and *C. diffusa* (Callaway and Aschehoug, 2000; Callaway and Vivanco, 2007; Thorpe *et al.*, 2009), *Fallopia × bohemia* (Murrell *et al.*, 2011), and *Acroptilon repens* (Stermitz *et al.*, 2003).

Management

The most direct way to manage allelo-chemicals is to add compounds that can sequester these allelochemicals, thus inhibiting their impact on soil microbes and native plants. Activated carbon, also known as activated charcoal, is highly absorptive due to its high density of micropores and sequestration of compounds through ionic bonding or adsorption (reviewed in Kulmatiski, 2011), which has resulted in its common use for chemical purification and pollutant removal from water and air. Additions of activated carbon have been effective in decreasing the negative impact of invasive species on native species in a number of systems (Callaway and Aschehoug, 2000; Ridenour and Callaway, 2001; Kulmatiski and Beard, 2006; Callaway and Vivanco, 2007; Lau et al., 2008; Thorpe et al., 2009; Kulmatiski, 2011). For example, the native grass Festuca idahoensis, when grown with the invasive C. maculosa, grew 85% larger with activated carbon than without (Ridenour and Callaway, 2001). It is important to note that activated carbon additions on their own are often not sufficient to decrease the abundance of invasive plants – clearing of invasive plants along with native seed planting is frequently required. In ex-arable fields in Washington State, USA, that were dominated by invasive plants for decades, the combination of clearing of invasive vegetation, a single application of activated carbon, and native seed additions shifted dominance from invasive to native plants, and this was maintained even after 6 years (Kulmatiski, 2011).

Allelochemicals can be highly species-specific in their impacts, which likely accounts for the fact that additions of activated carbon vary in their effectiveness in controlling invasive species, and may promote some, but not all native species (Lau et al., 2008; reviewed in Kulmatiski, 2011). Activated carbon additions also can increase the prevalence of some invasive species (Kulmatiski and Beard, 2006; Lau et al., 2008). Beyond the species-specific nature of activated carbon impacts, its use is far from straightforward because it not only sequesters allelochemicals, but also alters nutrient availability, rates of nutrient cycling, and the soil microbial community (Lau et al., 2008; Kulmatiski, 2011).

Allelochemicals generally are short-lived in the soil (hours to days) (Blair et al., 2005; Reigosa et al., 2006), suggesting that activated carbon may be most useful to minimize the effects of invaders currently at a site, or early in restoration, when it can sequester allelochemicals from newly invading individuals. To ameliorate potential longer term legacies of allelochemicals deposited through plant litter (Reigosa et al., 2006), best practices should include

removing all invasive plant material from a site.

Because the effects of allelochemicals are species-specific, another potential restoration approach is to plant native species that are not susceptible to these compounds (Perry et al., 2005; Alford et al., 2009). Plant species are being tested for innate resistance to the allelochemicals of the invasive C. maculosa. The establishment of these resistant species can prevent Centaurea from reinvading and may eventually facilitate the establishment of native species that are susceptible to these allelochemicals (Callaway and Aschehoug, 2000; Callaway and Vivanco, 2007; Thorpe et al., 2009).

The allelopathic effects of invasive species on native species may decrease with time, as native species adapt to these inputs (Callaway et al., 2005). Allelochemicals can have stronger impacts on heterospecific neighbors in their invaded ranges, compared to their home ranges (Bais et al., 2003; Callaway et al., 2008; Thorpe et al., 2009), suggesting that there has been ongoing selection for resistance in the home range. Over time in the new range, the inhibitory effects of allelochemicals may decrease as native species similarly evolve resistance to invasive species (Callaway et al., 2005; reviewed in Strayer et al., 2006). Breeding of resistant native plant genotypes may be a potential management approach. With time since invasion, the impacts of allelochemicals on the soil microbial community can also vary. Comparisons of sites that had been invaded by A. petiolata for 20–50 years, demonstrated that resistance of the microbial community to allelochemicals increased over time. In the longest invaded sites, Alliaria populations decreased allelochemical inputs, further decreasing overall impacts of the invasion on the microbial community (Lankau, 2011). These cases suggest that invasions that are facilitated by allelochemical inputs may be controlled over the course of 4 to 5 decades through the strong selection imposed by allelochemicals on the native plant and microbial communities. However, this selection may be at the cost of decreased diversity (e.g. Lankau, 2011).

Soil microbial community

The soil microbial community frequently mediates soil feedbacks associated with invasive plant species, but their specific effects can be difficult to predict. For example, the soil microbial community is altered by invasion of *Aegilops triuncialis* into an herbaceous serpentine community, leading to decreased growth and flowering time of one native forb, *Lasthenia californica*, but not other native species (Batten *et al.*, 2008). Similarly, in the Great Plains, the soil microbial community is altered by the invasion of *Agropyron cristatum*, *B. inermis*, and *Eu. esula*; each invader benefits from the changes it induces, but only a subset of native species are affected by the altered soil community of each invasive plant species (Jordan *et al.*, 2008). The lack of apparent generality, and thus unpredictability, of invasive species effects on soil microbial communities may be partly due to our poor understanding of the specific microbial taxa and mechanisms responsible for the observed feedbacks.

Our understanding of the mechanisms underlying soil microbial community feedbacks is most developed for pathogens and symbionts. Some invasive plant species are successful because they have escaped soil pathogens common in their native range, and this release from pathogens makes them more competitive against native species, which commonly experience negative feedbacks with the soil pathogen community (reviewed in Reinhardt and Callaway, 2006). However, invasive plants can exacerbate this feedback, because their leachates enhance the pathogens of native species (Mangla *et al.*, 2008). In other cases, invasive plants may be successful because the benefit obtained from local mycorrhizal mutualists is greater than the negative effects of pathogens in the new range (Klironomos, 2002).

Invasive plant species are usually colonized by local mycorrhizal fungi and can have direct effects on the composition and abundance of the mycorrhizal community that can feed back to the plant community (Hawkes *et al.*, 2006; Stinson *et al.*, 2006;

Callaway *et al.*, 2008; Wolfe *et al.*, 2008). As with the general soil microbial community, the strength and direction of feedbacks from mycorrhizal fungi are context-dependent, based on factors such as the identities of the invader and fungi, the ecosystem, and the identities and life stages of neighboring native plant species (reviewed in van der Heijden and Horton, 2009). For example, in California grasslands, USA, the invasive forb *Carduus pycnocephalus* grows best in soils without arbuscular mycorrhizal (AM) fungi and its growth decreases AM fungal densities in soil, resulting in reduced colonization of native roots and decreased growth of the native forb *Gnaphalium californicum* (Vogelsang *et al.*, 2004; Vogelsang and Bever, 2009). Other invaders, such as *C. maculosa*, appear to tap into existing native mycorrhizal networks, essentially parasitizing resources, which results in substantial growth benefits (Marler *et al.*, 1999; Callway *et al.*, 2003; Carey *et al.*, 2004). Invasive plants can also alter the composition of AM fungi mycorrhizae that colonize native plant roots. Invasion of annual grasses in California, USA (Hawkes *et al.*, 2006; Hausmann and Hawkes, 2009), as well as *C. maculosa* in Montana, USA (Mummey *et al.*, 2005), can shift the AM fungal community infecting native plant roots to substantially overlap with that of the exotic plants. While the mechanisms driving invasive plant effects on mycorrhizal communities are often unknown, in some cases, invasive plants that are less reliant on mycorrhizal fungi may release inhibitory compounds that can broadly reduce the abundance of mycorrhizal fungi in soil (Stinson *et al.*, 2006; Callaway *et al.*, 2008; Wolfe *et al.*, 2008). In other cases, invasive plants can associate with a subset of the mycorrhizal community, such as fungal generalists (Moora *et al.*, 2011) or those fungi most beneficial to the invader (Zhang *et al.*, 2010), which may promote the selected fungal taxa over others. In these cases, the network of mycorrhizal fungi supported by invasive plants may create a priority effect (reviewed in Hausmann and Hawkes, 2010).

Co-invasion by plants and their mycorrhizal fungi may also facilitate plant invasion success through positive feedbacks, such as with *Pinus* species and ectomycorrhizal fungi in New Zealand (Dickie *et al.*, 2010). Where ectomycorrhizal associates are spatially limited, the spread of exotic *Pinus* species can also be limited (Nuñez *et al.*, 2009). More than 200 species of ecto-mycorrhizal fungi have been introduced to new ranges worldwide; these fungi are largely associated with plantation forestry (Vellinga *et al.*, 2009) and thus the spread of the ectomycorrhizal fungi and their plant hosts may be linked in many cases.

Management

As described above, when invader-induced feedbacks are strong enough to prevent the original native species from persisting long enough to alter soil conditions, a multi-stage successional approach can be employed by initially planting species that are more tolerant of the invaded soil conditions. This is feasible because most plant–microbial interactions are species-specific. Once the initial plantings ameliorate the invaded soil legacies, the target native community can be reestablished as seeds or transplants (Jordan *et al.*, 2008). This plant-induced change to the microbial community may take time. While changes in plant species can impact some components of the microbial community within weeks to months, the microbes mediating plant–soil feedbacks can persist unchanged for at least a growing season (Kulmatiski and Beard, 2011).

A more aggressive approach would be to plant native species that culture soil pathogens that decrease the growth of invasive species (Knevel *et al.*, 2004). Finding and promoting such native species could be a key tool for disrupting invasive species' positive feedbacks within the soil community. Even without intervention, the strength of negative feedbacks on invasive species increases with time since establishment, suggesting that over the long term, the soil microbial community may decrease the dominance of invasive species (Diez *et al.*, 2010).

Another management option is to interfere with the plant inputs that shape the microbial community. Litter removal, or inputs of activated carbon to deactivate key plant metabolites have been effective in managing invasive species, but may also promote some invaders (reviewed in Kulmatiski, 2011). As described above, activated carbon can inhibit the impacts of allelochemicals on the microbial community. For example, in a case where a tropical invasive plant increases generalist soil pathogens, addition of activated carbon decreases pathogen spore numbers and increases native plant growth (Mangla *et al.*, 2008).

Few studies have assessed the impacts of disturbance regimes on invader–soil feed-backs. In Portuguese coastal dunes, fire decreased AM fungal colonization in all species, and rhizobial colonization in native, but not invasive legumes. Overall, fire enhanced invasive species performance by changing invader–soil biota feedbacks from neutral to positive, and native species feedbacks from negative to neutral (Carvalho *et al.*, 2010). The impacts of disturbance regimes on plant–soil feedbacks may be important to consider, because it may result in disturbance events that were meant to control invaders having the unintended consequence of strengthening of invader–soil feedbacks. While this example demonstrates that disturbance further strengthens invader feedbacks, disturbance may be effective in disrupting invader–soil feedbacks in other cases.

A number of studies have investigated the potential to inoculate invaded soils with desirable microbial communities (reviewed in Vessey, 2003; Schwartz *et al.*, 2006). On highly degraded soils with a depauperate soil community, inoculation with microbes, particularly mycorrhizal fungi, can enhance plant establishment unless soil conditions are too stressful (Kardol *et al.*, 2009). Where sites dominated by invasive plants have an intact soil community, inoculation can be more complicated because the inoculated microbial community may be inhibited by the microbes already present (Kardol *et al.*, 2009; Mummey *et al.*, 2009). However, in other cases, inoculation into intact communities can be effective. In the tallgrass prairie of the Central Plains, USA, for example, inoculation with AM fungi increased the cover of native grasses over weedy plants (Smith, M.R. *et al.*, 1998).

The effectiveness of microbial inoculation in controlling invasive plants is also complicated by the ecological specificity of interactions between plants and their microbial communities. Inoculation can increase or decrease plant growth, depending on the identity of inoculated microbes, the plant species, and the environmental conditions (reviewed in Harris, 2009; Mummey *et al.*, 2009), which supports the use of local microbes for inoculation efforts. The source of inocula can have a strong impact on restoration success. Most commercial inocula contain generalist AM fungi that may not support the native plant community and may decrease soil mycorrhizal diversity (reviewed in Harris, 2009). While generating native inoculum can be challenging, it can be critical for effective results. For example, on degraded shrublands in Spain, the biomass of plants was twice as high when inoculated with a mixture of indigenous AM fungi compared to inoculation with an exotic AM fungus (Requena *et al.*, 2001). Pre-inoculation of native seedlings with desirable AM fungi may further help to minimize the AM fungal taxa associated with exotic species (Mummey *et al.*, 2009).

Nitrogen

While many nutrients are critical in regulating plant growth, interactions between invasive plants and nitrogen (N) are particularly important because N is the most commonly limiting nutrient to plant growth in temperate terrestrial ecosystems, and as such, has strong impacts on plant species composition and diversity (Eviner and Chapin, 2003b; reviewed in Clark *et al.*, 2007; Suding *et al.*, 2008). On average, invasive compared to native plant species, enhance N availability through increases in decomposition and N mineralization rates (Ehrenfeld, 2003; Corbin and D'antonio, 2004; Liao *et al.*, 2008), although some invasive species decrease N availability, such as the invasion of *Ae. triuncialis* into grasslands of California, USA (Drenovsky and Batten, 2007), *Bromus tectorum* into western US shrublands (Bradley *et al.*, 2006), and *A. cristatum* into the northern Central Plains of the USA (Christian and Wilson, 1999). Whereas shifts in the soil microbial community tend to have species-specific impacts on plant growth, enhanced N availability often will increase the performance of most plants when grown alone in an invaded soil (Casper *et al.*, 2008). However, in mixed communities, increased soil N can shift plant community composition through selection for species that are more competitive (Clark *et al.*, 2007; Suding *et al.*, 2008). Of particular concern is that invader-induced increases in soil N availability will feed back to enhance invasion, because soils with high N availability are more susceptible to plant invasion (reviewed in Heneghan *et al.*, 2008; Suding *et al.*, 2008).

While the amount of soil N can be a key regulator of plant species composition, invasive species may also change the timing and location of N availability. For example, leaching from *Bromus tectorum* litter redistributes soil nitrate deep in the soil profile, where native grasses cannot access it, thus increasing N availability to *Bromus*. This enhances *Bromus* growth at the expense of the native grasses (Sperry *et al.*, 2006). *Bromus* also alters the timing of soil-N availability, with high soil-N availability occurring after the senescence of *Bromus* (Adair and Burke, 2010). Similarly, invasion of exotic grasses into Hawaiian woodlands greatly alters the seasonality of soil-N availability. Grass invasion shifts most net-N mineralization from the dry season to the wet season due to grass impacts on soil organic matter enhancing wet-season N cycling, and grass impacts on microclimate decreasing dry-season N cycling rates (Mack and D'Antonio, 2003). Invasive plants also can alter the form of N available. For example, in California grasslands, USA, invasive grasses increase the soil nitrifier population, and thus nitrification rates (Hawkes *et al.*, 2005). Conversely, the invasion of *Andropogon garanus* into Australian grasslands inhibits nitrification (Rossiter-Rachor *et al.*, 2009). While there are not clear examples of native versus invasive species performance being impacted by the form of N, the relative amount of N available as ammonium versus nitrate has been shown to alter competition between species (reviewed in Marschner, 1986; Crabtree and Bazzaz, 1993).

Management

Soil nitrogen can be removed by repeated disturbances, including burning, grazing, or mowing and removal of vegetation, and this decrease in N can cause a shift from dominance by competitive, weedy species, to a more diverse plant community (reviewed in Marrs, 1993; Walker *et al.*, 2004; Perry *et al.*, 2010). In many cases, these techniques are also used to directly decrease the prevalence of invaders (e.g. by removing invasive plants before they set seed), and while the timing of disturbance may have minimal impacts on N removal, it will be critical in influencing which plant species reestablish (Pollak and Kan, 1996; DiTomaso *et al.*, 2006; Holdredge and Bertness, 2011). In the long term, repeated disturbances can reduce N availability, but in the short term, N availability can be enhanced immediately after disturbance, and in some cases, this increase in N can be sustained during the next few disturbance cycles (reviewed in Perry *et al.*, 2010), making the system vulnerable to reinvasion if invasive species propagules are present.

In extremely high fertility sites, such as those that have been fertilized for years, it can take decades to adequately restore target N cycles and the plant community through grazing, burning, or mowing (reviewed in Walker *et al.*, 2004). In these extreme cases, topsoil removal (also known as sod-cutting) can rapidly remove accumulated nutrients and organic matter, as well as soil microbes and many invasive plant propagules in the seed bank (reviewed in Marrs, 1993; Walker *et al.*, 2004). Topsoil removal is the most effective method of quickly and reliably removing N (Perry *et al.*, 2010), but it also

removes the native seed bank and microbial community, which will need to be restored. While effective, topsoil removal can only be used in smaller restoration projects, and is limited to sites accessible to heavy machinery.

Additions of biologically available carbon, such as sawdust or sugar (as opposed to the more inert activated carbon), can fuel growth of soil microbes, thus sequestering N in microbial biomass. This approach has been effective in reducing a number of invasions, and seems to be particularly effective in inhibiting grasses (rather than forbs or shrubs), and in shifting dominance from invasive annual to native perennial species (reviewed in Perry *et al.*, 2010). However, its effectiveness in reducing soil-N availability and controlling invasive species is variable, and often short-lived. In some cases, adding carbon can actually enhance N availability and/or invasive species (Blumenthal *et al.*, 2003; Krueger-Mangold *et al.*, 2006; Corbin *et al.*, 2007; Eviner and Hawkes, 2008; reviewed in Alpert, 2010; Eviner *et al.*, 2010; Perry *et al.*, 2010; Kulmatiski *et al.*, 2011). The amount and type of carbon needed to sequester N can vary by species and site (Blumenthal *et al.*, 2003; Prober *et al.*, 2005). In some cases, the amount of carbon needed may be prohibitive due to expense and logistics, suggesting that it may be a tool appropriate to small, high-intensity restoration sites, but may not be feasible across large areas (Perry *et al.*, 2010). Even when it is effective in sequestering N, much of this N is re-released within a few months to a few years, so this technique is most often effective in conjunction with quickly restoring native plant species (reviewed in Perry *et al.*, 2010).

Planting native species that decrease N availability through high N uptake has decreased the prevalence of some invaders. In rangelands of northwestern USA, planting *Secale cereale* or the native perennial grass *Elymus elymoides* decreased available soil N, shifting competitive dominance from the invasive *C. maculosa* to the native late-seral species *Pseudoroegneria spicata* (Herron *et al.*, 2001). Effective management often requires a combination of approaches; using disturbance and/or carbon additions to temporarily decrease available N, in combination with fostering plants that can maintain low soil-N availability. For example, in Australian grasslands, the combination of burning, carbon additions, and seed additions of a native grass with high-N uptake (*Themeda triandra*) was required to effectively decrease weed cover and reduce soil nitrate to levels found on native-dominated sites (Prober and Lunt, 2009). Alternatively, if soil N can be adequately reduced by carbon additions or disturbance, low-N adapted plants can be introduced, and their low litter quality can feed back to maintain or further decrease low-N availability (reviewed in Perry et al., 2010).

In instances where invasive plant species may inhibit nitrification through allelochemicals, activated carbon may be effective in binding these allelochemicals and increasing nitrification rates (reviewed in Lau et al., 2008). In contrast, when invasive plant species enhance nitrification rates, commercial nitrification inhibitors can be used. These are commonly added to fertilized agricultural sites (Prasad and Power, 1995), and have been effective in decreasing some invasions, while enhancing native species (Young et al., 1997, 1998).

Soil salinity

A few invaders have been shown to increase soil salinity, thus decreasing the performance of native competitors. Examples include *Tamarix* species (Smith, S.D. et al., 1998; Ladenburger et al., 2006), *Carpobrotus edulis* (Kloot, 1983), and *Halogeton glomeratus* (Harper et al., 1996; Duda et al., 2003). Conversely, invasion of brackish marshes by *Phragmites australis* decreases salinity (cited in Ehrenfeld et al., 2005).

Management

Natural flooding and/or high rainfall can leach salts from soils in the short term, and can be used in conjunction with promoting native species tolerant of higher electrical conductivity levels. However, in many cases, longer term decreases in salinity will require restoration of historic flood regimes and/or ground water-table levels (reviewed in Ladenburger et al., 2006). Where invasive plants have redistributed salts to be concentrated at the soil surface, such as *Mesembryanthemum crystallinum* (Vivrette and Muller, 1977), topsoil removal may be required.

Disturbance as an important feedback pathway

While this chapter focuses on plant–soil feedbacks, invasive species can also greatly alter disturbance regimes to benefit themselves. For example, *B. tectorum* in the Great Basin, USA (Knick and Rotenberry, 1997), *T. caput-medusae* in the western USA (Davies and Svejcar, 2008), and invasive grasses in Hawaii, USA (D'Antonio and Vitousek, 1992), can increase fire frequency, thus enhancing their own growth at the expense of native species. *Brassica nigra* in California grasslands, USA, enhances herbivory of the native bunchgrass, *Nassella pulchra*, by small mammals, and this effect extends 30 m away from invaded patches (Orrock et al., 2008).

Challenges in Understanding and Managing Feedbacks

It is clear that invasive plant species can alter the soil in a way that benefits their own performance, and in these cases, their effective eradication may require inter-ference with invader–soil feedbacks. How-ever, the study of feedbacks is still a relatively new field, and effective management requires a better predictive ability of feedback mechanisms and their relative importance, specificity, context-dependence, and spatial and temporal patterns (reviewed in Ehrenfeld et al., 2005). Key challenges to understanding and managing feedbacks are discussed below.

1. Relative importance of different feedback mechanisms. Clearly, many mechanisms can drive plant–soil feedbacks, and these mechanisms vary in both how specific their impacts are for various native species and which management approaches will likely be effective. Although some broad management techniques are similar in method (e.g. planting transitional species that can tolerate invasive species' soil legacies), the selection of native species will depend on the underlying mechanisms. Individual invasive species may have multiple mechanisms driving soil feedbacks that must be considered when attempting to ameliorate invasive soil legacies. For example, *B. tectorum* alters the amount and distribution of available soil N (Sperry *et al.*, 2006), as well as disturbance regimes (D'Antonio and Vitousek, 1992), and microbial communities (Belnap and Phillips, 2001; Hawkes *et al.*, 2006). The relative importance of different feedback mechanisms likely varies with the specific invasive species, native species, and site conditions. Where multiple mechanisms are at play, selection of restoration approaches will require knowing if one key feedback mechanism can be targeted, or if management of each feedback pathway is required.

2. Specificity of feedback mechanisms. As reviewed in this chapter, many feedbacks depend on the identities of the invasive and native species. Soil feedbacks from one invader can impact a number of native species, while a second invader in the same ecosystem can have feedbacks that affect an entirely different set of native species. As we increase the number of well-developed case studies of invader feedbacks, we will improve our understanding of the types of native plants that are more sensitive to specific changes in the soil physical, chemical, and biotic environment.

3. Context-dependence of feedbacks. Many studies have shown that the impacts of plant species on soils vary with environmental conditions and the amount of time an invader has been present (reviewed in Ehrenfeld *et al.*, 2005; Strayer *et al.*, 2006; Eviner and Hawkes, 2008). For example, in North America, *B. tectorum* can increase rates of N cycling in cool deserts, and decrease N-cycling rates in warmer arid

grasslands (reviewed in Ehrenfeld *et al.*, 2005; Ehrenfeld, 2010). Similarly, the strength and magnitude of feedbacks are likely to vary across space and time, and depending on which species are interacting (reviewed in Bardgett *et al.*, 2005; Eviner *et al.*, 2010). While some species consistently generate negative soil feedbacks to conspecific species across sites, the direction and magnitude of feedbacks can differ by site for other species (Casper *et al.*, 2008). In a particularly interesting example from an annual-herb-dominated community in the UK, eight plant species significantly differed in their effects on soil properties, which then fed back to impact the relative growth of these species. N enrichment did not impact the effects of these species on soil properties, but the interaction of N enrichment with plant effects on soils greatly altered plant growth responses to species-specific changes to soils (Manning *et al.*, 2008).

Time is a particularly important driver of context-dependence of plant–soil feedbacks. Vulnerability to pathogens differs with life stage for a given plant species, and in the extent to which mycorrhizal fungi can be negative or beneficial (reviewed in Bardgett *et al.*, 2005; Casper and Castelli, 2007; van der Heijden and Horton, 2009). The length of time an invader has been at a site has large impacts on the extent to which it changes the soil (reviewed in Strayer *et al.*, 2006), and can thus alter feedbacks (reviewed in Bardgett *et al.*, 2005). For example, *B. tectorum* has its strongest positive feedback in its third generation on a given soil (Blank, 2010). Changes over time may be due to the accumulation of impacts (e.g. accumulation of soil organic matter) or shifts in the relative strengths of positive versus negative feedback pathways (e.g. soil symbionts versus pathogens), as occurs during succession (Kardol *et al.*, 2007).

Another key time-related concern is the persistence of invader effects on soils, even after invasive plant species have been removed from a site. When an invader has been at a site for decades to centuries, its impacts on soil microbes, organic matter, and nutrients can persist long after the invader has been removed (reviewed in Eviner and Hawkes, 2008; Eviner *et al.*, 2010). Even when an invader has been at a site for a short duration, its impacts on soil may persist long enough to interfere with native plant restoration (Grman and Suding, 2010).

4. Relative importance of feedbacks versus other drivers of invasion. There are many potential mechanisms driving invasions (Theoharides and Dukes, 2007), and a number of these may be operating simultaneously. It is critical to compare the relative importance of soil feedbacks to other mechanisms such as competition (Casper and Castelli, 2007), propagule pressure (Eppstein and Molofsky, 2007), and release from aboveground natural enemies such as herbivores and pathogens (Mitchell and Power, 2003; Agrawal *et al.*, 2005). For some invasive species, factors such as competition and climate are more important than soil-feedback effects (e.g. Yelenik and Levine, 2011). In other cases, invasive plant species dominance may be maintained by the combination of asymmetric competition generated through early germination and the negative feedbacks to native plants generated by soil legacies (Grman and Suding, 2010). More work will be required to understand the role of soil feedbacks relative to other mechanisms in invasive species success.

5. How prevalent does an invader need to be to induce feedbacks? It is often assumed that the impacts of a plant species on the soil are proportional to its biomass in the community (Grime, 1998; Parker *et al.*, 1999), but recent work has shown that some invasive species can have significant impacts on soils even when they are relatively rare. For example, in a river floodplain in New Zealand, non-native plants made up less than 3% of plant community biomass, but had significant impacts on soil carbon, microbial biomass, and microbial community structure (Peltzer *et al.*, 2009). Similarly, varying proportions of native and invasive plant litter demonstrated that the effects of litter of the invasive *Berberis*

thunbergii on the soil microbial community were not proportional to its relative abundance in the mixture (Elgersma and Ehrenfeld, 2011).

Summary

While there is still much to learn about the role of plant–soil feedbacks in exotic plant species invasions, they clearly do play an integral role in some systems, and must be addressed to restore resilient native communities. Despite the considerable variation in the effects of invasive species across space and time within a specific area, current tools for altering plant–soil feedbacks show considerable promise and will be improved with more case studies and collaborations between land managers and researchers. While few 'rules of thumb' for management are available from this emerging field, some general principles do apply:

- As with other mechanisms of invasion (e.g. high propagule availability), the most efficient management approach will be to quickly eradicate new infestations of invasive species, before they are able to alter soil conditions to benefit themselves.
- Testing potential approaches to manage plant–soil feedbacks (Table 7.1) without knowing the mechanism driving the feedback can be risky. The species-specific nature of most of these feedback mechanisms indicates that many of these management techniques have a chance to promote, rather than control invasive species. In cases where the mechanisms are not known, trials should be small-scale and well monitored, before they are applied to broader areas of invasion.
- The mechanisms driving plant–soil feedbacks, and the strength and direction of these feedbacks can change greatly over time since invasion, as well as site-to-site. Thus, even when management has successfully disrupted invader–soil feedbacks at one site, preliminary trials under different conditions (or at sites

with a very different length of time since invasion) should be undertaken.

• Restoration of sites that have been invaded for decades are likely to have strong soil legacies that may not be quickly reversed. In these cases, screening for native species which can tolerate the invader-cultured soil may be the best first stage of restoration, when little is known about the mechanisms driving the invader–soil feedbacks.

More concrete management recommendations will undoubtedly emerge in this rapidly developing field. Setting up field trials with control areas as comparisons, and follow-up monitoring of these trials, will increase the rate at which such 'rules of thumb' are available.

Acknowledgements

Thanks to Thomas Monaco and Roger Sheley for insightful comments that improved this chapter from its previous draft.

References

Adair, E.C. and Burke, I.C. (2010) Plant phenology and life span influence soil pool dynamics: *Bromus tectorum* invasion of perennial C3-C4 grass communities. *Plant and Soil* 355, 255–269.

Agrawal, A.A., Kotanen, P.M., Mitchell, C.E, Power, A.G., Godsoe, W. and Klironomos, J. (2005) Enemy release? An experiment with congeneric plant pairs and diverse above- and belowground enemies. *Ecology* 86, 2979–2989.

Alford, E.R., Vivanco, J.M. and Paschke, M.W. (2009) The effects of flavonoid allelochemicals from knapweeds on legume-rhizobia candidates for restoration. *Restoration Ecology* 17, 506–514.

Alpert, P. (2010) Amending invasion with carbon: after fifteen years, a partial success. *Rangelands* 32, 12–15.

Amatangelo, K.L., Dukes, J.S. and Field, C.B. (2008) Responses of a California annual grassland to litter manipulation. *Journal of Vegetation Science* 19, 605–612.

Bais, H.P., Vepachedu, R., Gilroy, S., Callaway, R.M. and Vivanco, J.M. (2003) Allelopathy and exotic plant invasion: from molecules and genes to species interactions. *Science* 301, 1377–1380.

Bais, H.P., Weir, T.L., Perry, L.G., Gilroy, S. and Vivanco, J.M. (2006) The role of root exudates in rhizosphere interactions with plants and other organisms. *Annual Review Plant Biology* 57, 233–266.

Bardgett, R.D., Bowman, W.D., Kaufmann, R.D. and Schmidt, S.K. (2005) A temporal approach to linking aboveground and belowground ecology. *Trends in Ecology and Evolution* 20, 634–641.

Batten, K.M., Scow, K.M. and Espeland, E.K. (2008) Soil microbial community associated with an invasive grass differentially impacts native plant performance. *Microbial Ecology* 55, 220–228.

Belnap, J. and Phillips, S.L. (2001) Soil biota in an ungrazed grassland: response to annual grass (*Bromus tectorum*) invasion. *Ecological Applications* 11, 1261–1275.

Blair, A.C., Hanson, B.D., Brunk, G.R., Marrs, R.A., Westra, P., Nissen, C.J. and Hufbauer, R.A. (2005) New techniques and findings in the study of a candidate allelochemical implicated in invasion success. *Ecology Letters* 8, 1039–1047.

Blank, R.R. (2010) Intraspecific and interspecific pair-wise seedling competition between exotic annual grasses and native perennials: plant-soil relationships. *Plant and Soil* 326, 331–343.

Blumenthal, D.M., Jordan, N.R. and Russelle, M.P. (2003) Soil carbon controls weeds and facilitates prairie restoration. *Ecological Applications* 13, 605–615.

Bradley, B.A., Houghton, R.A., Mustard, J.F. and Hamburg, S.P. (2006) Invasive grass reduces aboveground carbon stocks in shrublands of the Western US. *Global Change Biology* 12, 1815–1822.

Callaway, R.M. and Aschehoug, E.T. (2000) Invasive plants versus their new and old neighbors: a mechanism for exotic invasion. *Science* 290, 521–523.

Callaway, R.M. and Vivanco, J.M. (2007) Invasion of plants into native communities using the underground information superhighway. *Allelopathy Journal* 19, 143–151.

Callaway, R.M., Mahall, B.E., Wicks, C., Pankey, J. and Zabinski, C. (2003) Soil fungi and the effects of an invasive forb on grasses: neighbor identity matters. *Ecology* 84, 129–135.

Callaway, R.M., Ridenour, W.M., Laboski, T., Weir, T. and Vivanco, J.M. (2005) Natural selection for resistance to the allelopathic effects of invasive plants. *Journal of Ecology* 93, 576–583.

Callaway, R.M., Cipollini, D., Barto, K., Thelen, G.C., Hallett, S.C., Prati, D., Stinson, K. and Klironomos, J. (2008) Novel weapons: invasive plant suppresses fungal mutualists in America but not in its native Europe. *Ecology* 89, 1043–1065.

Carey, E.V., Marler, M.J. and Callaway, R.M. (2004) Mycorrhizae transfer carbon from a native grass to an invasive weed: evidence from stable isotopes and physiology. *Plant Ecology* 172, 133–141.

Carvalho, L.M., Antunes, P.M., Martins-Loucao, M.A. and Klironomos, J.N. (2010) Disturbance influences the outcome of plant-soil biota interactions in the invasive Acacia longifolia and in native species. *Oikos* 119, 1172–1180.

Casper, B.B. and Castelli, J.P. (2007) Evaluating plant-soil feedback together with competition in a serpentine grassland. *Ecology Letters* 10, 394–400.

Casper, B.B., Bentivenga, S.P., Ji, B., Doherty, J.H., Edenborn, H.M. and Gustafson, D.J. (2008) Plant-soil feedback: testing the generality with the same grasses in serpentine and prairie soils. *Ecology* 89, 2154–2164.

Christian, J.M. and Wilson, S.D. (1999) Long-term ecosystem impacts of an introduced grass in the Northern Great Plains. *Ecology* 80, 2397–2407.

Clark, C.M., Cleland, E.E., Collins, S.L., Fargione, J.E., Gough, L., Gross, K.L., Pennings, S.C., Suding, K.N. and Grace, J.B. (2007) Environmental and plant community determinants of species loss following nitrogen enrichment. *Ecology letters* 10, 596–607.

Corbin, J. and D'Antonio, C.M. (2004) Effects of invasive species on soil nitrogen cycling: implications for restoration. *Weed Technology* 18, 1464–1467.

Corbin, J.D., Dyer, A.R. and Seabloom, E.W. (2007) Competitive interactions. In: Stromberg, M., Corbin, J. and D'Antonio, C. (eds) *Ecology and Management of California Grasslands*. University of California Press, Berkeley, California, pp. 156–168.

Crabtree, R.C. and Bazzaz, F.A. (1993) Seedling response of four birch species to simulated nitrogen deposition: ammonium vs. nitrate. *Ecological Applications* 3, 315–321.

D'Antonio, C.M. and Vitousek, P.M. (1992) Biological invasion by exotic grasses, the grass/fire cycle and global change. *Annual Review of Ecology and Systematics* 23, 63–87.

Davies, K.W. and Svejcar, T.J. (2008) Comparison of medusahead-invaded and noninvaded Wyoming big sagebrush steppe in southeastern Oregon. *Rangeland Ecology and Management* 61, 623–629.

DiTomaso, J.M., Brooks, M.L., Allen, E.B., Minnich, R., Rice, P.M. and Kyser, G.B. (2006) Control of invasive weeds with prescribed burning. *Weed Technology* 20, 535–548.

Dickie, I.E., Bolstridge, N., Cooper, J.A. and Peltzer, D.A. (2010) Co-invasion by *Pinus* and its mycorrhizal fungi. *New Phytologist* 187, 475–484.

Diez, J.M., Dickie, I., Edwards, G., Hulme, P.E., Sullivan, J.J. and Duncan, R.P. (2010) Negative soil feedbacks accumulate over time for non-native plant species. *Ecology Letters* 13, 803–809.

Drenovsky, R.E. and Batten, K.M. (2007) Invasion by *Aegilops triuncialis* (barb goatgrass) slows carbon and nutrient cycling in a serpentine grassland. *Biological Invasions* 9, 107–116.

Duda, J.J., Freeman, D.C., Emlen, J.M., Belnap, J., Kitchen, S.G., Zak, J.C., Sobek, E., Tracy, M. and Montante, J. (2003) Differences in native soil ecology associated with invasion of the exotic annual chenopod, *Halogenton glomeratus*. *Biology and Fertility of Soils* 38, 72–77.

Ehrenfeld, J.G. (2003) Effects of exotic plant invasions on soil nutrient cycling processes. *Ecosystems* 6, 503–523.

Ehrenfeld, J.G. (2010) Ecosystem consequences of biological invasions. *Annual Review of Ecology, Evolution and Systematics* 41, 59–80.

Ehrenfeld, J.G., Ravit, B. and Elgersma, K. (2005) Feedback in the plant-soil system. *Annual Review of Environment and Resources* 30, 75–115.

Elgersma, K.J. and Ehrenfeld, J.G. (2011) Linear and non-linear impacts of a non-native plant invasion on soil microbial community structure and function. *Biological Invasions* 13, 757–768.

Eppinga, M.B., Kaproth, M.A., Collins, A.R. and Molofsky, J. (2011) Litter feedbacks, evolutionary change and exotic plant invasion. *Journal of Ecology* 99, 503–514.

Eppstein, M.J. and Molofsky, J. (2007) Invasiveness in plant communities with feedbacks. *Ecology Letters* 10, 253–263.

Eviner, V.T. and Chapin III, F.S. (2003a) Functional matrix: a conceptual framework for predicting multiple plant effects on ecosystem processes. *Annual reviews in Ecology, Evolution and Systematics* 34, 455–485.

Eviner, V.T. and Chapin III, F.S. (2003b) Biogeochemical interactions and biodiversity. In: Melillo, J.M., Field, C.B. and Moldan, B. (eds) *Interactions of the Major Biogeochemical Cycles: Global Change and Human Impacts*. Island Press, Washington, DC, pp. 151–173.

Eviner, V.T. and Hawkes, C.V. (2008) Embracing variability in the application of plant-soil interactions to the restoration of communities and ecosystems. *Restoration Ecology* 16, 713–729.

Eviner, V.T., Hoskinson, S.A. and Hawkes, C.V. (2010) Ecosystem impacts of exotic plants can feed back to increase invasion in western US rangelands. *Rangelands* 31, 21–31.

Facelli, J.M. and Pickett, S.T.A. (1991) Indirect effects of litter on woody seedlings subject to herb competition. *Oikos* 62, 129–138.

Farrer, E.C. and Goldberg, D.E. (2009) Litter drives ecosystem and plant community changes in cattail invasion. *Ecological Applications* 19, 398–412.

Flory, S.L. and Clay, K. (2010) Non-native grass invasion suppresses forest succession. *Oecologia* 164, 1029–1038.

Foster, B.L. and Gross, K.L. (1998) Species richness in a successional grassland: effects of nitrogen enrichment and plant litter. *Ecology* 79, 2593–2602.

Grime, J.P. (1979) *Plant Strategies and Vegetation Processes*. John Wiley, Chichester, UK.

Grime, J.P. (1998) Benefits of plant diversity to ecosystems: immediate, filter and founder effects. *Journal of Ecology* 86, 902–910.

Grman, E. and Suding, K.N. (2010) Within-year soil legacies contribute to strong priority effects of exotics on native California grassland communities. *Restoration Ecology* 18, 664–670.

Harper, K.T., Van Buren, R. and Kitchen, S.G. (1996) Invasion of alien annuals and ecological consequences in salt desert shrublands of western Utah. In: Barrow, J.R., McArthur, E.D., Sosebee, R.E. and Tausch, J.R. (eds) *Proceedings – Symposium on Shrubland Ecosystem Dynamics in a Changing Environment*. US Department of Agriculture, Forest Service, Intermountain Research Station, Ogden, Utah, pp. 58–65.

Harris, J. (2009) Soil microbial communities and restoration ecology: facilitators or followers? *Science* 325, 573–574.

Hausmann, T. and Hawkes, C.V. (2009) Plant neighborhood control of arbuscular mycorrhizal communities. *New Phytologist* 183, 1188–1200.

Hausmann, T. and Hawkes, C.V. (2010) Order of plant host establishment alters the composition of arbuscular mycorrhizal communities. *Ecology* 91, 2333–2343.

Hawkes, C.V. (2007) Are invaders moving targets? The generality and persistence of advantages in size, reproduction, and enemy release in invasive plant species with time since introduction. *American Naturalist* 170, 832–843.

Hawkes C.V., Wren, I.F., Herman, D.H. and Firestone, M.K. (2005) Plant invasion alters nitrogen cycling by modifying the soil nitrifying community. *Ecology Letters* 8, 976–985.

Hawkes, C.V., Belnap, J., D'Antonio, C. and Firestone, M.K. (2006) Arbuscular mycorrhizal assemblages in native plant roots change in the presence of invasive exotic grasses. *Plant and Soil* 281, 369–380.

Heneghan, L., Miller, S.P., Baer, S., Callaham Jr, M.A., Montgomery, J., Pavao-Zuckerman, M., Rhoades, C.C. and Richardson, S. (2008) Integrating soil ecological knowledge into restoration management. *Restoration Ecology* 16, 608–617.

Herron, G.J., Sheley, R.L., Maxwell, B.D. and Jacobsen, J.S. (2001) Influence of nutrient availability on the interaction between spotted knapweed and bluebunch wheatgrass. *Restoration Ecology* 9, 326–331.

Hobbs, R.J., Higgs, E. and Harris, J.A. (2009) Novel ecosystems: implications for conservation and restoration. *Trends in Ecology and Environment* 24, 599–609.

Holdredge, D. and Bertness, M.D. (2011) Litter legacy increases the competitive advantage of invasive *Phragmites australis* in New England wetlands. *Biological Invasions* 13, 423–433.

Jordan, N.R., Larson, D.L. and Huerd, S.C. (2008) Soil modification by invasive plants: effects on native and invasive species of mixed-grass prairies. *Biological Invasions* 10, 177–190.

Kardol, P., Bezemer, T.M. and van der Putten, W.H. (2006) Temporal variation in plant-soil feedback controls succession. *Ecology Letters* 9, 1080–1088.

Kardol, P., Cornips, N.J., van Kempen, M.M.L., Bakx-Schotman, J.M.T. and van der Putten, W.H. (2007) Microbe-mediated plant-soil feedback causes historical contingency effects in plant community assembly. *Ecological Monographs* 77, 147–162.

Kardol, P., Bezemer, T.M. and van der Putten, W.H. (2009) Soil organism and plant introductions in restoration of species-rich grassland communities. *Restoration Ecology* 17, 258–269.

Klironomos, J.N. (2002) Feedback with soil biota contributes to plant rarity and invasiveness in communities. *Nature* 417, 67–70.

Kloot, P.M. (1983) The role of common iceplant (*Mesembryanthemum crystallinum*) in the deterioration of medic pastures. *Australian Journal of Ecology* 8, 301–306.

Knevel, I.C., Lans, T., Menting, F.B.J., Hertling, U.M. and van der Putten, W.H. (2004) Release from native root herbivores and biotic resistance by soil pathogens in a new habitat both affect the alien *Ammophila arenaria* in South Africa. *Oecologia* 141, 502–510.

Knick, S.T. and Rotenberry, J.T. (1997) Landscape characteristics of disturbed shrubsteppe habitats in southwestern Idaho (USA). *Landscape Ecology* 12, 287–297.

Krueger-Mangold, J.M., Sheley, R.L. and Svejcar, T.J. (2006) Toward ecologically-based invasive plant management on rangeland. *Weed Science* 54, 597–605.

Kulmatiski, A. (2011) Changing soils to manage plant communities: Activated carbon as a restoration tool in ex-arable fields. *Restoration Ecology* 19, 102–110.

Kulmatiski, A. and Beard, K.H. (2006) Activated carbon as a restoration tool: potential for control of invasive plants in abandoned agricultural fields. *Restoration Ecology* 14, 251–257.

Kulmatiski, A. and Beard, K.H. (2011) Long-term plant growth legacies overwhelm short-term plant growth effects on soil microbial community structure. *Soil Biology and Biochemistry* 43, 823–830.

Kulmatiski, A. and Kardol, P. (2008) Getting plant-soil feedbacks out of the greenhouse: experimental and conceptual approaches. *Progress in Botany* 69, 449–472.

Kulmatiski, A., Beard, K.H. and Stark, J.M. (2004) Finding endemic soil-based controls for weed growth. *Weed Technology* 18, 1353–1358.

Kulmatiski, A., Beard, K.H., Stevens, J. and Cobbold, S.M. (2008) Plant-soil feedbacks: a meta-analytical review. *Ecology Letters* 11, 980–992.

Kulmatiski, A., Heavillin, J. and Beard, K.H. (2011) Testing predictions of a three-species plant-soil feedback model. *Journal of Ecology* 99, 542–550.

Lacey, J., Marlow, C. and Lane, J. (1989) Influence of spotted knapweed (*Centaurea maculosa*) on surface runoff and sediment yield. *Weed Technology* 3, 627–631.

Ladenburger, C.G., Hild, A.L., Kazmer, D.J. and Munn, L.C. (2006) Soil salinity patterns in Tamarix invasions in the Bighorn Basin, Wyoming, USA. *Journal of Arid Environments* 65, 111–128.

Lankau, R.A. (2011). Resistance and recovery of soil microbial communities in the face of *Alliaria petiolata* invasions. *New Phytologist* 189, 536–548.

Lau, J.A., Puliafico, K.P., Kopshever, J.A., Steltzer, H., Jarvis, E.P., Schwarzlander, M., Strauss, S.Y. and Hufbauer, R.A. (2008) Inference of allelopathy is complicated by effects of activated carbon on plant growth. *New Phytologist* 178, 412–423.

Lenz, T.I., Moyle-Croft, J.L. and Facelli, J.M. (2003) Direct and indirect effects of exotic annual grasses on species composition of a South Australian grassland. *Austral Ecology* 28, 23–32.

Leung, B., Drake, J.M. and Lodge, D.M. (2004) Predicting invasions: propagule pressure and the gravity of allee effects. *Ecology* 85, 1651–1660.

Liao, C., Peng, R., Luo, Y., Zhou, X., Wu, X., Fang, C., Chen, J. and Li, B. (2008) Altered ecosystem carbon and nitrogen cycles by plant invasion: a meta-analysis. *New Phytologist* 177, 706–714.

Lockwood, J.L., Cassey, P. and Blackburn, T. (2005) The role of propagule pressure in explaining species invasions. *Trends in Ecology and Evolution* 20, 223–228.

Mack, M.C. and D'Antonio, C.M. (2003) Exotic grasses alter controls over soil nitrogen dynamics in a Hawaiian woodland. *Ecological Applications* 13, 154–166.

Mangla, S., Inderjit and Callaway, R.M. (2008) Exotic invasive plant accumulates native soil pathogen which inhibit native plants. *Journal of Ecology* 96, 58–67.

Manning, P., Morrison, S.A., Bonkowski, M. and Bardgett, R.D. (2008) Nitrogen enrichment modifies plant community structure via changes to plant-soil feedback. *Oecologia* 157, 661–673.

Marler, M.J., Zabinski, C.A. and Callaway, R.M. (1999) Mycorrhizae indirectly enhance competitive effects of an invasive forb on a native bunchgrass. *Ecology* 80, 1180–1186.

Marrs, R.H. (1993) Soil fertility and nature conservation in Europe: theoretical considerations and practical management solutions. *Advances in Ecological Research* 24, 241–301.

Marschner, H. (1986) *Mineral Nutrition of Higher Plants*, 2nd edn. Academic Press, London, UK.

Mitchell, C.E. and Power, A.G. (2003) Release of invasive plants from fungal and viral pathogens. *Nature* 421, 625–627.

Moora, M., Berger, S., Davison, J., Öpik, M., Riccardo, B., Bruelheide, H., Kühn, I., Kunin, W.E., Metsis, M., Rortais, A., Vanatoa, A., Vanatoa, E., Stout, J.C., Truusa, M., Westphal, C., Zobel, M. and Walther, G.-R. (2011) Alien plants associate with widespread generalist arbuscular mycorrhizal fungal taxa: evidence from a continental-scale study using massively parallel 454 sequencing. *Journal of Biogeography* doi:10.1111/j.1365-2699.2011.02478.x.

Mummey, D.L., Rillig, M.C. and Holben, W.E. (2005) Neighboring plant influences on arbuscular mycorrhizal fungal community composition as assessed by T-RFLP analysis. *Plant and Soil* 271, 83–90.

Mummey, D.L., Antunes, P.M. and Rillig, M.C. (2009) Arbuscular mycorrhizal fungi pre-inoculant identity determines community composition in roots. *Soil Biology and Biochemistry* 41, 1173–1179.

Murrell, C., Gerber, E., Krebs, C., Parepa, M., Schaffner, U. and Bossdorf, O. (2011) Invasive knotweed affects native plants through allelopathy. *American Journal of Botany*, 98, 38–43.

Nillson, M.C. (1994) Separation of allelopathy and resource competition by the boreal dwarf shrub *Empetrum hermaphroditum* Hagerup. *Oecologia* 98, 1–7.

Nuñez, M.A., Horton, T.R. and Simberloff, D. (2009) Lack of belowground mutualisms hinders Pinaceae invasions. *Ecology* 90, 2352–2359.

Orrock, J.L., Witter, M.S. and Reichman, O.J. (2008) Apparent competition with an exotic plant reduces native plant establishment. *Ecology* 89, 1168–1174.

Parker, I., Simberloff, D., Lonsdale, W., Goodell, K., Wonham, M., Kareiva, P., Williamson, M., von Hollo, B., Moyle, P., Byers, J. and Goldwasser, L. (1999) Impact: toward a framework for understanding the ecological effects of invaders. *Biological Invasions* 1, 3–19.

Peltzer, D.A., Bellingham, P.J., Kurokawa, H., Walker, L.R., Wardle, D.A. and Yeates, G.W. (2009) Punching above their weight: low-biomass non-native plant species alter soil properties during primary succession. *Oikos* 118, 1001–1014.

Perry, L.G., Johnson, C., Alford, E.R., Vivanco, J.M. and Paschke, M.W. (2005) Screening of grassland plants for restoration after spotted knapweed invasion. *Restoration Ecology* 13, 725–735.

Perry, L.G., Blumenthal, D.M., Monaco, T.A., Paschke, M.W. and Redente, E.F. (2010) Immobilizing nitrogen to control plant invasion. *Oecologia* 163, 13–24.

Pollak, O. and Kan, T. (1996) The use of prescribed fire to control invasive exotic weeds at Jepson Prairie Preserve. In: Witham, C.W., Bauder, E.T., Belk, D., Ferren, W.R. and Ornduff, R. (eds) *Proceedings, Ecology, Conservation, and Management of Vernal Pool Ecosystems*. California Native Plant Society, Sacramento, California, pp. 241–249.

Prasad, R. and Power, J.F. (1995) Nitrification inhibitors for agriculture, health, and the environment. *Advances in Agronomy* 54, 233–281.

Prober, S.M. and Lunt, I.D. (2009) Restoration of *Themeda australis* swards suppresses soil nitrate and enhances ecological resistance to invasion by exotic annuals. *Biological Invasions* 11, 171–181.

Prober, S.M., Thiele, K.R., Lunt, I.D. and Koen, T.B. (2005) Restoring ecological function in temperate grassy woodlands: manipulating soil nutrients, exotic annuals and native perennial grasses through carbon supplements and spring burns. *Journal of Applied Ecology* 42, 1073–1085.

Reigosa, M.J., Pedrol, N. and Gonzalez, L. (2006) *Allelopathy: a Physiological Process with Ecological Implications*. Springer, Dordrecht, the Netherlands.

Reinhart, K.O. and Callaway, R.M. (2006) Soil biota and invasive plants. *New Phytologist* 170, 445–457.

Reinhart, K.O., Packer, A., Van der Putten, W.H. and Clay, K. (2003) Plant-soil biota interactions and spatial distribution of black cherry in its native and invasive ranges. *Ecology Letters* 6, 1046–1050.

Requena, N., Perez-Solis, E., Azcon-Aguilar, C., Jeffries, P. and Barea, J.M. (2001) Management of indigenous plant-microbe symbioses aids restoration in desertified ecosystems. *Applied and Environmental Microbiology* 67, 495–498.

Ridenour, W.M. and Callaway, R.M. (2001) The relative importance of allelopathy in interference: the effects of an invasive weed on a native bunchgrass. *Oecologia* 126, 444–450.

Rossiter-Rachor, N.A., Setterfield, S.A., Douglas, M.M., Hutley, L.B., Cook, G.D. and Schmidt, S. (2009) Invasive *Andropogon gayanus* (gamba grass) is an ecosystem transformer of nitrogen relations in Australian savanna. *Ecological Applications* 19, 1546–1560.

Schramm, J.W. and Ehrenfeld, J.G. (2010) Leaf litter and understory canopy shade limit the establishment, growth and reproduction of *Microstegium vimineum*. *Biological invasions* 9, 3195–3204.

Schwartz, M.W., Hoeksema, J.D., Gehring, C.A., Johnson, N.C., Klironomos, J.N., Abbott, L.K. and Pringle, A. (2006) The promise and the potential consequences of the global transport of mycorrhizal fungal inoculum. *Ecology Letters* 9, 501–515.

Seastedt, T.R., Hobbs, R.J. and Suding, K.N. (2008) Management of novel ecosystems: are novel approaches required? *Frontiers in Ecology and the Environment* 6, 547–553.

Sheley, R.L., Carpinelli, M.F. and Reever Morghan, K.J. (2007) Effects of Imazapic on target and nontarget vegetation during revegetation. *Weed Technology* 21, 1071–1081.

Simberloff, D. and Von Holle, B. (1999) Positive interaction of nonindigenous species: invasional meltdown? *Biological Invasions* 1, 21–32.

Smith, M.R., Charvat, I. and Jacobson, R.L. (1998)

Arbuscular mycorrhizae promote establishment of prairie species in a tallgrass prairie restoration. *Canadian Journal of Botany* 76, 1947–1954.

Smith, S.D., Devitt, D.A., Sala, A., Cleverly, J.R. and Busch, D.E. (1998) Water relations of riparian plants from warm desert regions. *Wetlands* 18, 687–696.

Sperry, L.J., Belnap, J. and Evans, R.D. (2006) *Bromus tectorum* invasion alters nitrogen dynamics in an undisturbed arid grassland ecosystem. *Ecology* 87, 603–615.

Stermitz, F.R., Bais, H.P., Foderaro, T.A. and Vivanco, J.M. (2003) 7,8-Benzoflavone: a phytotoxin from root exudates of invasive Russian knapweed. *Phytochemistry* 64, 493–497.

Stinson, K.A., Campbell, S.A., Powell, J.R., Wolfe, B.E., Callaway, R.M., Thelen, G.C., Hallet, S.G., Prati, D. and Klironomos, K.N. (2006) Invasive plant suppresses the growth of native tree seedlings by disrupting belowground mutualisms. *PLoS Biology* 4, e140.

Strayer, D.L., Eviner, V.T., Jeschke, J.M. and Pace, M.L. (2006) Understanding the long-term effects of species invasions. *Trends in Ecology and Evolution* 21, 645–651.

Suding, K.N., Gross, K.L. and Houseman, G.R. (2004) Alternative states and positive feedbacks in restoration ecology. *Trends in Ecology and Evolution* 19, 46–53.

Suding, K.N., Collins, S.L., Gough, L., Clark, C., Cleland, E.E., Gross, K.L., Milchunas, D.G. and Pennings, S. (2008) Functional and abundance based mechanisms explain diversity loss due to N fertilization. *Proceedings of the National Academy of Sciences* 102, 4387–4392.

Theoharides, K.A. and Dukes, J.S. (2007) Plant invasion across space and time: factors affecting nonindigenous species success during four stages of invasion. *New Phytologist* 176, 256–273.

Thorpe, A.S., Thelen, G.C., Diaconu, A. and Callaway, R.M. (2009) Root exudate is allelopathic in invaded community but not in native community: field evidence for the novel weapons hypothesis. *Journal of Ecology* 97, 641–645.

van der Heijden, M.G.A. and Horton, T.R. (2009) Socialism in soil? The importance of mycorrhizal fungal networks for facilitation in natural ecosystems. *Journal of Ecology* 97, 1139–1150.

van der Putten, W.H., Kowalchuk, G.A., Brinkman, E.P., Doodeman, G.T.A., van der Kaaij, R.M., Kamp, A.F.D., Menting, F.B.J. and Veenedaal, E.M. (2007) Soil feedback of exotic savanna grass relates to pathogen absence and mycorrhizal selectivity. *Ecology* 88, 978–988.

van der Putten, W.H., Bardgett, R.D., de Ruiter, P.C., Hol, W.H.G., Meyer, K.M., Bezemer, T.M., Bradford, M.A., Christensen, S., Eppinga, M.B., Fukami, T., Hemerik, L., Molofsky, J., Schadler, M., Scherber, C., Strauss, S.Y., Vos, M. and Wardle, D.A. (2009) Empirical and theoretical challenges in aboveground-belowground ecology. *Oecologia* 161, 1–14.

Vellinga, E.C., Wolfe, B.E. and Pringle, A. (2009) Global patterns of ectomycorrhizal introductions. *New Phytologist* 181, 960–973.

Vessey, J.K. (2003) Plant growth promoting rhizobacteria as biofertilizers. *Plant and Soil* 255, 571–586.

Vivrette, N.J. and Muller, C.H. (1977) Mechanism of invasion and dominance of coastal grassland by *Mesembryanthemum crystallinum*. *Ecological Monographs* 47, 301–318.

Vogelsang, K.M., Bever, J.D., Griswold, M. and Schultz, P.A. (2004) The use of mycorrhizal fungi in erosion control applications. Final Report for Caltrans. Contract No. 65A0070. California Department of Transportation, Sacramento, California.

Vogelsang, K.M. and Bever, J.D. (2009) Mycorrhizal densities decline in association with nonnative plants and contribute to plant invasion. *Ecology* 90, 399–407.

Walker, K.J., Stevens, P.A., Stevens, D.P., Mountford, J.O., Manchester, S.J. and Pywell, R.F. (2004) The restoration and re-creation of species rich lowland grassland on land formerly managed for intensive agriculture in the UK. *Biological Conservation* 119, 1–18.

Wardle, D.A., Nicholson, K.S. and Rahman, A. (1993) Influence of plant age on the allelopathic potential of nodding thistle (*Carduus nutans* L.) against pasture grasses and legumes. *Weed Research* 33, 69–78.

Wardle, D.A., Nicholson, K.S., Ahmed, M. and Rahman, A. (1994) Interference effects of the invasive plant *Carduus nutans* L. against nitrogen fixation ability of *Trifolium repens* L. *Plant and Soil* 163, 287–297.

Wardle, D.A., Nilsson, M.-C., Gallet, C. and Zackrisson, O. (1998) An ecosystem-level perspective of allelopathy. *Biological Review of the Cambridge Philosophical Society* 73, 305–319.

Wolfe, B.E., Rodgers, V.L., Stinson, K.A. and Pringle, A. (2008) The invasive plant *Alliaria petiolata* (garlic mustard) inhibits ectomycorrhizal fungi in its introduced range. *Journal of Ecology* 96, 777–783.

Wolkovich, E.M., Bolger, D.T. and Cottingham, K.L. (2009) Invasive grass litter facilitates native shrubs through abiotic effects. *Journal of Vegetation Science* 20, 1121–1132.

Yelenik, S.G. and Levine, J.M. (2011) The role of plant-soil feedbacks in driving native-species recovery. *Ecology* 92, 66–74.

Young, J.A., Clements, C.D. and Blank, R.R. (1997) Influence of nitrogen on antelope bitterbrush seedling establishment. *Journal of Range Management* 50, 536–540.

Young, J.A., Trent, J.D., Blank, R.R. and Palmquist, D.E. (1998) Nitrogen interactions with medusahead (*Taeniatherum caput-medusae* ssp. *asperum*) seedbanks. *Weed Science* 46, 191–195.

Young, T.A., Evans, R.A. and Kay, B.L. (1971) Germination of caryopses of annual grasses in simulated litter. *Agronomy Journal* 63, 551–555.

Zhang, Q., Yang, R., Tang, J., Yang, H., Hu, S. and Chen, X. (2010) Positive feedback between mycorrhizal fungi and plants influences plant invasion success and resistance to invasion. *PLoS ONE* 5, e12380.

8

Species Performance: the Relationship Between Nutrient Availability, Life History Traits, and Stress

Jeremy J. James

US Department of Agriculture, Eastern Oregon Agricultural Research Center, USA

Introduction

Differences in species performance (i.e. how a species captures and utilizes resources to maintain and increase population size) influence the rate and direction of plant community change (Sheley *et al.*, 2006). Species performance is determined by a number of interacting factors. This includes resource supply rates, physiological traits that determine how a species affects and responds to the environment, life history traits that determine patterns of birth, mortality, and growth of individuals in a population, as well as abiotic and biotic stressors such as herbivory or drought. Researchers and land managers long have recognized that our ability to predict and manage the spread of invasive species directly depends on our understanding of the processes that differentially impact the performance of invasive and native plants. In nutrient-poor systems across the globe, increases in nutrient availability increase the susceptibility of ecosystems to invasion (Mack *et al.*, 2000). This general and widespread response indicates that increases in nutrient availability are likely favoring the performance of invasive species over native species.

Having recognized this, restoration practitioners have long considered soil nutrient management to be a fundamental component to restoring invasive plant-infested systems. However, the role soil nutrient availability plays in our ability to restore invasive plant dominated systems is not as straightforward as it initially appears. The objectives of this chapter are to: (i) examine current paradigms and assumptions about how nutrient availability influences the relative performance of invasive and native plants; (ii) develop the argument that the influence of nutrient availability on species performance is mediated by life history traits and stressors such as drought and herbivory; and (iii) describe how these principles that determine species performance in different nutrient environments can be used to develop more effective invasive plant management strategies.

Current Paradigms of Nutrient Availability and Invasion

Background

Theoretical and empirical work has established a positive relationship between resource availability and habitat invasibility. For example, in nutrient-poor systems dominated by slow-growing species, invasions have commonly been attributed to increased soil nitrogen (N) availability following disturbance (McLendon and

Redente, 1991; Kolb *et al.*, 2002; Brooks, 2003). Due to their faster growth rates and ability to rapidly take up N, fast-growing invasive species are thought to be more competitive than slow-growing native species in high N soils (Norton *et al.*, 2007; Mazzola *et al.*, 2008; Mackown *et al.*, 2009; Perry *et al.*, 2010). In contrast, slow-growing native species, with their greater investment in belowground structures and ability to recycle and store N (Chapin, 1980; Fargione and Tilman, 2002), are thought to be favored under low N conditions. These observations have led to the logical assumption that managing soils for low nutrient availability will facilitate restoration of invasive plant-dominated systems.

Based on the positive relationship between nutrient availability and habitat invasibility, land managers and researchers have been applying a range of techniques to lower soil N prior to native plant restoration, including the use of cover crops, carbon amendments, and topsoil removal (Perry *et al.*, 2010). Following decades of manipulations, however, the effect of soil N management on the relative performance of native and invasive species has been mixed. For example, work in some grassland systems has shown that reductions in soil N availability may reduce the growth of invasive species and facilitate the establishment of native species (Morgan, 1994; Blumenthal *et al.*, 2003). Likewise, reduction in soil N availability to invasive annual grass-dominated coastal prairie and old-fields in North America and pasture in Australia lowered soil N availability and facilitated reestablishment of perennial grasses (Alpert *et al.*, 2000; Paschke *et al.*, 2000; Prober *et al.*, 2005). While these studies did not identify how decreasing soil N facilitated establishment of native species, these results have been largely attributed to an overall lower N requirement of native plants or to changes in competitive relationships among species as N availability declined.

A number of other studies, however, have not observed the expected effect of lowering soil N on invasive and native species performance. For example, lowering soil N in invasive plant-dominated coastal and interior grasslands, as well as sagebrush steppe communities in North America, did not facilitate reestablishment of perennial grasses (Morghan and Seastedt, 1999; Corbin and D'Antonio, 2004; Huddleston and Young, 2005; Mazzola *et al.*, 2008). The discrepancies between these studies and the current paradigm of soil N effects on invasive and native plant performance have been explained a number of ways including lack of long-term effects of treatments on soil N, the type of treatments used to reduce soil N, and environmental conditions such as drought (Blumenthal, 2009; Brunson *et al.*, 2010; Perry *et al.*, 2010). While some variation among studies could be due to these factors, the large discrepancies in overall conclusions reached in these research efforts suggests the current conceptual framework for understanding the role nutrient management plays in our ability to restore invasive plant-dominated systems is incomplete.

Reconsideration of the nutrient availability–invasibility paradigm

The development of the current nutrient management paradigm is largely centered on the assumptions that traits such as fast growth make invasive species better 'adapted' to high nutrient environments than native species while slow growth makes native species better 'adapted' to low nutrient environments than invasive species. However, over the last decade or so, there have been major advances in understanding the plant physiological and morphological traits underpinning the different ecological strategies displayed by native and invasive species as well as the ecological trade-offs associated with these traits (Weiher and Keddy, 1999; Diaz *et al.*, 2004; Wright *et al.*, 2004). These insights are instrumental for determining how our ability to predict nutrient availability effects on species performance can be improved.

For example, current plant trait frameworks suggest plant ecological strategies may be more clearly differentiated based on how a species balances the trade-off between resource acquisition and resource conservation, rather than growth potential (Weiher and Keddy, 1999; Diaz et al., 2004; Wright et al., 2004). Under this framework, some species maximize resource conservation by making mechanical and chemical investments in tissue that increase tissue lifespan and decrease tissue loss due to herbivory or environmental stress (Lambers and Poorter, 1992; Westoby et al., 2002; Wright et al., 2004). While these investments decrease relative growth rate (RGR), these traits increase mean nutrient residence time, allowing a greater duration of return on nutrients captured and greater nutrient conservation (Berendse and Aerts, 1987). Over the long term, these traits are expected to be advantageous in low nutrient soils (Berendse, 1994; Aerts, 1999). On the other hand, some species make thin, short-lived tissue. While this decreases tissue life span and increases tissue susceptibility to herbivory or environmental stress, it allows these species to construct more absorptive root and leaf surface area per unit of biomass allocated to tissue, resulting in more rapid resource capture (Lambers and Poorter, 1992). These traits are expected to provide an advantage in nutrient-rich soils where resource capture is favored over resource conservation.

This research suggesting tissue construction strategy (thin, short-lived versus thick, long-lived) influences the balance between resource acquisition and resource conservation and, consequently, differentiates the ecological strategies of plants, has major implications for predicting how we expect soil N management to impact restoration efforts. Specifically, because certain invasive species invest little in tissue protection and defense, over the short term they may achieve greater growth rates than native species even in low nutrient soils (Burns, 2004; Garcia-Serrano et al., 2005; James, 2008). In addition, the advantages native species have in low nutrient soil, including greater nutrient mean residence time and greater protection from environmental stress or herbivory, are only going to be manifested through time and

when environmental stress or herbivory pressure is high. Therefore, in a restoration scenario where both native and invasive species are recruiting from the seed bank, invasive species may maintain an initial growth advantage even in low nutrient soils. If this initial growth advantage allows invasive species to interfere with the growth and survival of native species, then native species will have little opportunity to capitalize on traits that ultimately will provide them an advantage in nutrient-poor soils. Given this potential and the variable results seen in soil N manipulation studies, it may be helpful to examine a quantitative synthesis of the literature and determine if the balance of evidence supports the current paradigm or suggests a need to further refine a predictive framework for the role soil nutrient management plays in restoring invaded systems.

Evaluation of the nutrient availability–invasibility paradigm

A recent study used meta-analysis to evaluate the degree to which soil N management differentially impacts growth and competitive ability of invasive annual and native perennial grasses (James et al., 2011). Meta-analysis is a useful tool to quantitatively synthesize results of independent studies (Gurevitch and Hedges, 1993). In the experiments analyzed in this meta-analysis, all species were started from seed and thus provide useful insight into how soil conditions may impact initial growth and competitive interactions and set the stage for long-term plant community changes in a restoration setting where both native and invasive species are starting as seed. Here three hypotheses were tested: (i) decreasing soil N availability has a greater negative effect on biomass and tiller production of annual compared to perennial grass seedlings; (ii) however, seedlings of fast-growing species, including invasive annual grasses, maintain a higher RGR than seedlings of slow-growing species in low N environments, allowing invasive annual grasses to construct more biomass and tillers than perennials in the short-term; and (iii) as a result, lowering soil N availability will not alter the competitive advantage that annual grass seedlings have over perennial grass seedlings.

To test these hypotheses, 70 comparisons were extracted from 25 studies and effect size was calculated as the natural log response ratio (ln RR) where:

$$\ln RR = \ln \frac{T_z}{T_b} = \ln T_z - \ln T_b \qquad (8.1)$$

and T_z and T_b are means of the response variable for two different treatment groups (Hedges *et al.*, 1999). The parameter ln RR estimates the proportional difference between treatment groups. An ln RR of zero indicates the response variables do not vary between groups, and a positive or negative ln RR indicates the response variable is larger for the T_z or T_b group, respectively.

Figures 8.1–8.3 show the results of this quantitative synthesis. Figure 8.1 shows the effect of soil-N availability on the difference in (a) biomass, (b) number of tillers, and (c) relative growth rate (RGR) of invasive annual grasses and native perennial grasses growing in low and high N environments. Confidence intervals on biomass differences (annual – perennial) for high and low N do not overlap (Fig. 8.1a). This supports the first hypothesis that reducing soil N decreased annual grass biomass by a greater proportion than perennial grass biomass. The confidence interval on the RGR difference (fast growing species – slow growing species) is greater than 0 for low N (Fig. 8.1c). Therefore, fast growing invasive annual grasses maintained higher RGR under low and high N compared to slow-growing perennial grasses. As a result of this higher RGR, invasive annual grasses produced more biomass and tillers than perennial grasses when N availability was low (Fig. 8.1a, b). This supports the second hypothesis that seedlings of fast-growing species, including invasive annual grasses, maintain a higher RGR than seedlings of slow-growing species in low N environments, allowing invasive annual grasses to construct more biomass and tillers than perennials in the short term.

Consistent with the greater growth rate, biomass and tiller production by invasive annuals compared to native perennials in low N soil, and in support of the third hypothesis, lowering soil N availability did not alter the competitive advantage invasive annual grass seedlings have over native perennial grass seedlings. While all point estimates of competition parameters were negative, indicating plants competed under low and high N, confidence intervals for low and high N overlapped substantially (Fig. 8.2), suggesting N availability had little effect on competition intensity. Moreover, the competition studies showed that annual

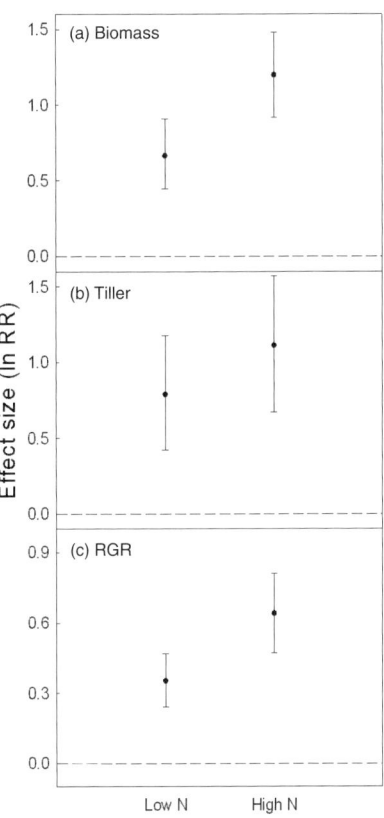

Fig. 8.1. Most likely parameter values (dots) and 95% confidence intervals (bars) on log response ratios (ln RR) where: (a) ln RR = ln(perennial grams per plant) – ln(annual grams per plant); (b) lnRR = ln(perennial tillers per plant) – ln(annual tillers per plant); and (c) ln RR = ln(RGR of fast growing species) – ln(RGR of slow growing species).

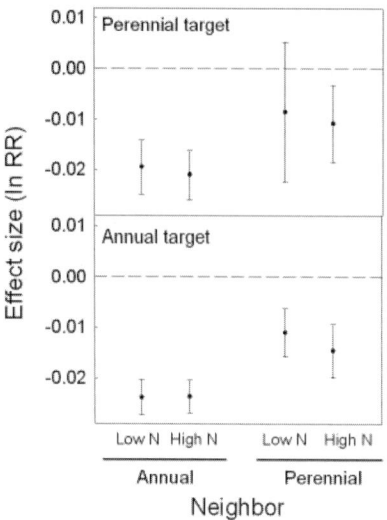

Fig. 8.2. Most likely parameter values (dots) and 95% confidence intervals (bars) on log response ratios (ln RR) from addition series studies included in the meta-analysis. More negative values indicate more competitive effects of neighbors on target plant biomass.

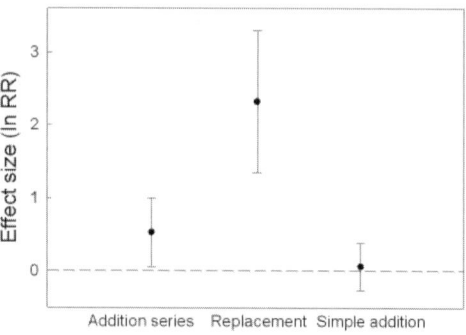

Addition series   Replacement   Simple addition

Fig. 8.3. Most likely parameter values (dots) and 95% confidence intervals (bars) on log response ratios (ln RR). The ln RR describes effects of nitrogen availability on biomass production of perennial grasses growing with invasive annual grass neighbors. Specifically, ln RR = ln(biomass of perennial plants grown with annual grasses under high N) – ln(biomass of perennial plants grown with annual grasses under low N). Separate estimates are provided for the three types of competition designs analyzed in the meta-analysis (addition series, replacement series, and simple addition).

grass neighbors had a stronger competitive effect on both annual and perennial targets compared to perennial neighbors and that the competitive effects of annual neighbors on perennials did not decrease with lower N availability (Fig. 8.2). In addition, perennial targets competing with annual grass neighbors did not incur a net cost in biomass production in high-N environments (Fig. 8.3). Point estimates and confidence intervals describing the difference in perennial plant biomass when perennial plants were grown with annual neighbors under high N compared to when perennial plants were grown with annual grass neighbors under low N were positive or overlapped zero in the different types of competition designs used. This indicates perennials produced either more biomass when grown with annual grass competitors under high N compared to when grown with annual grass competitors under low N, or that there was no detectable difference in biomass produced under high and low N.

The current soil N management framework rests on the assumption that because

increasing N availability facilitates invasion, decreasing N availability should facilitate restoration of systems dominated by invasive plants (McLendon and Redente, 1992; Alpert et al., 2000; Blumenthal et al., 2003). The meta-analysis described here rejects this assumption for invasive annual grasses. In addition, because many invasive plant species display traits that favor resource acquisition over resource conservation compared to their native counterparts (Grotkopp et al., 2002; James and Drenovsky, 2007; Leishman et al., 2007), these findings likely apply to other invasion scenarios. This means it is very likely that when invasive and native species are establishing from seed, soil nutrient management alone will not directly facilitate establishment of native species. While this conclusion is supported by an array of studies showing soil N management failed to facilitate restoration of invasive plant-dominated systems (e.g. Corbin and D'Antonio, 2004; Huddleston and Young, 2005; Mazzola et al., 2008), this conclusion is at odds with studies that have demonstrated that soil N management facilitates restoration of systems infested by a range of invasive plants (e.g. Alpert *et al.*, 2000; Paschke *et al.*, 2000; Prober *et al.*, 2005). An important question then is how can we use these results and current understanding of plant life history trade-offs to reconcile these discrepancies and improve our ability to predict how nutrient availability impacts species performance?

Influence of Nutrient Management on Species Performance: the Role of Life History, Drought, and Herbivory

Advances in understanding traits underlying variation in plant ecological strategies have been made. These advancements can improve our ability to understand and predict how soil nutrient management may influence restoration outcomes. Potential growth rate long has been a key trait used to characterize different plant ecological strategies (e.g. Grime, 1977; Goldberg and Landa, 1991; Loehle, 2000). When grown under optimum conditions, species that dominate nutrient-rich sites typically have a higher RGR than species that dominate nutrient-poor sites (Lambers and Poorter, 1992). While a high RGR allows rapid growth and resource capture, the ecological advantage a low RGR provides is less clear. Early interpretations focused on advantages directly due to low RGR, such as the potential for low RGR species to function closer to their physiological optimum in infertile soils compared to high RGR species (Grime and Hunt, 1975; Chapin, 1980). The current consensus, however, argues that traits associated with low RGR, not RGR itself, have been the target of selection in nutrient-poor systems (Lambers *et al.*, 1998; Aerts, 1999). Specific leaf area (SLA) is the principle trait influencing RGR variation among species (Lambers and Poorter, 1992). Morphological and chemical adjustments that increase leaf tissue toughness and protection from abiotic stress or herbivores decrease SLA and consequently, RGR (Poorter *et al.*, 2009). While a low SLA indirectly reduces growth and resource capture rates, it increases the ability to conserve resources because low SLA species tend to have longer leaf lifespan than high SLA species (Westoby *et al.*, 2002). A high SLA, on the other hand, allows quicker return on nutrients and dry matter invested in leaves, resulting in rapid resource capture but poor resource conservation due to shorter leaf lifespan. Variation in SLA among species, therefore, represents a key axis describing different plant ecological strategies that range between rapid resource capture and effective resource conservation (Diaz *et al.*, 2004; Wright *et al.*, 2004; Leishman *et al.*, 2007).

While we would expect resource conservation strategies to become increasingly important as soil nutrient availability declines, favoring low SLA species, the meta-analysis demonstrated that fast-growing, high SLA species, including invasive annual grasses, maintained greater growth rates in low-N soils than slow-growing species. These differences in initial RGR were reflected in the greater biomass and tiller production by invasive species in low-N soils compared to native species. High-SLA species also may produce thinner and less dense root tissue compared to low-SLA species, resulting in a lower specific root length (SRL) (Elberse and Berendse, 1993). Construction costs per unit tissue dry weight are similar among species (Poorter and Bergkotte, 1992). As a result the thinner or less dense root and leaf tissue generally produced by invasive compared to native species allows invasive species to maintain a greater rate of return on biomass allocated to leaves and roots and ultimately grow faster than native species. At the seedling stage, this difference in tissue production cost allows invasive species to preempt more belowground resources than native species in nutrient-poor soils (James, 2008). Therefore, because of these differences in tissue construction strategies, at the seedling stage, slow-growing native species have no direct advantage in terms of initial growth in nutrient-poor soils compared to invasive species. If restoration outcomes are largely determined during the seedling establishment phase, then soil nutrient management alone likely will not influence restoration outcomes.

Restoration outcomes, however, also are influenced by longer-term processes, not just processes associated with the seedling establishment phase. Internal plant nutrient recycling, for example, contributes significantly to the long-term performance of natives in low nutrient soils (Killingbeck and Whitford, 1996). Plants have two primary sources of nutrients: soil nutrient pools and nutrients retained within the plant. The reliance on these two different nutrient sources depends on a variety of factors, including internal plant nutrient recycling, soil nutrient availability, and life history. During tissue senescence, biomolecules are broken down and the mineral nutrients are translocated to storage tissues, such as stems and roots. Internal nutrient recycling buffers plants against variation in soil nutrient availability (Aerts and Chapin, 2000), impacts plant fitness (May and Killingbeck, 1992), and may be less costly than acquiring and assimilating soil nutrients (Wright and Westoby, 2003). Although the amount of nutrients resorbed (i.e. realized resorption) may vary annually, the maximum amount of nutrients that can be resorbed physiologically (i.e. potential resorption) is considered an evolved 'target value' (Killingbeck, 2004). As such, potential resorption is greater in plants from low-nutrient habitats compared to plants that dominate nutrient-rich systems (Killingbeck, 2004). If native species resorb more nutrients from dying tissue in nutrient-poor soils than invasive species, then over time natives will accumulate larger nutrient pools, which should place them at a competitive advantage.

In addition to nutrient recycling, native plants also conserve nutrients by investing heavily in structural support for leaf and root tissue. While these investments decrease SLA and SRL and lower rates of growth and resource capture, they increase tissue life span. As a consequence, these traits result in longer mean nutrient residence time and a lower relative nutrient requirement for native compared to invasive species (Berendse and Aerts, 1987; Berendse

et al., 2007). Similar to nutrient recycling traits, because these traits deflect some biomass away from growth functions (e.g. creation of leaf or root surface areas for photosynthesis and nutrient uptake) toward nutrient conservation functions, these traits place native seedlings at an initial disadvantage in nutrient-poor soils. Modeling and empirical work, however, demonstrate that over time, these traits allow slow-growing species to accumulate more biomass and have faster population growth rates than fast-growing species in nutrient-poor soils (Berendse et al., 1992; Berendse, 1994; Ryser, 1996).

The outcome of competition at the seedling stage, however, is not solely based on resource capture differences, and other abiotic and biotic factors likely will be important in influencing the outcome of competition between invasive and native species in nutrient-poor soils. In addition to increasing tissue life span and nutrient residence time, the thicker, longer-lived leaf and root tissue constructed by native compared to invasive species can decrease tissue loss from drought. Structural investments in leaf and root cells and tissue xylem increase tissue density and are an opportunity cost in terms of investment of biomass in surface area for resource capture. These investments, however, allow plants to maintain physiological function as soils dry compared to plants that invest less in these structures. These differences in drought tolerance may heavily influence competition outcomes in environments with low and variable precipitation inputs (Goldberg and Novoplansky, 1997; Chesson et al., 2004). Likewise, mechanical investments in tissue, that decrease susceptibility to drought, also can decrease tissue palatability, making these plants less susceptible to attack by generalist invertebrate herbivores. For example, Buckland and Grime (2000) showed that in field microcosms, fast-growing species dominated low, moderate, and high fertility soils in the absence of a generalist invertebrate herbivore. When a generalist herbivore was added, the abundance of slow-growing species increased in low-fertility soils but not in high-fertility soils. Thus, population dynamics of invertebrate herbivores that target fast-growing species with less structural investment in tissue, may play an important role in determining outcomes of competition and assembly in nutrient-poor soils (Fraser and Grime, 1999; Olofsson, 2001; Burt-Smith *et al.*, 2003).

Designing Ecologically Based Invasive Plant Management Strategies Based on Principles of Species Performance in Low and High Nutrient Environments

Theory, empirical work, and the results of the meta-analysis described here all support the notion that increases in nutrient availability favor invasive species' performance over native species (Huenneke *et al.*, 1990; Stohlgren *et al.*, 1999; Davis *et al.*, 2000). There also is evidence that invasive species, once established may accelerate nutrient cycles and maintain high levels of nutrient availability, ensuring resource conditions continue to favor dominance of invasive species (Liao *et al.*, 2008). From these results, it is clear that invasive-plant managers should carefully examine how management decisions and natural processes impact nutrient availability. Management decisions and natural processes that increase nutrient availability will favor the performance of invasive over native species. Therefore, maintaining low nutrient environments is a central goal of ecologically based invasive plant management programs.

However, the meta-analysis and our current understanding of traits underlying variation in plant ecological strategies suggest nutrient management alone will not be sufficient for facilitating establishment of native species in invasive plant-dominated systems. By considering the meta-analysis and the trait-based frameworks of plant ecological strategies we can formulate some principles and predictions about how and when soil N management will positively influence restoration outcomes. Three principles emerge:

1. Managing environments for low nutrient availability will favor resource conservation over resource capture by plants, favoring desired species over invasive species.
2. Initial establishment of desired species needs to be successfully managed to realize any benefit of nutrient management.
3. When desired species are establishing from seed, a sufficient amount of stress that inhibits the performance of invasive species will need to be applied to realize any benefit of nutrient management.

To understand how these principles can be incorporated into a practical invasive-plant management program it may be useful to consider two contrasting restoration scenarios. In both scenarios there is a significant invasive plant seed bank, but in scenario one, most native plants must be seeded, while in scenario two, a significant number of native plants are still present at the site.

If native species must be seeded, they cannot capitalize on key traits that provide them an advantage in low nutrient soils, such as the ability to recycle stored nutrients or their greater mean nutrient residence time. Managing soils for low nutrients will not provide a growth advantage to native compared to invasive species at the seedling stage. If herbivore pressure or abiotic stress is pronounced, the lower SLA and SRL of native seedlings may be advantageous due to greater investment in lignin or other compounds that increase tissue toughness. On the other hand, if stresses like drought or herbivory do not occur, or cannot be applied in a manner that preferentially impacts invasives (e.g. livestock grazing at inappropriate times), the performance of invasive species will have to be controlled via other means (e.g. herbicide) for a long enough period to give native species an opportunity to establish and realize the benefits of their resource conservation traits. If this is achieved, managers should expect a long-term benefit to soil nutrient management. If, however, the initial performance of invasive species is not managed, managers should not expect to receive a benefit from nutrient management.

For the second restoration scenario, in which there is a significant amount of native plants established or if factors such as dispersal heterogeneity or low annual grass propagule pools allow perennial grasses to survive the first growing season (e.g. DiVittorio et al., 2007), mangers should expect soil nutrient management to have a direct positive impact on restoration efforts. Here, established plants have an opportunity to begin to capitalize on their resource conservation traits and do not have to compete with invasive species during the seedling stage. The greater size and greater nutrient conservation ability should rapidly place natives at a competitive advantage in low nutrient soils (e.g. Lulow, 2006; Abraham et al., 2009).

Conclusion

By considering our current understanding of traits underlying variation in plant ecological strategies, the meta-analysis presented here, and the two contrasting restoration scenarios outlined above, it is possible to improve prediction and management of native and invasive plant performance. In situations in which a remnant native plant community exists or there is sufficient time for seeded species to establish, management to reduce nutrient availability likely will have a direct positive effect on relative performance of native compared to invasive species, resulting in long-term positive changes in plant community dynamics. In the more common and complex situation in which both native and invasive species recruit from seed, soil nutrient management will not have a direct effect on restoration outcomes. Instead, management strategies or natural processes that reduce invasive species performance over the short-term and allow seeded species to establish will need to be applied before any soil nutrient management benefit is realized. If invasive plant propagule pools and performance can be successfully managed the first year, then low-nutrient soils should begin to place native species at a competitive advantage, and this advantage should increase through time.

References

Abraham, J.K., Corbin, J.D. and D'antonio, C.M. (2009) California native and exotic perennial grasses differ in their response to soil nitrogen, exotic annual grass density, and order of emergence. Plant Ecology 201, 445–456.

Aerts, R. (1999) Interspecific competition in natural plant communities: mechanisms, trade-offs and plant-soil feedbacks. Journal of Experimental Botany 50, 29–37.

Aerts, R. and Chapin, F.S. (2000) The mineral nutrition of wild plants revisited: a re-evaluation of processes and patterns. Advances in Ecological Research 30, 1–67.

Alpert, P., Alpert, P. and Maron, J.L. (2000) Carbon addition as a countermeasure against biological invasion by plants. Biological Invasions 2, 33–40.

Berendse, F. (1994) Competition between plant populations at low and high nutrient supplies. Oikos 71, 253–260.

Berendse, F. and Aerts, R. (1987) Nitrogen-use-efficiency: a biologically meaningful definition? Functional Ecology 1, 293–296.

Berendse, F., Elberse, W.T. and Geerts, R. (1992) Competition and nitrogen loss from plants in grassland ecosystems. Ecology 73, 46–53.

Berendse, F., De Kroon, R.H. and Braakhekke, W.B. (2007) Acquisition, use, and loss of nutrients. In: Pugnaire, F.I. and Valladares, F. (eds) Functional Plant Ecology. CRC Press, Boca Raton, Florida, pp. 259–285.

Blumenthal, D.M. (2009) Carbon addition interacts with water availability to reduce invasive forb establishment in a semi-arid grassland. Biological Invasions 11, 1281–1290.

Blumenthal, D.M., Jordan, N.R. and Russelle, M.P. (2003) Soil carbon addition controls weeds and facilitates prairie restoration. Ecological Applications 13, 605–615.

Brooks, M.L. (2003) Effects of increased soil nitrogen on the dominance of alien annual plants in the Mojave Desert. Journal of Applied Ecology 40, 344–353.

Brunson, J.L., Pyke, D.A. and Perakis, S.S. (2010) Yield responses of ruderal plants to sucrose in invasive-dominated sagebrush steppe of the Northern Great Basin. Restoration Ecology 18, 304–312.

Buckland, S.M. and Grime, J.P. (2000) The effects of trophic structure and soil fertility on the assembly of plant communities: a microcosm experiment. Oikos 91, 336–352.

Burns, J.H. (2004) A comparison of invasive and non-invasive dayflowers (Commelinaceae) across experimental nutrient and water gradients. *Diversity and Distributions* 10, 387–397.

Burt-Smith, G.S., Grime, J.P. and Tilman, D. (2003) Seedling resistance to herbivory as a predictor of relative abundance in a synthesised prairie community. *Oikos* 101, 345–353.

Chapin, F.S. (1980) The mineral nutrition of wild plants. *Annual Review of Ecology and Systematics* 11, 233–260.

Chesson, P., Gebauer, R.L.E., Schwinning, S., Huntly, N., Wiegand, K., Ernest, M.S.K., Sher, A., Novoplansky, A. and Weltzin, J.F. (2004) Resource pulses, species interactions, and diversity maintenance in arid and semi-arid environments. *Oecologia* 141, 236–253.

Corbin, J.D. and D'Antonio, C.M. (2004) Can carbon addition increase competitiveness of native grasses? A case study from California. *Restoration Ecology* 12, 36–43.

Davis, M.A., Grime, J.P. and Thompson, K. (2000) Fluctuating resources in plant communities: a general theory of invasibility. *Journal of Ecology* 88, 528–534.

Diaz, S., Hodgson, J.G., Thompson, K., Cabido, M., Cornelissen, J.H.C., Jalili, A., Montserrat-Marti, G., Grime, J.P., Zarrinkamar, F., Asri, Y., Band, S.R., Basconcelo, S., Castro-Diez, P., Funes, G., Hamzehee, B., Khoshnevi, M., Perez-Harguindeguy, N., Perez-Rontome, M.C., Shirvany, F.A., Vendramini, F., Yazdani, S., Abbas-Azimi, R., Bogaard, A., Boustani, S., Charles, M., Dehghan, M., De Torres-Espuny, L., Falczuk, V., Guerrero-Campo, J., Hynd, A., Jones, G., Kowsary, E., Kazemi-Saeed, F., Maestro-Martinez, M., Romo-Diez, A., Shaw, S., Siavash, B., Villar-Salvador, P. and Zak, M.R. (2004) The plant traits that drive ecosystems: evidence from three continents. *Journal of Vegetation Science* 15, 295–304.

DiVittorio, C.T., Corbin, J.D. and D'antonio, C.M. (2007) Spatial and temporal patterns of seed dispersal: an important determinant of grassland invasion. *Ecological Applications* 17, 311–316.

Elberse, W.T. and Berendse, F. (1993) A comparative study of the growth and morphology of eight grass species from habitats with different nutrient availabilities. *Functional Ecology* 7, 223–229.

Fargione, J. and Tilman, D. (2002) Competition and coexistence in terrestrial plants. In: Somm, U. and Worm, B. (eds) *Competition and Coexistence in Terrestrial Plants*. Springer-Verlag, Berlin, pp. 165–206.

Fraser, L.H. and Grime, J.P. (1999) Interacting effects of herbivory and fertility on a synthesized plant community. *Journal of Ecology* 87, 514–525.

Garcia-Serrano, H., Escarre, J., Garnier, E. and Sans, X.F. (2005) A comparative growth analysis between alien invader and native *Senecio* species with distinct distribution ranges. *Ecoscience* 12, 35–43.

Goldberg, D.E. and Landa, K. (1991) Competitive effect and response – hierarchies and correlated traits in the early stages of competition. *Journal of Ecology* 79, 1013–1030.

Goldberg, D.E. and Novoplansky, A. (1997) On the relative importance of competition in unproductive environments. *Journal of Ecology* 85, 409–418.

Grime, J.P. (1977) Evidence for existence of three primary strategies in plants and its relevance to ecological and evolutionary theory. *American Naturalist* 111, 1169–1194.

Grime, J.P. and Hunt, R. (1975) Relative growth rate: its range and adaptive significance in a local flora. *Journal of Ecology* 63, 393–422.

Grotkopp, E., Rejmanek, M. and Rost, T.L. (2002) Toward a causal explanation of plant invasiveness: seedling growth and life-history strategies of 29 pine (*Pinus*) species. *American Naturalist* 159, 396–419.

Gurevitch, J. and Hedges, L.V. (1993) Meta-analysis: combining the results of independent experiments. In: Scheiner, S.M. and Gurevitch, J. (eds) *Design and Analysis of Ecological Experiments*. Chapman and Hall, New York, pp. 347–370.

Hedges, L.V., Gurevitch, J. and Curtis, P.S. (1999) The meta-analysis of response ratios in experimental ecology. *Ecology* 80, 1150–1156.

Huddleston, R.T. and Young, T.P. (2005) Weed control and soil amendment effects on restoration plantings in an Oregon grassland. *Western North American Naturalist* 65, 507–515.

Huenneke, L.F., Hamburg, S.P., Koide, R., Mooney, H.A. and Vitousek, P.M. (1990) Effects of soil resources on plant invasion and community structure in Californian serpentine grassland. *Ecology* 71, 478–491.

James, J.J. (2008) Effect of soil nitrogen stress on the relative growth rate of annual and perennial grasses in the Intermountain West. *Plant and Soil* 310, 201–210.

James, J.J. and Drenovsky, R.E. (2007) A basis for relative growth rate differences between native and invasive forb seedlings. *Rangeland Ecology and Management* 60, 395–400.

James, J.J., Drenovsky, R.E., Monaco, T.A. and Rinella, M.J. (2011) Managing soil nitrogen to restore annual grass-infested plant communities: effective strategy or incomplete framework? *Ecological Applications* 21, 490–502.

Killingbeck, K.T. (2004) Nutrient resorption. In: Noonden, L.D. (ed.) *Plant Cell Death Processes.* Elsevier, Amsterdam, the Netherlands, pp. 215–226.

Killingbeck, K.T. and Whitford, W.G. (1996) High foliar nitrogen in desert shrubs: an important ecosystem trait or defective desert doctrine? *Ecology* 77, 1728–1737.

Kolb, A., Alpert, P., Enters, D. and Holzapfel, C. (2002) Patterns of invasion within a grassland community. *Journal of Ecology* 90, 871–881.

Lambers, H. and Poorter, H. (1992) Inherent variation in growth rate between higher plants: a search for physiological causes and ecological consequences. *Advances in Ecological Research* 23, 187–261.

Lambers, H., Chapin, F.S. and Pons, T.L. 1998. *Plant Physiological Ecology.* Springer-Verlag, New York.

Leishman, M.R., Haslehurst, T., Ares, A. and Baruch, Z. (2007) Leaf trait relationships of native and invasive plants: community- and global-scale comparisons. *New Phytologist* 176, 635–643.

Liao, C.Z., Peng, R.H., Luo, Y.Q., Zhou, X.H., Wu, X.W., Fang, C.M., Chen, J.K. and Li, B. (2008) Altered ecosystem carbon and nitrogen cycles by plant invasion: a meta-analysis. *New Phytologist* 177, 706–714.

Loehle, C. (2000) Strategy space and the disturbance spectrum: a life-history model for tree species coexistence. *American Naturalist* 156, 14–33.

Lulow, M.E. (2006) Invasion by non-native annual grasses: the importance of species biomass, composition, and time among California native grasses of the Central Valley. *Restoration Ecology* 14, 616–626.

Mack, R.N., Simberloff, D., Lonsdale, W.M., Evans, H., Clout, M. and Bazzaz, F.A. (2000) Biotic invasions: causes, epidemiology, global consequences, and control. *Ecological Applications* 10, 689–710.

Mackown, C.T., Jones, T.A., Johnson, D.A., Monaco, T.A. and Redinbaugh, M.G. (2009) Nitrogen uptake by perennial and invasive annual grass seedlings: nitrogen form effects. *Soil Science Society of America Journal* 73, 1864–1870.

May, J.D. and Killingbeck, K.T. (1992) Effects of preventing nutrient resorption on plant fitness and foliar nutrient dynamics. *Ecology* 73, 1868–1878.

Mazzola, M.B., Allcock, K.G., Chambers, J.C., Blank, R.R., Schupp, E.W., Doescher, P.S. and Nowak, R.S. (2008) Effects of nitrogen availability and cheatgrass competition on the establishment of Vavilov Siberian Wheatgrass. *Rangeland Ecology and Management* 61, 475–484.

McLendon, T. and Redente, E.F. (1991) Nitrogen and phosphorus effects on secondary succession dynamics on a semiarid sagebrush site. *Ecology* 72, 2016–2024.

McLendon, T. and Redente, E.F. (1992) Effects of nitrogen limitation on species replacement dynamics during early secondary succession on a semiarid sagebrush site. *Oecologia* 91, 312–317.

Morgan, J.P. (1994) Soil impoverishment: a little-known technique holds potential for establishing prairie. *Restoration Notes* 12, 55–56.

Morghan, K.J.R. and Seastedt, T.R. (1999) Effects of soil nitrogen reduction on nonnative plants in restored grasslands. *Restoration Ecology* 7, 51–55.

Norton, J.B., Norton, U. and Monaco, T.A. (2007) Mediterranean annual grasses in western North America: kids in a candy store. *Plant and Soil* 298, 1–5.

Olofsson, J. (2001) Influence of herbivory and abiotic factors on the distribution of tall forbs along a productivity gradient: a transplantation experiment. *Oikos* 94, 351–357.

Paschke, M.W., McLendon, T. and Redente, E.F. (2000) Nitrogen availability and old-field succession in a shortgrass steppe. *Ecosystems* 3, 144–158.

Perry, L.G., Blumenthal, D.M., Monaco, T.A., Paschke, M.W. and Redente, E.F. (2010) Immobilizing nitrogen to control plant invasion. *Oecologia* 163, 13–24.

Poorter, H. and Bergkotte, M. (1992) Chemical composition of 24 wild species differing in relative growth rate. *Plant Cell and Environment* 15, 221–229.

Poorter, H., Niinemets, U., Poorter, L., Wright, I.J. and Villar, R. (2009) Causes and consequences of variation in leaf mass per area (LMA): a meta-analysis *New Phytologist* 183, 1222–1222.

Prober, S.M., Thiele, K.R., Lunt, I.D. and Koen, T.B. (2005) Restoring ecological function in temperate grassy woodlands: manipulating soil nutrients, exotic annuals and native perennial grasses through carbon supplements and spring burns. *Journal of Applied Ecology* 42, 1073–1085.

Ryser, P. (1996) The importance of tissue density for growth and life span of leaves and roots: a comparison of five ecologically contrasting grasses. *Functional Ecology* 10, 717–723.

Sheley, R.L., Mangold, J.M. and Anderson, J.L. (2006) Potential for successional theory to guide restoration of invasive-plant-dominated rangeland. *Ecological Monographs* 76, 365–379.

Stohlgren, T.J., Binkley, D., Chong, G.W., Kalkhan, M.A., Schell, L.D., Bull, K.A., Otsuki, Y., Newman, G., Baskin, M. and Son, Y. (1999) Exotic species invade hot spots of native plant diversity. *Ecological Monographs* 69, 25–46.

Weiher, J.B. and Keddy, P.A. (1999) *Ecological Assembly Rules: Perspectives, advances, retreats.* Cambridge University Press, Cambridge, UK.

Westoby, M., Falster, D.S., Moles, A.T., Vesk, P.A. and Wright, I.J. (2002) Plant ecological strategies: Some leading dimensions of variation between species. *Annual Review of Ecology and Systematics* 33, 125–159.

Wright, I.J. and Westoby, M. (2003) Nutrient concentration, resorption and lifespan: leaf traits of Australian sclerophyll species. *Functional Ecology* 17, 10–19.

Wright, I.J., Reich, P.B., Westoby, M., Ackerly, D.D., Baruch, Z., Bongers, F., Cavender-Bares, J., Chapin, T., Cornelissen, J.H.C., Diemer, M., Flexas, J., Garnier, E., Groom, P.K., Gulias, J., Hikosaka, K., Lamont, B.B., Lee, T., Lee, W., Lusk, C. and Midgley, J.J. (2004) The worldwide leaf economics spectrum. *Nature* 428, 821–827.

9

Reducing Invasive Plant Performance: a Precursor to Restoration

Joseph M. DiTomaso[1] and Jacob N. Barney[2]

[1] Department of Plant Sciences, University of California, USA
[2] Department of Plant Pathology, Virginia Tech, USA

Introduction

Most non-native plants in natural areas do not out-compete native species or cause significant impacts to ecosystem function (Rejmánek, 2000, 2011; Smith and Knapp, 2001). It has been estimated that <10% of invasive species that have established and persist in natural areas actually transform the ecosystem by changing the character, condition, form, or nature of an area (Richardson et al., 2000). There are many theories and reviews on why species become invasive, including release from natural enemies and herbivores in their native range (Keane and Crawley, 2002; Daehler, 2003), improved competitive ability through a shift in allocation from defense to growth (Blossey and Nötzold, 1995), and the development of novel growth or functional forms in invasive species that have competitive advantages over native species (Alpert et al., 2000). While these general theories may contribute to invasiveness in non-native species, specific traits related to reproduction, establishment, growth, competitive ability, susceptibility to herbivory and pathogens, and dispersal can all contribute to the success of invasive species in specific habitats (Smith and Knapp, 2001; Van Clef and Stiles, 2001; Ehrenfeld, 2004).

In general, most successful invasive plants tend to have greater adaptive qualities, often referred to as phenotypic plasticity, than native species. Plants with greater adaptive qualities are expected to maintain higher fitness and perform better in a wider range of habitats and environments (Rejmánek and Richardson, 1996; Alpert et al., 2000; Rejmánek, 2011), particularly in disturbed environments where conditions are in frequent flux (Claridge and Franklin, 2002; Daehler, 2003). These same traits that allow a species to be a successful invader outside its native range also account for widespread distribution in its native range (Richardson and Pyšek, 2006). However, it is important to recognize that increased adaptive qualities in non-native invaders do not indicate a performance advantage over native species within any single, defined environment (Daehler, 2003).

Phenotypic adaptation not only enhances the invasive potential of non-native species, but also increases the challenge of manipulating their performance in management and restoration programs. Long-term sustainable management programs will require persistent competition from desirable species or increased environmental stress to reduce the abundance and performance of invasive species. This chapter will discuss the various ecological processes operating at the plant and ecosystem level associated with the invasibility of natural ecosystems, and the management tools that can manipulate these processes to achieve a more desirable plant community.

© CAB International 2012. *Invasive Plant Ecology and Management: Linking Processes to Practice* (eds T.A. Monaco and R.L. Sheley)

Key Ecological Processes

Resource acquisition and plant–soil resource interactions

Resources that are most limiting in many environments include light, water, and soil nutrients, particularly nitrogen and phosphorus. Plant communities that are considered to be most susceptible to invasion are those with high availability of unused resources (Davis *et al.*, 2000; Smith and Knapp, 2001). This generally occurs in communities where the invasive species does not encounter intense competition for resources from resident native or desirable species (Richardson and Pyšek, 2006). For example, species-poor communities, often caused by natural or human-related disturbance, can have more free resources compared to undisturbed species-rich communities. Although not always the case (Alpert *et al.*, 2000; Maron and Marler, 2008), invasive species grow more quickly and are more successful than native species when resources are high and stress is low, but native species often compete better with invasive species when resource availability is low and stress is high (Alpert *et al.*, 2000). Thus, the level of environmental stress can greatly impact the relative performance of native or desirable species compared to invasive plants.

When increased resources are available, frequently through disturbance, invasive species are generally able to use them more quickly and efficiently than native species (Funk and Vitousek, 2007; Sheley *et al.*, 2010; Rejmánek, 2011). Disturbance, whether natural or human-related, can vary seasonally, causing resource pulses, defined by high resource fluctuation and availability. These pulses in resource availability can be a key factor controlling invasibility of ecosystems (Davis *et al.*, 2000; Chambers *et al.*, 2007), particularly when invasive species propagule pressure to the system is high (Richardson and Pyšek, 2006).

Light

Light can be the principal limiting factor in many environments, particularly when the vegetation is multilayered, such as forests and riparian areas. An increase in light availability can facilitate rapid invasion of many woody species. For example, in an eastern US forest community, three invasive woody species, Chinese tallow tree (*Triadica sebifera* (L.) Small), Amur honeysuckle (*Lonicera maackii* (Rupr.) Herder), and oriental bittersweet (*Celastrus orbiculatus* Thunb.), often form a seed and seedling bank under shaded conditions of an established forest. Once light becomes available through natural disturbance, the seedlings of invasive species are poised to rapidly respond and quickly occupy the canopy of the site (Jones and McLeod, 1989; Luken and Goessling, 1995; Marks and Gardescu, 1998). In contrast, the native species *Celastrus scandens* L. does not have a persistent seed or seedling bank, and is outcompeted for light by its congeneric invasive relative or other invasive species (Van Clef and Stiles, 2001).

Even in grassland communities, light can play a significant role in altering the regeneration and growth of native grass and forb species (Gerlach *et al.*, 1998). In California grasslands, seedlings of invasive forbs, including yellow starthistle (*Centaurea solstitialis* L.) and artichoke thistle (*Cynara cardunculus* L.), are sensitive to low light and do not effectively compete for light in well established dense grasslands (DiTomaso *et al.*, 2003; White and Holt, 2005). However, grazing or mowing can increase light penetration resulting in rapid establishment and dominance of invasive forbs in these ecosystems.

Once established, invasive species can dominate plant canopies, thus preventing light penetration to native species below. This is most notable for many climbing vines that grow over adjacent vegetation, such as kudzu (*Pueraria montana* (Lour.) Merr. var. *lobata* (Willd.) Maesen & S. Almeida), bittervine (*Mikania micrantha* Kunth), Japanese climbing fern (*Lygodium japonicum* L.), and old world climbing fern (*Lygodium myriophyllum* L.) (Bryson and Carter, 2004; Brooks *et al.*, 2008). In a comparative study with native woody species in riparian and foothill habitats of California, USA, invasive species nearly always reduced canopy openness to a higher level compared to the native woody vegetation (Sedna and Rejmánek, unpublished data). The level of openness under the invasive shrub Scotch broom (*Cytisus scoparius* (L.) Link) was significantly lower compared to the native manzanita shrub (*Arctostaphylos viscida* Parry) during the growing season. In addition, light levels under fig trees (*Ficus carica* L.) were an order of magnitude lower than the native woodlands, which were expressed in the virtual absence of other species below its canopy. Similar results were also found for other invasive species, including giant reed (*Arundo donax* L.), tree-of-heaven (*Ailanthus altissima* (Mill.) Swingle), and Himalaya blackberry (*Rubus armeniacus* Focke), which all dramatically reduce native species richness and diversity in their understory compared to native woody communities.

Soil nutrients

Although phosphorus is important in plant growth and performance, nitrogen availability plays a greater role in deter-mining species dominance in most communities, including mesic heathland (Aerts and Berendse, 1988), abandoned cropland (Wilson and Gerry, 1995), shrub-steppe rangeland in North America (Paschke *et al.*, 2000) and Australia (Snyman, 2002), arid shrublands (Milberg *et al.*, 1999; Monaco *et al.*, 2003), and wetlands (Daehler, 2003). In the western USA, studies have shown that the addition of mineral forms of nitrogen to disturbed rangelands increases the relative abundance of the invasive annual grass downy brome (*Bromus tectorum* L.) compared with native perennial species (Paschke *et al.*, 2000). The negative effect of nitrogen addition to grassland and rangeland communities is exacerbated by altered disturbance regimes caused by human activities (Daehler, 2003). Typically, nutrient enrichment shifts species composition to fewer relatively fast growing species, primarily annual grasses and robust perennial species (Alpert *et al.*, 2000).

While high nitrogen conditions rarely favor native species, reductions in available nitrogen often increase the relative performance and competitive ability of native species, particularly perennial grasses (Alpert *et al.*, 2000; Paschke *et al.*, 2000; Daehler, 2003). This is not always the case, however. Under low nitrogen conditions, invasive annual rangeland grasses acquired more nitrate than the associated native perennial grasses, resulting in equal or greater invasiveness compared to higher nitrogen conditions (Monaco *et al.*, 2003). In some cases, perennial invasive species with high plant uptake rates, such as musk thistle (*Carduus nutans* L.), can cause long-term declines in soil nitrogen availability. Because its seedlings can tolerate low nitrogen levels it creates an environment that favors its own establishment (Wardle *et al.*, 1994). This and other cases illustrate the potential for some invasive plants to impact nutrient cycling to their advantage in a positive plant–soil feedback cycle (Drenovsky and Batten, 2007).

Young *et al.* (1998) showed that fertilizing rangelands with either nitrate (NO_3^-) or ammonium (NH_4^+), enhanced seedling emergence and establishment of the invasive annual grass medusahead (*Taeniatherum caput-medusae* (L.) Nevski). However, Monaco *et al.* (2003) did not find a significant difference in either seedling growth or nitrogen allocation between invasive annual grasses and native perennial grasses to the two forms of nitrogen (NO_3^- and NH_4^+). Thus, the form of nitrogen may give a competitive advantage to either native or invasive species, depending on the specific response of the species within the com-munity.

Invasive nitrogen-fixing species, mainly members of the Fabaceae, can alter soil chemistry, soil enzyme activity, and possibly change the coupling of nutrient cycles within the ecosystem (Caldwell, 2006). It has been proposed that natural nutrient enrichment by nitrogen-fixing shrubs can promote establishment of other invasive species, particularly annual species, whose propagules are present (Alpert *et al.*, 2000).

Water

In arid and semi-arid ecosystems, water is the most important resource driving ecosystem processes and community composition (Chambers *et al.*, 2007). In general, increased water addition to mesic and xeric plant communities, primarily through precipitation, will increase susceptibility to invasion (Maron and Marler, 2008). In these same communities, drought stress generally limits invasibility, in part, due to the lower drought tolerance of certain invasive species compared to many native species (Alpert *et al.*, 2000; Daehler, 2003), which have evolved in that ecosystem. For example, diffuse knapweed (*Centaurea diffusa* Lam.) and yellow starthistle, like many other invasive thistles, produce roots that penetrate deep into the soil profile, allowing them to utilize late-season, deep-soil moisture (Kiemnec *et al.*, 2003; Enloe *et al.*, 2004). In cool and wet conditions, root penetration into the deep soil profile was greater for diffuse knapweed compared to the native perennial grass bluebunch wheatgrass (*Pseudoroegneria spicata* (Pursh) A. Löve), but under warm and dry conditions, there was no difference between the species (Kiemnec *et al.*, 2003). Under these drier conditions, the perennial grass was more competitive for both nutrients and water, as diffuse knapweed was unable to utilize the late season resources.

Some invasive species are more drought-tolerant than associated native species (Alpert *et al.*, 2000). For example, the invasion success of buffelgrass (*Pennisetum ciliare* (L.) Link) is partially due to its ability to emerge following relatively low precipitation levels under desert conditions (Ward *et al.*, 2006). In Montana, spotted knapweed (*Centaurea stoebe* L. ssp. *micranthos* (Gugler) Hayek) maintained greater water potentials despite greater transpiration compared to established perennial grasses (Hill *et al.*, 2006). This adaptation to drought-prone areas was attributed to its deep root system, which provided access to more consistent deep-soil moisture. In riparian areas in southwestern USA, human activities (i.e. dam building,

flow diversions) have altered the stream flow characteristics of many river systems, reducing water availability, and promoting the spread of the more drought-tolerant invasive shrub saltcedar (*Tamarix ramosissima* Ledeb.) (Cleverly *et al.*, 1997).

The timing of water use can also differ between invasive and native species. In arid environments that have few native annual grasses, non-native winter annual species such as downy brome, medusahead, and red brome (*Bromus rubens* L.) can complete their life cycle during short periods when water availability is high, prior to the dry season (Alpert *et al.*, 2000). In more mesic environments, non-native winter annual grasses can extract surface water early in the growing season, before many native forbs or perennial grasses are active (Young *et al.*, 2009). In comparison, late-season, deep-rooted, invasive forbs utilize late-season moisture (Young *et al.*, 2010a, b). This combination of early season, shallow, water use by non-native winter annual grasses and late season, water use by deep-taprooted perennials can explain why grassland communities in California, USA, resist establishment of native species, which tend to largely extract soil water in the middle of the growing season (Kulmatiski *et al.*, 2006).

Invasive species, such as yellow starthistle, can create locally dry conditions and maintain a significantly drier soil profile than either an annual or perennial grass community (DiTomaso *et al.*, 2003; Enloe *et al.*, 2004; Gerlach, 2004). When infestations are dense, soil moisture in the following season does not recharge after subnormal or even normal winter and spring precipitation (DiTomaso *et al.*, 2003).

Reproduction and dispersal

Pollination, seed production, and propagule dispersal play a fundamental role in the maintenance of natural communities and in the potential invasion of non-native plant species (Traveset and Richardson, 2006). Short generation times, from germination to reproduction, and high seed production are characteristics of many invasive species that accomplish rapid dispersal and the buildup of large seed banks (Bryson and Carter, 2004; Rejmánek, 2011). Ideally, the mating system of a successful invader would include the ability to self-pollinate when infestations are small, but later, when populations are large and pollinators are not limited, to increase genetic diversity and polymorphism by outcrossing (Rejmánek, 2011).

Pollination

Unlike many native species that have co-evolved with a suite of both specialist and generalist pollinators, invasive species generally attract only pollination generalists in their introduced range (Bartomeus et al., 2008), particularly those favored and introduced by humans (e.g. honeybees). Such insects can displace specialist pollinators and can often have negative effects on native species recruitment (Traveset and Richardson, 2006). In addition, invasive species could competitively affect pollination of desirable plants, including wind-pol-linated (through reduction in pollen quality) and animal-pollinated species (through reduction in pollen quantity and quality) (Brown and Mitchell, 2001). This could occur through pollen swamping, which can reduce seed set through stigma or stylar clogging. Examples of pollinator competition include the invasive purple loosestrife (Lythrum salicaria L.), which was shown to sub-stantially reduce the seed set of the native congener Lythrum alatum Pursh (Brown and Mitchell, 2001), and the invasive police-man's helmet (Impatiens glandulifera Royle), which halved the seed set of the native Stachys palustris L. (Chittka and Schürkens, 2001).

When invasive species populate areas with congeneric native species, there is the possibility for hybridization. These hybrids can dilute the native species gene pool and potentially lead to more vigorous invasive hybrids (Brown and Mitchell, 2001). For example, hybrid vigor between the native cordgrass (Spartina foliosa Trin.) and the invasive smooth cordgrass (Spartina alterniflora Loisel.) has dramatically impacted the San Francisco Bay estuary, USA, as the hybrid is more vigorous than either parent (Anttila et al., 1998; Ayres et al., 2004).

Seeds and other propagules

The seed banks of most perennial herbaceous species, especially grasses, are typically small due to variable seed production and short-lived seeds (Chambers et al., 2007). By comparison, large long-lived seed banks can account for the persistence of many invasive species (Alpert et al., 2000; Bryson and Carter, 2004; Krueger-Mangold et al., 2006). Even species that reproduce by vegetative means, such as giant reed, can develop rhizomes that remain viable for at least 16 weeks after isolation from the parent plant (Boose and Holt, 1999; Decruyenaere and Holt, 2001). However, long-lived seed banks are not always the case with invasive species and often depend on the specific habitat. Many invasive woody or perennial grass species of riparian areas (e.g. Tamarix spp. and Cortaderia spp.) have small seeds with short-lived seed banks (DiTomaso, 1998; Drewitz and DiTomaso, 2004). This is an adaptation to broad dispersal and rapid germination on very wet mineral substrates (Rejmánek, 2011).

Propagule pressure

Propagule pressure from either seeds or reproductive vegetative fragments is influenced by many characteristics, including dispersal agents, the degree of habitat fragmentation, and human activities (Alpert et al., 2000). The amount of propagule pressure in an area is more important than biological characteristics among species that account for invasive potential, such as adaptation to disturbance and efficient use of resources (Rejmánek, 2011). For example, species that are highly competitive and widely invasive through a variety of characteristics may not become established due to limitations in dispersal of propagules, but less invasive species may dominate an area if their propagule pressure is extremely high (Lonsdale, 1999). Once invasive species become well established and dominant, the propagule pool of desirable species may be much lower proportionally compared to the invasive, thus compromising site rehabilitation efforts (Krueger-Mangold et al., 2006).

Propagule dispersal

The seeds of plants, including invasive species, are primarily dispersed long distances by humans, animals, water, and wind. Plants have many different adaptations or preadaptations for dispersal by these vectors (Rejmánek, 2011). Overall, however, invasive species have a seed dispersal advantage over many native species (Daehler, 2003). One reason for this advantage is because human dispersal, either intentional or accidental, is more common in invasive plants than in native species, and is often the most significant driver of invasions (Bryson and Carter, 2004; Richardson and Pyšek, 2006). For example, in California, USA, 63% of the most invasive 200 species were introduced and spread intentionally as ornamentals, or for other purposes such as erosion control, livestock forage, or agricultural products (Bossard et al., 2006). Additionally, in a global evaluation, non-native species increased proportionally in nature reserves with higher number of visitors (Lonsdale, 1999). Dispersal of invasive plant seeds and propagules can occur via many routes, including direct movement through planting for ornamental, soil erosion, shading, forage, or crop use, or indirect movement as contaminants of soil, vehicles, equipment, or clothing.

In addition to human dispersal, seed dispersal by vertebrate animals is responsible for the success of many invaders into both disturbed and undisturbed habitats (Rejmánek, 2011). Mammals, including both livestock and wildlife, can transport seeds or fruit on their fur, by hoarding seeds in caches, or following ingestion (Davies and Sheley, 2007). Fish and even insects, particularly ants, can also disperse some species with edible seeds or fruit, but these vectors typically only move seeds a short distance. Many native species have tight relationships with specific frugivore dispersers and are not generally known to be dispersed long-distance by a single animal vector (Meisenburg and Fox, 2002). In contrast, invasive species typically attract more generalist vertebrate dispersers. Among the animal seed vectors, birds are the most important vertebrate responsible for long-distance dispersal of invasive species (Rejmánek, 2011).

There are two mechanisms by which animals disperse fruits or seeds (Meisenburg and Fox, 2002). The first is through ectozoochoric fruit or seed, which attach to animals with awns, hooks, barbs, or sticky secretions. The second is by endozoochoric fruit, which provide a food source for animals. For invasive species that rely on endozoochory for their dispersal, it is nearly always assured that a suitable vertebrate disperser will be available (Meisenburg and Fox, 2002). Fruits can possess an edible mesocarp or endosperm, or may contain an edible large elaiosome or a fleshy aril that surrounds the seed (Meisenburg and Fox, 2002; Rejmánek, 2011). However, not all animals that eat fruit disperse the seed. Many animals can actually be seed predators that destroy seed viability. In Florida, animal ingestion of endozoochoric fruits was the most common method of dispersal, representing half of the invasive species evaluated (Meisenburg and Fox, 2002).

The dominance of an animal-dispersed, invasive species can also impact the animal community within an area. For example, invasive gorse (Ulex europaeus L.) replaced New Zealand plant communities formerly dominated by the native kanuka (Kunzea ericoides (A. Rich.) Joy Thomps.). This shift in dominance has also changed the proportion of seed-dispersing mammals and birds in the invaded region (Williams and Karl, 2002).

Water and wind dispersal can greatly impact the speed and opportunity for invasion of non-native species. Buoyant seeds, particularly those with winged appendages (e.g. Rumex spp., Sagittaria spp., Sesbania punicea (Cab.) Benth.), have the potential to be moved long distances by water (Rejmánek, 2011). Seed or fruit movement in the atmosphere by wind is facilitated by pappi, or plumed or winged appendages. For example, cogongrass (*Imperata cylindrica* (L.) Beauv.) seeds can be dispersed by wind at distances up to 25 km (Bryson and Carter, 2004). Wind can also disperse plants through a tumbleweed action, where the entire aboveground shoot (e.g. *Salsola tragus* L., *Bassia scoparius* (L.) A.J. Scott) or inflorescence (e.g. *Panicum capillare* L.) breaks off from the stem and is dispersed by tumbling (Davies and Sheley, 2007).

Although many invasive plants have well-developed, long-distance dispersal mechanisms, a number of species shed their seeds in close proximity to their parent plant. These species either lack a facilitated dispersal feature or disperse seeds through a mechanism that only moves seeds a short distance (Cousens and Mortimer, 1995). For example, some fruit can disperse seeds a few meters through explosive dehiscence of the fruit. Ballistic dispersal is often important in local expansion, but does not assist in long-distance propagule movement at the landscape or regional scale.

Invasibility of ecosystems

Invasion success not only depends on the processes directly influencing invasive species, but also the processes operating within the plant community (Barney and Whitlow, 2008). Invasibility is defined as the intrinsic susceptibility of an area to invasion (Richardson, 2001), and depends on a number of factors, including identity and characteristics of the invading species, evolutionary history, species diversity, community structure, the strength of interactions between species, propagule pressure, assemblages of predators or pathogens, disturbance and stress, resource availability, and growing conditions (Lonsdale, 1999; Alpert et al., 2000; Daehler, 2003; Barney and Whitlow, 2008). In particular, disturbance, habitat and species diversity, resource availability and open niches, and propagule pressure are critical aspects that contribute to the invasibility of an ecosystem.

Disturbance

Disturbance is often cited as an important precursor for invasion of an ecosystem (Symstad, 2000). While invasions can take place without disturbance, events that kill or damage extant organisms, or reduce their biomass or performance, can strongly affect habitat invasibility (Alpert et al., 2000). There are several types of natural and human-related disturbances that can affect vegetation. These include grazing, fire, flood, and drought, which can occur either regularly or episodically at high or low intensity and at a landscape level or more locally (James et al., 2010). Such events reduce the level of competition among species and often lower competitive stress by increasing resource availability. An increase in resource availability can occur when plant uptake rates decline due to suppression of resident vegetation, or when resource supply increases at a rate faster than the resident vegetation can sequester it, which often follows burning (Davis et al., 2000). This can open niches within ecosystems and increase safe site availability for the successful dispersal and establishment of invasive species.

In many cases, disturbance is a prerequisite for invasion (Smith and Knapp, 2001). Disturbance associated with human activities generally gives invaders the largest performance advantage, particularly when the disturbance intensity and intervals are very different from the natural disturbance regime (Daehler, 2003). In general, an increase in disturbance frequency and intensity tends to favor invasive species (James et al., 2010). As disturbance intensity increases, nutrient-cycling rates increase, and the ability of resident vegetation to sequester nutrients decreases (James et al., 2010). This results in enhanced nutrient availability. Some habitats with a long history of human disturbance may be less invasible because the resident vegetation would have already been selected to perform well under a frequent disturbance regime (Alpert et al., 2000).

While invasions of non-native species are most common in agricultural or urban sites, where human disturbance is regular, they are least common in natural areas with low rates of disturbance, such as deserts and savannas (Lonsdale, 1999; Daehler, 2003). When disturbance frequency is maintained at a low rate, species, particularly native ones, which tolerate stressful conditions (i.e. limiting resources), will continue to dominate (Krueger-Mangold et al., 2006). Non-native species that are efficient competitors for limiting resources in undisturbed plant communities will often be the most successful invaders and the worst weeds of natural ecosystems (Rejmánek, 2011).

Habitat and species diversity

A number of studies have shown that more diverse plant communities resist invasions and use resources more completely (Case, 1990; Tilman, 1997; Maron and Marler, 2008). However, other studies demonstrate that communities richer in native species have more invasive species (Stohlgren et al., 1998, 2003; Lonsdale, 1999). In these cases, non-native species can more easily invade areas with high resource availability, which are often riparian areas or islands, and more available resources also promotes increased native plant diversity (Stohlgren et al., 1998; Alpert et al., 2000; Richardson and Pyšek, 2006). From these two concepts, it can be concluded that either unexploited available resources or near complete exploitation of resources can occur in both low and high diversity ecosystems, such that invasibility is not necessarily directly related to species richness (Bulleri et al., 2008). More importantly, habitat diversity or heterogeneity within a large scale ecosystem may provide more opportunity for suitable habitat for both native and invasive species (Stohlgren et al., 2003; Richardson and Pyšek, 2006; Bulleri et al., 2008). Although loss of species diversity alone may not affect community invasibility, communities with fewer species are often more susceptible to invasion (Dukes, 2001).

Resource availability and open niches

Invasion occurs when the growth form and traits of an invader are favored by the local conditions of the specific habitat, and the propagule pressure is adequate to allow establishment (Tilman, 1997; Alpert et al., 2000; Davis et al., 2000; Rejmánek et al., 2005). Thus, even in areas considered relatively invasion-resistant, introduction of an appropriately matched species at the right time can lead to invasion. For example, the Mojave and Sonoran deserts are two highly stressful, low resource environments in North America. Although these ecosystems have few non-native plants, they have recently been invaded by African mustard (Brassica tournefortii Gouan) and buffelgrass (Rejmánek et al., 2005). Successful invasion occurs when the fitness difference between the resident vegetation and the non-native species favors dominance of the latter or when niche differences (i.e. resource acquisition characteristics) are large enough to allow the latter to establish despite lower population fitness (Rejmánek, 2011). Conversely, invasion resistance occurs when the fitness of the resident vegetation is greater than the non-native species or when niche overlap is large between the resident vegetation and the latter. In many cases, there is little difference in fitness between the invasive and the resident vegetation and niche overlap is small. In these situations, the invader coexists with the resident vegetation and does not cause detectable displacement of the desired species (Tilman, 1997; Rejmánek, 2011).

Niche occupation is generally related to functional diversity within a site. Higher functional diversity, often associated with species richness, can reduce resource availability and provide greater resistance to invasion compared to areas with low functional diversity (Symstad, 2000; Dukes, 2001). However, even in areas with high functional diversity, fluctuations in resource availability can occur from year to year, or within a year. These episodic fluctuations or events can dramatically increase resources and greatly enhance invasion success (Davis et al., 2000; Rejmánek et al., 2005; Chambers et al., 2007).

Although relatively uncommon, the existence of one species in an ecosystem, native or invasive, can facilitate the success of other invasive species by creating an unused niche. For example, the establishment of invasive ruderal annual species in California coastal prairies, USA, is enhanced by the native nitrogen-fixing shrub *Lupinus arboreus* Sims (Maron and Connors, 1996). As another more specific interaction, the establishment of invasive dandelion (*Taraxacum officinale* F.H. Wigg.) is facilitated by the presence of cushions formed by the native plant *Azorella monantha* Clos in the alpine zone of the Chilean Andes (Cavieres *et al.*, 2008). These cushions increased nutrient availability, which was speculated to benefit dandelion establishment.

Propagule pressure

While the invasibility of an environment by a non-native species depends on its characteristics and the susceptibility of the environment to invasion, the invasion process itself depends on conditions of resource enrichment or release and the timely availability of invading propagules (Lonsdale, 1999; Davis *et al.*, 2000). Propagule pressure is often a strong determinant of habitat invasibility (Von Holle and Simberloff, 2005). In plant communities that are considered less resistant to invasion, fewer propagules are needed for a newly introduced species to establish and invasion rates are relatively fast. In comparison, habitats with high invasion resistance require much higher rates of propagule supply to overwhelm the ecological resistance of the site (Rejmánek *et al.*, 2005; Von Holle and Simberloff, 2005; Richardson and Pyšek, 2006). For example, the resident species richness of experimental riparian plots in California, USA, did not have as great an effect on invader establishment as did the number of invader propagules added to the plots (Levine, 2000). In another study in South Africa, the intensity of propagule pressure, rather than any environmental factor, best explained the invasion of three woody invasive species (Rouget and Richardson, 2003). Once the invader population establishes and matures,

its seed bank increases in proportion to that of the native or desirable vegetation. This leads to further decline in the resident species habitat as a result of seed swamping by the invasive species (Daehler, 2003).

Managing plant performance and ecosystem processes

A tremendous body of work exists determining both species characteristics that contribute to successful invasion, and habitat characteristics that facilitate susceptibility to invasion. However, few studies have bridged the chasm with empirical studies on the interactions between species and habitats (Barney and Whitlow, 2008). Heger and Trepl (2003) proposed a conceptual model of the 'key-lock approach' that considers the features of both the invasive species and the habitat into a relational model. They contend that species with specific characteristics (keys) fit into specific habitats (locks) that end in successful invasion. Hence, not all habitats are equitably invasible, and there are no 'invasive characters.' Rather, a specific combination is required to facilitate successful invasion.

The future of identifying mechanisms along the invasion pathway will be founded in integrated studies of species, habitats, and the role of propagule pressure (Richardson and Pyšek, 2006; Barney and Whitlow, 2008). Understanding the complex interactions of species traits contributing to invasion, and the 'weakness' of the receiving habitat can directly influence our ability to successfully manage the invasion. According to the principles of ecologically-based invasive plant management (EBIPM), there are three general causes of succession: (i) site availability; (ii) species availability; and (iii) relative species performance (James *et al.*, 2010; Sheley *et al.*, 2010). The underlying cause of invasion must first be identified to allow land managers to appropriately manipulate one or more of these factors to guide ecosystem change and modify or repair the ecological processes that favor a more desired community or vegetation trajectory.

As previously discussed, invasibility of a particular environment and invasiveness of individual non-native species can depend on specific disturbance regimes, propagule characteristics, resource availabilities, and community composition that favor the performance of the invader. However, these same processes can be manipulated and controlled to influence the performance of both invasive and desired species. Through these activities it can be possible to direct successional transitions that promote a more desired plant community (James et al., 2010; Sheley et al., 2010). A critical element in this process is to initially identify the abiotic (i.e. disturbance factors, climate, resource availability) or biotic (i.e. competing vegetation) filters or stressors that contribute to invasion. With this understanding it is then possible to manipulate these through modification in disturbance intervals and intensities, resource availability, hydrological regimes, management strategies, and propagule pressure, which can also be tied to revegetation efforts (Krueger-Mangold et al., 2006).

Disturbance regimes

Many ecosystems can be managed to favor a particular set of species or community composition by manipulating the type, interval, and intensity of disturbance. While certain disturbance regimes often lead to plant invasions, they can conversely be intentionally influenced to direct a community toward a more preferred vegetation state by providing appropriate germination conditions and sites, maintaining recruitment timeframes for resident species, reducing invasive species performance, and maximizing growth and reproductive performance of native or desired species (Daehler, 2003; Krueger-Mangold et al., 2006; James et al., 2010; Sheley et al., 2010). In addition, a particular disturbance regime can be used to suppress or manage an invasive species, and these types of strategies have been commonly employed by land managers (DiTomaso, 2000; DiTomaso et al., 2010).

In general, small scale, less frequent, and low intensity disturbances tend to favor establishment and growth of desirable species (James et al., 2010), while large scale, more frequent and higher intensity disturbances favor invasion by ruderal species (Alpert et al., 2000). Because nearly all native communities have evolved under a particular historic disturbance regime, restoring a more natural disturbance regime will generally favor native plants (Daehler, 2003). This is not always the case, however, as there are other disturbance regimes that can also favor native over specific invasive species. In contrast, suppressing natural disturbance regimes can also increase invasion (Alpert et al., 2000). It is important to recognize that even when natural disturbance regimes are established, a large number of invasive species within or near the community will increase the probability that some of these species will tolerate the historic disturbance regime (Daehler, 2003).

Disturbances that can alter the competitive relationships between native and invasive species include fire, grazing, mechanical damage (e.g. mowing or tillage), and alterations in hydrological regimes. Different types of disturbance can have different effects in the same habitat (Alpert et al., 2000). For example, maintaining historical fire regimes often favors native over invasive species (Daehler, 2003). In the Great Basin region of western USA, burning increased the availability of soil water and nitrate. While this often leads to an increase in downy brome populations when native perennial species are in low abundance, the opposite occurred when there was a high population of native perennial species. In this case, the native perennial species responded positively to the increased availability of water and nitrate and limited the increase in downy brome (Chambers et al., 2007). As another example, altering the historic flood regimes along rivers in southwestern USA is thought to have led to the establishment and spread of invasive saltcedar (Sher et al., 2000; Bay and Sher, 2008). By restoring the natural flood regime, it may be possible to reestablish the native woody vegetation while suppress further saltcedar encroachment.

Propagule pressure

Recruitment limitations are a critical aspect of long-term ecosystem management (Tilman, 1997). A particular species, whether invasive or desirable, will have a greater chance of successfully establishing and colonizing a site when its propagule numbers are greater than competing species occupying a similar niche (Von Holle and Simberloff, 2005). Thus, shifting the relative abundance towards desirable species will generally require manipulating and managing the availability, frequency, and abundance of desirable species propagules relative to that of invaders over time (Davies and Sheley, 2007; James et al., 2010; Sheley et al., 2010). Manipulating propagules can occur by preventing seed recruitment and dispersal of invasive plants from adjacent areas or by enhancing desirable plant recruitment, dispersal, or abundance of desirable species via intentional seed introduction.

Managing recruitment of invasive plant propagules, primarily seeds, will differ depending on the dispersal vector responsible for introduction. In many areas, introduction of new propagules is through specific corridors, such as roads, trails, streams, and rivers, and winter-feeding areas for livestock and wildlife (Krueger-Mangold et al., 2006). These source sites should be identified and prioritized for strategic management to reduce the risk of reintroduction or introducing new invasive species (Davies and Sheley, 2007). For example, water corridors can serve as very effective dispersal vectors. As such, it may be critical to prevent introduction of invasive species seeds or vegetative propagules into water by prioritizing the management of their populations near water's edge.

Preventing or limiting movement of invasive species can be achieved by managing propagules along the primary vectors or corridors by several control options, but most often by mechanical or chemical methods.

Minimizing unnecessary disturbance, using weed-free hay and uncontaminated soil in construction activities, and revegetating with high purity seed can also help prevent unintended introductions. Physical barriers or properly timed grazing can reduce the risk of moving invasive species propagules from one area to another. This is particularly true for invasive species with seeds possessing awns, hooks, or barbs that facilitate their dispersal (Davies and Sheley, 2007). Although manipulation in grazing timing and intensity is more practical and easily accomplished with livestock than wildlife, establishing barriers, such as fences or enclosures, can also influence wildlife grazing. For invasive plants that are wind dispersed, it may be possible to promote taller vegetation that not only reduces wind velocities but also physically intercepts seeds within these sites to form a barrier to long-distance dispersal to uninfested area (Davies and Sheley, 2007). The dispersal of seeds through tumbling action in wind can be limited with fencing along the perimeter of the restored area (Baker et al., 2010).

It may not be completely necessary to prevent dispersal and new recruitment of invasive plants to an area, provided that desirable species disperse and successfully germinate first (James et al., 2010). Early germination timing and establishment of desirable species can greatly favor their success or, conversely, delayed dispersal of invasive species can limit their success and promote plant community change to a more desired vegetation state.

Manipulating soil resources

The competitive relationships and relative performance of desirable versus invasive species can shift dramatically under a particular set of resource conditions. Typically, invasive species dominate over native and other desirable non-native species when resource availability, including light, nutrients, and water, are high (Daehler, 2003). Presumably, a return to the natural resource levels would favor native species that had evolved under historic conditions compared to non-native invasive species.

This assumes there is an adequate propagule pressure of desirable species and that disturbance levels are not too dissimilar from natural conditions.

In many natural environments, historical resource levels were low, and native species often evolved under relatively high stress conditions. Increased disturbance and resource availability has favored invasive over native species (James et al., 2010). Environmental manipulations that lower resource availability, particularly nitrogen, and increase stress can shift the community dynamics of an area in favor of native species (Alpert et al., 2000; Krueger-Mangold et al., 2006). In addition to manipulating total resources available to the ecosystem, conditions that favor native or other desirable plants may also be achieved by altering the intensity of the stress (i.e. heat of fire and level of grazing pressure) or the timing of resource availability (Daehler, 2003). For example, moderate prolonged resource stress favors desired species over invasive species compared with short-duration and more intense stress (James et al., 2010).

Although possible, manipulating resource availability is often difficult to accomplish and generally impractical at the landscape level. Reducing light availability can be accomplished by revegetating with taller herbaceous perennial or woody desirable species. Invasive species, such as yellow starthistle and jubatagrass (Cortaderia jubata (Lem.) Stapf.), are suppressed by reduced light and occur only infrequently under the canopy of taller vegetation (DiTomaso et al., 2003; Stanton and DiTomaso, 2004). In contrast, the performance of the invasive perennial vine Cape ivy (Delairea odorata Lem.) is very poor under high light conditions and can be managed by increasing light intensity (Robison et al., 2011). Similarly, the availability and flow of water can be manipulated in some riparian areas where dams or the flow of rivers and streams are controlled, and it has been proposed that manipulating hydrological regimes may be effective for enhancing the establishment of native woody vegetation while managing invasive saltcedar (Sher et al., 2000; Bay and Sher, 2008).

Soil nutrient availability can also be manipulated to favor certain desirable species. Strategies to reduce nitrogen availability can give an advantage to native species and reduce the competitive ability of invasive species, as well as prevent invasion and establishment of new invasive propagules (Alpert et al., 2000). There are several ways of manipulating nitrogen availability in invaded ecosystems. In some cases nitrogen can be added to the system to enhance native or desirable plants. For example, increasing the proportion of nitrogen as ammonium or urea in fertilizer mixtures can restrict germination or growth of ammonium- or urea-sensitive weeds and enhance native species (DiTomaso, 1995; Monaco et al., 2003). However, this requires a thorough understanding of the response of both desirable and invasive species to various forms of nitrogen. Alternatively, fertilizer applications during the time when nutrient uptake is maximal in desirable plant roots can increase their competitive ability against invasive species that utilize nutrients later in the year (DiTomaso, 1995). Another method to selectively manipulate the timing of nutrient use in desired species is through seed priming. Seed priming is a technique where seeds are sown after they are partially hydrated and germination has begun, but emergence of the radicle has not yet occurred (Baskin and Baskin, 1998). This strategy can ensure resource preemption and improve the coupling of nutrient cycling between plants and soils (Krueger-Mangold et al., 2006).

Other nutrient manipulations are designed to reduce soil nitrogen availability. For example, fires promoted by invasive annual grasses can temporarily increase available nitrogen but will decrease total nitrogen in the system through volatilization (Daehler, 2003). Over time and through several fire cycles in areas without a preponderance of nitrogen-fixers, a gradual reduction in nitrogen could promote a shift away from annual grass dominance back toward a late seral native community. Other more immediate methods of reducing available soil nitrogen can include carbon amendments to the soil in the form of sawdust, mulch, or sucrose (Corbin and

D'Antonio, 2004; White and Holt, 2005). These carbon sources reduce mineral nitrogen availability by promoting immobilization by soil microbial populations, which can effectively decrease invasive plant performance and favor native perennial species (Perry *et al.*, 2010). For example, by experimentally reducing nitrogen with sucrose addition over a 4-year period in Colorado, USA, plots dominated by downy brome were converted to a community that closely resembled the natural late-seral, shortgrass steppe vegetation (Paschke *et al.*, 2000).

Management strategies for reducing species performance of invasive plants

Management strategies should, ideally, target the most vulnerable or susceptible stages in the life cycle of the invasive plant while minimizing the negative effects on native or desirable species. Thus, management strategies that directly influence key ecological processes, including altering disturbance regimes, manipulating resource availability, limiting invasive seed dispersal and propagule pressure, and enhancing desirable species germination, seedling establishment, survivorship, growth, and reproduction (Table 9.1). The timing and techniques used to achieve this objective are critical to successful invasive plant management. For example, rapid germination and establishment of the invasive perennial species buffelgrass and crimson fountaingrass (*Pennisetum setaceum* (Forsk.) Chiov.) occur soon after precipitation events in many arid environments of the world. Monitoring areas of seedling development and initiating management programs within a short period following rainfall events can be an effective strategy for preventing subsequent establishment of large infestations (Ward *et al.*, 2006; Rahlao *et al.*, 2010).

While soil cultivation is a disturbance that generally increases the dominance of many invasive plants in grasslands and other natural areas (Stromberg and Griffin, 1996), timely light disking and soil imprinting can be used to damage annual invasive species and promote the rapid establishment of native plants in revegetation projects (Sheley *et al.*, 2010). Other mechanical techniques, such as mowing, can be used in a timely manner to facilitate the growth of native species and control non-native species. For example, annual mowing of the invasive perennial tall oatgrass (*Arrhenatherum elatius* (L.) P. Beauv ex J. Presl & C. Presl) during its flowering phase, in late spring or early summer, reduced its cover and biomass while increasing flowering and growth of native perennial grasses in an Oregon, USA, prairie (Wilson and Clark, 2001).

Under some situations, restrictions in grazing may be necessary to allow desirable vegetation to recover following management of invasive species, although this is not always possible in areas occupied by livestock. In contrast, restoration of historic grazing regimes can often promote native species diversity (Daehler, 2003). In rangelands, it is possible to adjust the intensity, duration, and timing of livestock grazing to favor particular species (Larson and Kiemnec, 2005). In addition, certain classes of livestock preferentially graze on specific plant life-forms or species. Sheep, for example, usually prefer forbs over grasses and shrubs, cattle generally graze on grasses, and goats often favor shrubs (Popay and Field, 1996; Krueger-Mangold *et al.*, 2006). In sensitive riparian areas or grasslands, for example, goats have been used to graze on spiny brush species, including wild blackberries (*Rubus* spp.), sweet roses (*Rosa* spp.), and matagouri (*Discaria toumatou* Raoul) without damaging the desired native vegetation or forage species (Dellow *et al.*, 1987; Holgate and Weir, 1987; Cossens *et al.*, 1989). By manipulating timing and intensity of grazing, defoliation will primarily target invasive species to give a competitive advantage to other plants in the community (Krueger-Mangold *et al.*, 2006). As an example, when medusahead was grazed by sheep at the 'boot' stage, just prior to exposure of the inflorescences, its cover was reduced by over 86%, while native forb cover significantly increased (DiTomaso *et al.*, 2008).

Table 9.1. Linkages between ecosystem processes and management strategies.

Process	Characteristic	Factors leading to invasiveness		Management strategy
		Plant	Ecosystem	
Disturbance (shift from natural regime)	No historical occurrence	Increases available resource (i.e. light, nutrient, water) or provides pulse in resource availability for use by invasive species	Physical damage to desirable species Reduced competitive ability o desirable species	• Eliminate cause of disturbance (i.e. grazing, burning, mechanical or hydrological disturbance) • Restore desirable fire or hydrology regime (i.e. occurrence, frequency, timing, and intensity) • Modify grazing practices through livestock type, frequency, duration, timing, and intensity • Specific mechanical control techniques to favor desirable species (e.g. mowing or tillage) • Selective herbicides or application rate, timing, or technology to increase stress on invasive species
	Frequency		Reduction in propagule production in desired plants	
	Intensity		Higher level of physical damage to desirable species Reduced competitive ability of desirable species Changes in nutrient cycling to favor invasive species	
Propagule pressure	Pollination	Generalist pollinators increase seed production of invasive species Pollen swamping of desirable species or congeners Hybrid vigor creates more aggressive invasive species or diluted native plant gene pool		• Prevent flowering of invasive species
	Seed or vegetative production	High fecundity of invasive species Long-lived seed banks or vegetative propagules		• Increase propagule numbers for desirable species (e.g. recruitment of reseeding) • Prevent seed production or recruitment of invasive species • Use of mowing, timely grazing or selective herbicides to prevent seed production
	Reproduction timing	Short generation times from germination to seed production in invasive species	Timely susceptibility of ecosystem to invasive species propagules	• Timely seeding of desirable species, transplanting or seed priming to achieve early germination and establishment
	Dispersal	Natural long-distance dispersal (i.e. water, wind, animals) Human facilitated dispersal (i.e. livestock, vehicles and equipment, soil, clothing, ornamental, use, forage or crop, erosion or flood control) Propagule build-up of invasive species	Changes in animal communities that facilitate dispersal of desirable species	• Management of invasive species with long-distance dispersal methods in areas in close proximity to restoration site or along transport corridors

Resource availability	Nutrients, light, water	Efficient resource capture though physiological mechanisms (i.e. high water use efficiency, photosynthetic rates or nutrient uptake) or morphological characteristic (i.e. deep roots or climbing habit)	High availability of unused resource year-round or seasonally	• Addition of nitrogen forms that favor desirable species • Reduce nutrient availability through carbon amendments or by removing nitrogen-fixers or altering fire occurrence • Timely addition of nutrients to enhance desirable vegetation • Planting desired vegetation to reduce canopy light penetration
Community composition	Species diversity		Species present do not provide full resource acquisition year-round or seasonally Invasive nitrogen-fixers or other resource provider can promote further invasions	• Revegetation with functionally similar species to effectively compete with invader • Diverse niche occupation by appropriate revegetation species mixes • Seeding in phases over multiple seasons or years to maximize potential for ideal establishment conditions • Timely selective control options to manage invasive species seedling establishment • Early to late seral planting of desirable species to accelerate successional community shift • Timely seeding to follow exposure to stress (i.e. grazing or fire) or management (i.e. herbicide, mowing, or tillage) • Ensure proper seedbed preparation to facilitate germination and establishment of seeded species

Much like grazing, properly timed prescribed burning can also manage invasive species and select for more desired native or desirable species. In California, USA, early summer prescribed burning significantly reduced the cover of yellow starthistle, medusahead (Fig. 9.1), and barb goatgrass (*Aegilops triuncialis* L.), and also dramatically increased the cover of native perennial grasses by at least ten-fold (DiTomaso *et al.*, 1999, 2001, 2006; Kyser *et al.*, 2008). Periodic burning may be an important tool to maintain healthy perennial grass or fire-tolerant shrub communities even after other control tools have been used to reduce invasive plant populations.

Early-seral species are favored by higher available soil nitrogen and can prevent the establishment of desirable late-seral native species through intense competition during the establishment phase. As a means to manipulate soil nutrient levels, Herron *et al.* (2001) showed that soil nitrogen could be reduced by planting an ephemeral cover crop (*Secale cereale* L.) or the mid-seral native bottlebrush squirreltail (*Elymus elymoides* (Raf.) Swezey). These species accelerated the successional change from a weedy plant community dominated by spotted knapweed toward a more desired plant community.

Herbicides are widely used tools for control of invasive plants in many ecosystems. Many herbicides selectively control specific groups or even taxa of species, while having little effect on others (Fig. 9.2). Choosing the proper herbicide, applying it at the correct rate and at the most appropriate timing, and using application technology that maximizes its effectiveness and selectivity can give successful invasive plant control and minimize damage to non-target species (Krueger-Mangold *et al.*, 2006).

Revegetation manipulations

Manipulating propagule pressure, disturbance, and soil resources with management tools is critical to the restoration of invaded natural ecosystems. However, it is just as important to know if and when revegetation efforts are necessary and to understand the revegetation methods and techniques that favor successful management of invasive species and restoration of a desired plant community.

The important question of whether or not it is necessary to actively revegetate invaded ecosystems takes on greater importance when monetary resources are limited. In some situations, not using active

Fig. 9.1. Prescribed burn for control of medusahead.

Fig. 9.2. Aerial application of the herbicide clopyralid for the selective control of yellow starthistle in a California, USA, grassland.

management practices and allowing the ecosystem to passively recover may be enough (Holl and Aide, 2010). However, passive restoration often takes several years, and many agencies or land managers feel the pressure to accelerate the process through expensive active revegetation efforts. For example, passive restoration of forest systems often requires more than 40 years to recover without revegetation efforts (Jones and Schmitz, 2009). In some cases, removing grazing and reducing the frequency of fire allows recovery of extensive areas of tropical dry forest, but even these practices require decades (Janzen, 2002). In ecosystems that are heavily invaded, where the level of degradation becomes more intense and the native plant seed bank is reduced, active restoration that includes revegetation efforts and stress manipulations through herbicide use, tillage, periodic flooding, prescribed burning, or timely strategic grazing are often necessary to recover certain ecosystem functions (Alpert *et al.*, 2000; Holl and Aide, 2010).

In most situations, revegetation efforts include establishing propagules of perennial species, either herbaceous or woody. This is particularly true for riparian, forests, and woodland areas, but also for grasslands and

rangelands (Chambers *et al.*, 2007). Choosing the most appropriate species for revegetation is among the most critical steps to a successful restoration program. Species-rich seed mixes can increase the likelihood of survival of some species under varying stressful environmental conditions and also provide more opportunities for occupying complementary functional guilds with invaders (Krueger-Mangold *et al.*, 2006). Introducing or maintaining high functional diversity can more fully utilize available resources, both spatially and temporally. This will increase the competitive influence of the desired species on the invasive species and provide a greater level of resistance to reinvasion or dominance of an invasive species (Davies *et al.*, 2007; James *et al.*, 2010).

In addition to choosing appropriate species mixes, it is critical to consider the relative propagule pressure between the desired and invasive species within a site. This requires both management of the invasive species seed bank (control strategies) and seeding at high enough rates to offset invasive species seed numbers. When this is not possible, it may be necessary to transplant seedlings or young plants of desired species into restoration sites (Kulmatiski *et al.*, 2006). In this case, transplants must be able to establish root systems quickly to capture resource before the invasive species.

Timing of revegetation efforts can also determine the success of a restoration program. In some cases, seeding should be conducted soon after management practices have been applied. This ensures rapid establishment of the desired plant canopy and subsequent suppression of invasive species. In other cases, seeding should be conducted when predicted weather conditions are most conducive to successful germination and establishment of the desired species (Kulmatiski *et al.*, 2006). Although more expensive, seeding can also be conducted in phases throughout multiple seasons or years to increase the chances of encountering environmental conditions that maximize seedling establishment and survival (Krueger-Mangold *et al.*, 2006). As another alternative, seeding different species over a period of time, termed 'assisted succession,' can increase the long-term success of a restoration program through assisted succession (Cox and Anderson, 2004). In this situation, species known to effectively compete with the invasive species are initially seeded. This is followed by subsequent seeding of other native species that provide added ecosystem values. For example, establishment of crested wheatgrass (*Agropyron cristatum* (L.) Gaertn.) in the Intermountain Region of western USA initially decreased downy brome cover, and through 'assisted succession' eventually led to the successful restoration of native sagebrush-grassland steppe (Cox and Anderson, 2004).

Proper seedbed preparation can also facilitate better germination and establishment of seeded desirable species. For example, large depressions trapped and retained more seeds and moisture in arid environments, resulting in high seedling emergence compared with smaller depressions (Chambers, 2000). To further increase successful germination of desired species, it is possible to treat the seed (seed coating) with products that decrease

susceptibility to pests and pathogens or alleviate abiotic stress (Madsen *et al.*, 2009).

Conclusions

Reducing invasive plant performance depends on an understanding of the characteristics and processes that make a plant invasive, as well as the aspects of an ecosystem that facilitate or enhance the invasion process. With a foundational knowledge of these concepts, it is possible to employ a number of tools to enhance the competitive ability of native and other desirable species, while minimizing the performance of invasive species. The various methods of manipulating propagule pressure of invasive plants, altering disturbance regimes to favor desirable species, modifying resource availability to support a more desired plant community, and employing management tools that selectively damage or suppress invasive species are critical aspects to a successful long-term passive or active restoration program. When active revegetation is necessary in a restoration effort, there are additional techniques that can further enhance the probability of successfully achieving a desired outcome. The practices should be incorporated into an EBIPM program to maximize the probability of successfully converting degraded landscapes into healthy functional ecosystems.

References

Aerts, R. and Berendse, F. (1988) The effect of increased nutrient availability on vegetation dynamics in wet heathlands. *Vegetation* 76, 63–69.

Alpert, P., Bone, E. and Holzapfel, C. (2000) Invasiveness, invasibility and the role of environmental stress in the spread of non-native plants. *Perspectives in Plant Ecology, Evolution, and Systematics* 3, 52–66.

Anttila, C.K., Daehler, C.C., Rank, N.E. and Strong, D.R. (1998) Greater male fitness of a rare invader (*Spartina alterniflora*, Poaceae) threatens a common native (*Spartina foliosa*) with hybridization. *American Journal of Botany* 85, 1597–1601.

Ayres, D.R., Smith, D.L., Zaremba, K., Klohr, S. and Strong, D.R. (2004) Spread of exotic cordgrasses and hybrids (*Spartina* sp.) in the tidal marshes of San Francisco Bay, California, USA. *Biological Invasions* 6, 221–231.

Baker, D.V., Withrow, J.R., Brown, C.S. and Beck, K.G. (2010) Tumbling: use of diffuse knapweed (*Centaurea diffusa*) to examine an understudied dispersal mechanism. *Invasive Plant Science and Management* 3, 301–309.

Barney, J.N. and Whitlow, T.H. (2008) A unifying framework for biological invasions: the state factor model. *Biological Invasions* 10, 59–272.

Bartomeus, I., Vila, M. and Santamaria, L. (2008) Contrasting effects of invasive plants in plant-pollinator networks. *Oecologia* 155, 761–770.

Baskin, C.C. and Baskin, J.M. (1998) Ecology of seed dormancy and germination in grasses. In: Cheplick, G.P. (ed.) *Population Biology of Grasses*. Cambridge University Press, New York, pp. 30–83.

Bay, R.F. and Sher, A.A. (2008) Success of active revegetation after *Tamarix* removal in riparian ecosystems of the southwestern United States: a quantitative assessment of past restoration projects. *Restoration Ecology* 16, 113–128.

Blossey, B. and Nötzold, R. (1995) Evolution of increased competitive ability in invasive nonindigenous plants – a hypothesis. *Journal of Ecology* 83, 887–889.

Boose, A.B. and Holt, J.S. (1999) Environmental effects on asexual reproduction in *Arundo donax*. *Weed Research* 39, 117–127.

Bossard, C., Brooks, M., DiTomaso, J.M., Randall, J., Roye, C., Sigg, J., Stanton, A. and Warner, P. (2006) *California Invasive Plant Inventory*. California Invasive Plant Council, Berkeley, California, Publ. #2006-02, 39 pp.

Brooks, S.J., Panetta, F.D. and Galway, K.E. (2008) Progress towards the eradication of mikania vine (*Mikania micrantha*) and limnocharis (*Limnocharis flava*) in Northern Australia. *Invasive Plant Science and Management* 1, 296–303.

Brown, B.J. and Mitchell, R.J. (2001) Competition for pollination: effects of pollen of an invasive plant on seed set of a native congener. *Oecologia* 129, 43–49.

Bryson, C.T. and Carter, R. (2004) Biology of pathways for invasive weeds. *Weed Technology* 18, 1216–1220.

Bulleri, F., Bruno, J.F. and Benedetti-Cecchi, L. (2008) Beyond competition: incorporating positive interactions between species to predict ecosystem invasibility. *PloS Biology* 6, 1136–1140.

Caldwell, B.A. (2006) Effects of invasive scotch broom on soil properties in a Pacific coastal prairie soil. *Applied Soil Ecology* 32, 149–152.

Case, T.J. (1990) Invasion resistance arises in strongly interacting species-rich model competitive systems. *Proceedings of the National Academy of Sciences, USA* 87, 9610–9614.

Cavieres, L.A., Quiroz, C.L. and Molina-Montenegro, M.A. (2008) Facilitation of the non-native *Taraxacum officinale* by native nurse cushion species in the high Andes of central Chile; are there differences between nurses? *Functional Ecology* 22, 148–156.

Chambers, J.C. (2000) Seed movements and seedling fates in disturbed sagebrush steppe ecosystems: implications for restoration. *Ecological Applications* 10, 1400–1413.

Chambers, J.C., Roundy, B.A., Blank, R.R., Meyer, S.E. and Whittaker, A. (2007) What makes Great Basin sagebrush ecosystems invasible by *Bromus tectorum*? *Ecological Monographs* 77, 117–145.

Chittka, L. and Schürkens, S. (2001) Successful invasion of a floral market. *Nature* 411, 653–653.

Claridge, K. and Franklin, S.B. (2002) Compensation and plasticity in an invasive plant species. *Biological Invasions* 4, 339–347.

Cleverly, J.R., Smith, S.D., Sala, A. and Devitt, D.A. (1997) Invasive capacity of *Tamarix ramosissima* in a Mojave Desert floodplain: the role of drought. *Oecologia* 111, 12–18.

Corbin, J.D. and D'Antonio, C.M. (2004) Can carbon addition increase competitiveness of native grasses? A case study from California. *Restoration Ecology* 12, 36–43.

Cossens, G.G., Mitchell, R.B. and Crossan, G.S. (1989) Matagouri, hawkweed and purple fuzzweed control with sheep, goats and legumes in the New Zealand tussock grassland. *Proceedings of the Brighton Crop Protection Conference, Weeds*. Brighton, UK, pp. 879–884.

Cousens, R. and Mortimer, M. (1995) *Dynamics of Weed Populations*. Cambridge University Press, New York.

Cox, R.D. and Anderson, V.J. (2004) Increasing native diversity of cheatgrass-dominated rangeland through assisted succession. *Rangeland Ecology & Management* 57, 203–210.

Daehler, C.C. (2003) Performance comparisons of co-occurring native and alien invasive plants: implications for conservation and restoration. *Annual Review of Ecology, Evolution, and Systematics* 34, 183–211.

Davies, K.W. and Sheley, R.L. (2007) A conceptual framework for preventing the spatial dispersal of invasive plants. *Weed Science* 55, 178–184.

Davis, M.A., Grime, J.P. and Thompson, K. (2000) Fluctuating resources in plant communities: a general theory of invasibility. *Journal of Ecology* 88, 528–534.

Decruyenaere, J.G. and Holt, J.S. (2001) Seasonality of clonal propagation in giant reed. *Weed Science* 49, 760–767.

Dellow, J.J., Mitchell, T., Johnston, W., Hennessey, G. and Gray, P. (1987) Large area blackberry, *Rubus fruticosus* agg., control using grazing goats. *Proceedings of the 8th Australian Weeds Conference*, Sydney, Australia, 70 pp.

DiTomaso, J.M. (1995) Approaches for improving crop competitiveness through the manipulation of fertilization strategies. *Weed Science* 43, 491–497.

DiTomaso, J.M. (1998) Impact, biology, and ecology of saltcedar (*Tamarix* spp.) in the southwestern United States. *Weed Technology* 12, 236-336.

DiTomaso, J.M. (2000) Invasive weeds in range-lands: species, impacts and management. *Weed Science* 48, 255–265.

DiTomaso, J.M., Kyser, G.B. and Hastings, M.S. (1999) Prescribed burning for control of yellow starthistle (*Centaurea solstitialis*) and enhanced native plant diversity. *Weed Science* 47, 233–242.

DiTomaso, J.M., Heise, K.L., Kyser, G.B., Merenlender, A.M. and Keiffer, R.J. (2001) Carefully timed burning can control barb goatgrass. *California Agriculture* 55(6), 47–53.

DiTomaso, J.M., Kyser, G.B. and Pirosko, C.B. (2003) Effect of light and density on yellow starthistle (*Centaurea solstitialis*) root growth and soil moisture use. *Weed Science* 51, 334–341.

DiTomaso, J.M., Brooks, M.L., Allen, E.B., Minnich, R., Rice, R.M. and Kyser, G.B. (2006) Control of invasive weeds with prescribed burning. *Weed Technology* 20, 535–548.

DiTomaso, J.M., Kyser, G.B., George, M.R., Doran, M.P. and Laca, E.A. (2008) Control of medusahead using timely sheep grazing. *Invasive Plant Science and Management* 1, 241–247.

DiTomaso, J.M., Masters, R.A. and Peterson, V.F. (2010) Rangeland invasive plant management. *Rangelands* 32, 43–47.

Drenovsky, R.E. and Batten, K.M. (2007) Invasion by *Aegilops triuncialis* (barb goatgrass) slows carbon and nutrient cycling in a serpentine grassland. *Biological Invasions* 9, 107–116.

Drewitz, J.J. and DiTomaso, J.M. (2004) Seed biology of jubatagrass (*Cortaderia jubata*). *Weed Science* 52, 525–530.

Dukes, J.S. (2001) Biodiversity and invasibility in grassland microcosms. *Oecologia* 126, 563–568.

Ehrenfeld, J.G. (2004) Implications of invasive species for belowground community and nutrient processes. *Weed Technology* 18, 1232–1235.

Enloe, S.F., DiTomaso, J.M., Orloff, S.B. and Drake, D.J. (2004) Soil water dynamics differ among rangeland plant communities dominated by yellow starthistle (*Centaurea solstitialis*), annual grasses, or perennial grasses. *Weed Science* 52, 929–935.

Funk, J.L. and Vitousek, P.M. (2007) Resource-use efficiency and plant invasion in low-resource systems. *Nature* 446, 1079–1081.

Gerlach, J.D. Jr (2004) The impacts of serial land-use changes and biological invasions on soil water resources in California, USA. *Journal of Arid Environments* 57, 365–379.

Gerlach, J.D., Dyer, A. and Rice, K. (1998) Grassland and foothill woodland ecosystems of the Central Valley. *Fremontia* 26, 39–43.

Heger, T. and Trepl, L. (2003) Predicting biological invasions. *Biological Invasions* 5, 313–321.

Herron, G.J., Sheley, R.L., Maxwell, B.D. and Jacobsen, J.S. (2001) Influence of nutrient availability on the interaction between *Centaurea maculosa* and *Pseudoroegneria spicata*. *Ecological Restoration* 9, 326–331.

Hill, J.P., Germino, M.J., Wraith, J.M., Olson, B.E. and Swan, M.B. (2006) Advantages in water relations contribute to greater photosynthesis in *Centaurea maculosa* compared with established grasses. *International Journal of Plant Sciences* 167, 269–277.

Holgate, G.L. and Weir, D.A. (1987) Sweet brier control with goats. *Proceedings of the New Zealand Grassland Association* 48, 157–161.

Holl, K.D. and Aide, T.M. (2010) When and where to actively restore ecosystems? *Forest Ecology and Management* 261, 1558–1563.

James, J.J., Smith, B.S., Vasquez, E.A. and Sheley, R.L. (2010) Principles for ecologically based invasive plant management. *Invasive Plant Science and Management* 3, 229–239.

Janzen, D.H. (2002) Tropical dry forest: area de Conservacion Guanacaste, northwestern Costa Rica. In: Perrow, M.R. and Davy, A.J. (eds) *Handbook of Ecological Restoration*. Cambridge University Press, Cambridge, UK, pp. 559–583.

Jones, H.P. and Schmitz, O.J. (2009) Rapid recovery of damaged ecosystems. *PLoS One* 4, e3653, doi:10.1371/journal.pone.0005653.

Jones, R.H. and McLeod, K.W. (1989) Shade tolerance in seedlings of Chinese tallow tree, American sycamore, and cherrybark oak. *Bulletin of the Torrey Botanical Club* 116, 371–377.

Keane, R.M. and Crawley, M.J. (2002) Exotic plant

invasions and the enemy release hypothesis. *Trends in Ecology and Evolution* 17, 164–170.

Kiemnec, G., Larson, L.L. and Grammon, A. (2003) Diffuse knapweed and bluebunch wheatgrass seedling growth under stress. *Journal of Range Management* 56, 65–67.

Krueger-Mangold, J.M., Sheley, R.L. and Svejcar, T.J. (2006) Toward ecologically-based invasive plant management on rangeland. *Weed Science* 54, 597–605.

Kulmatiski, A., Beard, K.H. and Stark, J.M. (2006) Exotic plant communities shift water-use timing in a shrub-steppe ecosystem. *Plant and Soil* 288, 271–284.

Kyser, G.B., Doran, M.P., McDougald, N.K., Orloff, S.B., Vargas, R.N., Wilson, R.G. and DiTomaso, J.M. (2008) Site characteristics determine the success of prescribed burning for medusahead (*Taeniatherum caput-medusae*) control. *Invasive Plant Science and Management* 1, 376–384.

Larson, L. and Kiemnec, G. (2005) Germination of two noxious range weeds under water and salt stress with variable light regimes. *Weed Technology* 19, 197–200.

Levine, J.M. (2000) Species diversity and biological invasions: relating local process to community pattern. *Science* 288, 852–854.

Lonsdale, W.M. (1999) Global patterns of plant invasions and the concept of invasibility. *Ecology* 80, 1522–1536.

Luken, J.O. and Goessling, N. (1995) Seedling distribution and potential persistence of the exotic shrub *Lonicera maackii* in fragmented forests. *American Midland Naturalist* 133, 124–130.

Madsen, M.D., Petersen, S.L., Roundy, B.A., Taylor, A.G. and Hopkins, B.G. (2009) Innovative use of seed coating technologies for the restoration of soil wettability and perennial grasses on burned semi-arid rangelands. *Water Resources Research* 1, 283–286.

Marks, P.L. and Gardescu, S. (1998) A case study of sugar maple (*Acer saccharum*) as a forest seedling bank species. *Journal of the Torrey Botanical Society* 125, 287–296.

Maron, J.L. and Connors, P.G. (1996) A native nitrogen-fixing shrub facilitates weed invasion. *Oecologia* 105, 302–312.

Maron, J.L. and Marler, M. (2008) Effects of native species diversity and resource additions on invader impacts. *The American Naturalist* 172 (suppl.), S18–S33.

Meisenburg, M.J. and Fox, A.M. (2002) What role do birds play in dispersal of invasive plants? *Wildland Weeds* 5, 8–14.

Milberg, P., Lamont, B.B. and Perez-Fernandez, M.A. (1999) Survival and growth of native and exotic composites in response to a nutrient gradient. *Plant Ecology* 145, 125–132.

Monaco, T.A., Johnson, D.A., Norton, J.M., Jones, T.A., Connors, K.J., Norton, J.B. and Redinbaugh, M.B. (2003) Contrasting responses of Intermountain West grasses to soil nitrogen. *Journal of Range Management* 56, 282–290.

Paschke, M.W., McLendon, T. and Redente, E.F. (2000) Nitrogen availability and old-field succession in a shortgrass steppe. *Ecosystems* 3, 144–158.

Perry, L.G., Blumenthal, D.M., Monaco, T.A., Paschke, M.W. and Redente, E.F. (2010) Immobilizing nitrogen to control plant invasion. *Oecologia* 163, 13–24.

Popay, I. and Field, R. (1996) Grazing animals as weed control agents. *Weed Technology* 10, 217–231.

Rahlao, S.J., Esler, K.J., Milton, S.J. and Barnard, P. (2010) Nutrient addition and moisture promote the invasiveness of crimson fountaingrass (*Pennisetum setaceum*). *Weed Science* 58, 154–159.

Rejmánek, M. (2000) Invasive plants: approaches and predictions. *Austral Ecology* 25, 497–506.

Rejmánek, M. (2011) Invasiveness. In: Simberloff, D. and Rejmánek, M. (eds) *Encyclopedia of Biological Invasions*. University of California Press, Berkeley, California, pp. 379–385.

Rejmánek, M. and Richardson, D.M. (1996) What attributes make some plant species more invasive? *Ecology* 77, 1655–1661.

Rejmánek, M., Richardson, D.M. and Pyšek, P. (2005) Plant invasions and invasibility of plant communities. In: van der Maarel, E. (ed.) *Vegetation Ecology*. Blackwell Publishing, Malden, Massachusetts, pp. 332–355.

Richardson, D.M. (2001) Plant invasions. In: Levin, S. (ed.) *Encyclopedia of Biodiversity*. Academic Press, San Diego, California, pp. 677–688.

Richardson, D.M. and Pyšek, P. (2006) Plant invasions: merging the concepts of species invasiveness and community invasibility. *Progress in Physical Geography* 30, 409–431.

Richardson, D.M., Pyšek, P., Rejmánek, M., Barbour, M.G., Panetta, F.D. and West, C.J. (2000) Naturalization and invasion of alien plants: concepts and definitions. *Diversity and Distributions* 6, 93–107.

Robison, R.A., Kyser, G.B., Rice, K.J. and DiTomaso, J.M. (2011) Light intensity is a limiting factor to the inland expansion of Cape ivy (*Delairea odorata*). *Biological Invasions* 13, 35–44.

Rouget, M. and Richardson, D.M. (2003) Inferring process from pattern in plant invasions: a semi-mechanistic model incorporating propagule pressure and environmental factors. *American Naturalist* 162, 713–724.

Sheley, R., James, J., Smith, B. and Vasquez, E. (2010) Applying ecologically based invasive-plant management. *Rangeland Ecology and Management* 63, 605–613.

Sher, A.A., Marshall, D.L. and Gilbert, S.A. (2000) Competition between native *Populus deltoides* and invasive *Tamarix ramosissima* and the implications for reestablishing flooding disturbance. *Conservation Biology* 14, 1744–1754.

Smith, M.D. and Knapp, A.K. (2001) Physiological and morphological traits of exotic, invasive exotic, and native plant species in tallgrass prairie. *International Journal of Plant Science* 162, 785–792.

Snyman, H.A. (2002) Short-term response of rangeland botanical composition and productivity to fertilization (N and P) in a semi-arid climate of South Africa. *Journal of Arid Environments* 50, 167–183.

Stanton, A.E. and DiTomaso, J.M. (2004) Growth response of *Cortaderia selloana* and *Cortaderia jubata* (Poaceae) seedlings to temperature, light, and water. *Madroño* 51, 312–321.

Stohlgren, T.J., Bull, K.A., Otsuki, Y., Villa, C.A. and Lee, M. (1998) Riparian zones as havens for exotic plant species in the central grasslands. *Plant Ecology* 138, 113–125.

Stohlgren, T.J., Barnett, D.T. and Kartesz, J.T. (2003) The rich get richer: patterns of plant invasions in the United States. *Frontiers of Ecology and the Environment* 1, 11–14.

Stromberg, M.R. and Griffin, J.R. (1996) Long-term patterns in coastal California grasslands in relation to cultivation, gophers, and grazing. *Ecological Applications* 6, 1189–1211.

Symstad, M.J. (2000) A test of the effects of functional group richness and composition on grassland invasibility. *Ecology* 81, 99–109.

Tilman, D. (1997) Community invasibility, recruitment limitation, and grassland biodiversity. *Ecology* 78, 81–92.

Traveset, A. and Richardson, D.M. (2006) Biological invasions as disruptors of plant reproductive mutualisms. *Trends in Ecology and Evolution* 21, 208–216.

Van Clef, M. and Stiles, E.W. (2001) Seed longevity in three pairs of native and non-native congeners: assessing invasive potential. *Northeastern Naturalist* 8, 301–310.

Von Holle, B. and Simberloff, D. (2005) Ecological resistance to biological invasion overwhelmed by propagule pressure. *Ecology* 86, 3212–3218.

Ward, J.P., Smith, S.E. and McClaran, M.P. (2006) Water requirements for emergence of buffelgrass (*Pennisetum ciliare*). *Weed Science* 54, 720–725.

Wardle, D.A., Nicholson, K.S., Ahmed, M. and Rahman, A. (1994) Interference effects of the invasive plant *Carduus nutans* L. against the nitrogen fixation ability of *Trifolium repens* L. *Plant and Soil* 163, 287–297.

White, V.A. and Holt, J.S. (2005) Competition of artichoke thistle (*Cynara cardunculus*) with native and exotic grassland species. *Weed Science* 53, 826–833.

Williams, P.A. and Karl, B.J. (2002) Birds and small mammals in kanuka (*Kunzea ericoides*) and gorse (*Ulex europaeus*) scrub and the resulting seed rain and seedling dynamics. *New Zealand Journal of Ecology* 26, 31–41.

Wilson, M.V. and Clark, D.L. (2001) Controlling invasive *Arrhenatherum elatius* and promoting native prairie grasses through mowing. *Applied Vegetation Science* 4, 129–138.

Wilson, S.D. and Gerry, A.K. (1995) Strategies for mixed grass prairie restoration: herbicide, tilling, and nitrogen manipulation. *Restoration Ecology* 3, 290–298.

Young, J.A., Trent, J.D., Blank, R.R. and Palmquist, D.E. (1998) Nitrogen interactions with medusahead (*Taeniatherum caput-medusae* ssp. *asperum*) seedbanks. *Weed Science* 46, 191–195.

Young, S.L., Barney, J.N., Kyser, G.B., Jones, T.S. and DiTomaso, J.M. (2009) Functionally similar species confer greater resistance to invasion: implications for grassland restoration. *Restoration Ecology* 17, 884–892.

Young, S.L., Barney, J.N., Kyser, G.B., Claassen, V.P. and DiTomaso, J.M. (2010a) Spatio-temporal relationship between water depletion and root distribution patterns of *Centaurea solstitialis* and two native perennials. *Restoration Ecology* 18 (suppl. 2), 323–333.

Young, S.L., Barney, J.N., Kyser, G.B., Claassen, V.P. and DiTomaso, J.M. (2010b) The role of light and soil moisture in plant community resistance to invasion by yellow starthistle (*Centaurea solstitialis*). *Restoration Ecology* 19, 599–606.

10

Revegetation: Using Current Technologies and Ecological Knowledge to Manage Site Availability, Species Availability, and Species Performance

Jane M. Mangold

Department of Land Resources and Environmental Sciences, Montana State University, USA

Introduction

For sites severely degraded by exotic invasive plants, simply controlling the weed to release desirable plants from competition may not be adequate. Introducing propagules of desired species through revegetation may be required. Revegetation is a resource-intensive venture that often results in less than optimum outcomes. Successional management (Pickett *et al.*, 1987; Sheley *et al.*, 1996; Krueger-Mangold *et al.*, 2006), in which site availability, species availability, and species performance are manipulated to direct plant communities from an undesirable state to a desirable state, may serve as a useful framework for assessing site conditions, choosing exotic plant control methods and revegetation strategies, and planning follow-up management (Fig. 10.1). Designing revegetation programs that are based on our best understanding of the primarily ecological processes responsible for plant community dynamics at a given site, may initiate outcomes that more fully meet management objectives. This chapter will briefly address the ecological theory for revegetation. Then, current revegetation strategies, limitations of those strategies, and new approaches for revegetation will be described in the context of site availability, species availability, and species performance.

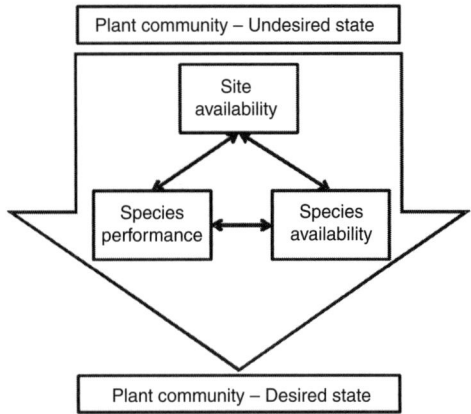

Fig. 10.1. Successional management can serve as an ecological framework for planning and implementing revegetation of weed-infested plant communities. In successional management, site availability, species availability, and species performance are manipulated to direct plant communities from an undesirable state to a desirable state (adapted from Pickett *et al.*, 1987).

Ecological Theory Behind the Necessity for Revegetation

Over the past several decades, increased education and awareness concerning the impacts of invasive plants has resulted in increased efforts to control them. When invasive plants are controlled via herbicides, grazing, mowing, biocontrol, or other methods, open niches are created in the plant community. Desirable species, released from the competitive effects of the invasive plant, often respond to the increase in site availability, and re-occupy the site (Sheley *et al.*, 2000). However, in plant communities that have been dominated by invasive plants for some time, desirable species may be exceedingly rare or even completely absent from existing vegetation and the seed bank. If invasive plants are controlled, but propagules of desirable species are not present to occupy open niches, invasive plants are likely to reestablish (Laufenberg *et al.*, 2005; Mangold *et al.*, 2007a; Reinhardt Adams and Galatowitsch, 2008). In some cases the same invasive species reestablishes, but in other cases a different, but no less troublesome invasive plant becomes dominant. For example, at two sites in western Montana, USA, the root weevil *Cyphocleonus achates* (Fahraeus) drastically decreased spotted knapweed (*Centaurea stoebe* L.) populations, but the invasive annual grass cheatgrass (*Bromus tectorum* L.) comprised 50–90% of the replacement vegetation (Story *et al.*, 2006).

Ecological processes such as nutrient cycling (Ehrenfeld, 2003; Rodgers *et al.*, 2008), carbon storage (Ehrenfeld, 2003), and fire regimes (Brooks, 2008) can be altered in an invasive plant-dominated system compared to the original system. In such cases, multiple and incremental actions must be undertaken to amend ecological processes so that the system is capable of supporting native vegetation and/or desirable introduced vegetation. Active revegetation will likely be a necessary action to suppress reestablishment of invasive plants and initiate the reestablishment of systems dominated by desirable vegetation that meets management objectives (Borman *et al.*, 1991; Lym and Tober, 1997; DiTomaso, 2000).

While revegetation of invasive plant-infested rangeland sounds ideal in theory, it is challenging in practice and often results in invasive plant species remaining dominant (D'Antonio and Meyerson, 2002). One major challenge of revegetating invasive plant-infested areas is overcoming the abundance of invasive plant propagules and lack of abundance of desired species propagules. Even when invasive plants are controlled, they often re-occupy the site due to a sizeable seed bank that responds quickly to increased site availability created by the disturbance of control itself (Hobbs and Huenneke, 1992; Reinhardt Adams and Galatowitsch, 2008). Revegetation to address site and species availability must be integrated with control efforts to address species performance with the goal of hindering performance of weedy species while establishing desirable species and promoting their performance in subsequent years.

Current Strategies for Revegetating Invasive-Plant Dominated Plant Communities

Site availability

Disturbance is the primary ecological process that regulates site availability and is commonly defined by the size, severity, and patchiness of the disturbance, time intervals between disturbance, and pre-disturbance history (Pickett *et al.*, 1987). Disturbance typically results in an increase in resource availability due to a decline in resource use by plants that were killed by the disturbance and an increase in resource supply rates through plant decomposition, thereby increasing susceptibility to invasion (Davis *et al.*, 2000). Disturbance is commonly viewed negatively by land managers because it contributes to weed invasion (Lozon and MacIsaac, 1997). However, land managers can design disturbances to manipulate site availability to meet revegetation needs. Designed disturbance has primarily been accomplished via mechanical means (i.e. tilling, harrowing, chaining, disking, plowing), fire, and herbicides, and is often part of controlling weedy plants at the site. The primary goal of designed disturbance is to prepare a seedbed that is conducive to seedling establishment, but a secondary objective is to control weedy species so that desired species can establish and grow with minimal competition from weeds for soil resources and light. Prescribed fire (Kyser and DiTomaso, 2002; Emery and Gross, 2005; Mangold *et al.*, 2007a) and herbicides (Sheley *et al.*, 2001, 2007; Monaco *et al.*, 2005; Davison and Smith, 2007) are more commonly used for controlling weedy species, while mechanical means are typically used for seedbed preparation (Sheley *et al.*, 1999; Thompson *et al.*, 2006; Fansler and Mangold, 2011).

Site availability, as modified through designed disturbance, should strive to create safe sites for desired species. While the precise conditions that might define a safe site for a given species are not known (but see Mangold *et al.*, 2007b), it is generally assumed that a safe site meets the light, water, and nutrient needs of a species so that a seed or vegetative propagule can germinate, emerge, and establish (Harper *et al.*, 1965). As mentioned above, mechanical means are often employed to create a seedbed that combines a certain degree of tilth and firmness to increase soil–seed contact (Sheley *et al.*, 2008). In some cases a soil imprinter is employed to create surface depressions that accumulate moisture, organic matter, and seeds (Dixon, 1988). Mechanically preparing a seedbed, however, can result in proliferation of annual weedy species, some of which can be invasive (J.M. Mangold, unpublished results). It is well established that disturbance results in an increase in nutrient availability, which invasive plants often capitalize on to a greater degree than native, desired species (Huenneke *et al.*, 1990). Therefore, care should be taken to minimize disturbance as much as possible while still placing desired species propagules in a microenvironment conducive to their growth. In regard to disturbance frequency, severity, and size, desired species will be favored when disturbances are less frequent, less severe, and smaller scale.

Because natural and direct human disturbances are known to promote invasive plants (Davis *et al.*, 2000; D'Antonio and Meyerson, 2002), aspects of designed disturbance can be manipulated to promote the abundance of desired vegetation over invasive plants. For example, fire has been used in fire-tolerant native plant communities to reduce the seed bank of the invasive forb yellow starthistle (*Centaurea solstitialis* L.) (Hastings and DiTomaso, 1996). However, properly manipulating disturbance will require further research and understanding of what constitutes a safe site for a suite of desired species. As site availability is better addressed in the future, it will be important to consider both the natural and human-directed disturbance regime (size, severity, frequency, patchiness, and history) at a site, how they might be contributing to invasion, and how they might be manipulated during revegetation to promote desired species (D'Antonio and Meyerson, 2002; Isselin-Nondedeu *et al.*, 2006; Watts, 2010).

Seeding method focuses on placing seeds in a safe site, although it can be used to modify species availability as well. The most common seeding methods are broadcast seeding and drill seeding. Choice of seeding method depends on factors like land accessibility, topography, seedbed characteristics, and economic constraints. Broadcast seeding is used where terrain is inaccessible to machinery due to steepness, rockiness, and/or remoteness. Large areas may be quickly and effectively seeded by plane or helicopter, while smaller areas can be seeded with a hand spreader (Sheley *et al.*, 2008). A few of the major drawbacks of broadcast seeding are the difficulty in preparing seedbeds and the inability to directly place seeds of desired species into probable safe sites. In general, broadcast seeding is deficient in providing sufficient seed-to-soil contact that many species require for adequate germination, emergence, and establishment. Because of the lack of adequate seed-to-soil contact with broadcast seeding, careful attention to site availability through seedbed preparation is necessary (see discussion above). When direct seedbed preparation is not possible, fire is believed to provide an adequate seedbed, and sites are often broadcast seeded following wildfires or prescribed burns. In the absence of fire and when seedbeds cannot be prepared, seeds are typically broadcasted at rates two to three times higher than the recommended drill rate (Sheley *et al.*, 2008).

If the land is accessible to machinery and is not too rocky, drill seeding using a no-till drill is the preferred method of seeding (Monsen and Stevens, 2004). A no-till drill is a tractor-pulled machine that opens a furrow in the soil, drops seeds in the furrow at a specified rate and depth, and rolls the furrow closed (Fig. 10.2). The benefit of this method of seeding is that the seed is placed into a predefined safe site (depth), thus increasing desired species establishment. Both site availability (i.e. safe sites created with drill) and species availability (i.e. seeding rate) are controlled. There are some drawbacks to drill seeding, which include: (i) plants develop in rows resembling a crop; (ii) seeds can lodge in the drill and/or separate in the seed box based on weight and size; and (iii) drill furrows can cause soil erosion from water flow, but there are fairly simple techniques for overcoming these, and most land managers are willing to negotiate the drawbacks in return for the improvement in species establishment (Sheley *et al.*, 2008).

Species availability

Selecting species to address species availability is often defined by management objectives for the site. For example, if livestock grazing is the primary management goal at a site, revegetating with competitive, grazing adapted, high production forage grasses may be appropriate. Alternatively, the primary management goal may be habitat for specific wildlife species, which will require the establishment of native forbs, grasses, and shrubs (USDI USFWS, 2001). In addition to management objectives, species are typically selected by matching environmental conditions at the site such as soil type, precipitation, temperature, and elevation, with species preferences and requirements. Ease of establishment is also important as seedlings of planted, desired species will be competing with weed seedlings emerging from the seed bank. Species that germinate quickly and have high growth rates have been good candidates for revegetating invasive plant-infested sites (Sheley *et al.*, 2008).

Historically, a narrow suite of species (relative to total number of naturally occurring species) have been commercially

Fig. 10.2. Rangeland no-till drill seeder. Photo courtesy of Jane Mangold, November 2009.

available for large-scale restoration of landscapes degraded by invasive plants, overgrazing, wildfire, etc.; therefore, many degraded plant communities were seeded with introduced, competitive forage grasses. For example, to support managed grazed systems, drought- and grazing-tolerant introduced species such as crested wheatgrass (*Agropyron cristatum* L.), smooth brome (*Bromus inermis* Leyss.), and timothy (*Phleum pretense* L.) were widely established in North America. The use of native species is being increasingly recommended to restore functional and diverse plant communities while maintaining ecological stability and reducing the risk of seeding an aggressive or invasive species (USDI and USDA, 2002). Fortunately, plant material availability continues to increase, likely due to the increasing recognition that many drastically disturbed landscapes will require revegetation for successful restoration; availability of native species is especially improving due to extensive research efforts (Shaw and Pellant, 2010) and increasing demand (USDI and USDA, 2002; Shaw *et al.*, 2005; O'Driscoll, 2007).

Properly distributing seeds for revegetation is heavily dependent on seeding rates to facilitate colonization. Generally, seeding rates should attempt to increase the frequency of desired propagules relative to weedy propagules. Typically a seeding rate of 215 to 550 viable seeds m^{-2} is desired (Sheley *et al.*, 2008). However, when revegetating invasive plant-infested sites, seeding rates two to three times higher are recommended (Sheley *et al.*, 2008). One study suggested seedling establishment could be further improved by increasing seeding rates to 5 and 25 times the recommended rate when revegetating grasslands infested by invasive forbs like spotted knapweed (*Centaurea stoebe* L.) (Sheley *et al.*, 1999). Ultimately, seeding rate should be determined by simultaneously reviewing processes that influence species and site availability. For example, if an overabundance of safe sites are present, then seeding rates should be high. Conversely, inundating the site with propagules where there are not adequate safe sites to

accommodate them will result in lower than expected establishment. Ideally, there should be adequate propagules of desired species to fill all available safe sites that are naturally occurring and/or created through designed disturbance (Satterthwaite, 2007).

Species performance

There are many well-established methods for controlling the performance of invasive plants in an effort to reduce their dominance in plant communities. Because several of these methods are discussed extensively in other chapters of this book (see DiTomaso and Barney, Chapter 9, this volume), they will not be discussed in detail here. However, it is commonly believed that performance of invasive plants *must be* controlled, sometimes for multiple years, in order for successful establishment of more desirable vegetation through revegetation (Berger, 1993; Wilson and Ingersoll, 2004; Fansler and Mangold, 2011). Invasive plant control can be achieved with a variety of mechanical, cultural, biological, and chemical methods, but control will be highest when methods address key ecological processes that are regulating plant community dynamics at a specific site. For example, if continual soil disturbance is present at a site, then mechanical control of invasive plants would probably not be the best option, and efforts should focus on the least disruptive method while working to minimize disturbance (Watts, 2010). In general, the performance of invasive plants should be addressed in an effort to reduce the production of propagules and decrease interference between invasive and desired plants.

After the extensive efforts implemented to prepare a site through designed disturbance, select species, and seed species at the appropriate rate and with the appropriate method, follow-up management must be employed. Follow-up management should be designed to bolster the performance of seeded species and hinder the performance of invasive plants, and often takes the form of herbicide applications to control reinvading weeds, deferring grazing to protect recently seeded species, and additional seeding to supplement areas of poor initial establishment. Unfortunately, follow-up management is often overlooked (D'Antonio and Meyerson, 2002). Because seeded species mortality can be high during the first year or two following seeding (Jessop and Anderson, 2007; Fansler and Mangold, 2011) and reinvading weeds can be prevalent (Mangold *et al.*, 2007a; Pokorny *et al.*, 2010), monitoring to assess desired species establishment compared to weedy species establishment is critical. Additional seeding and/or invasive plant control may be necessary to further move the plant community along a trajectory to a desired state. The site must be managed into the future to prevent reinvasion and dominance by invasive plants. In many cases, management regimes that led to dominance by invasive plants prior to restoration must be changed so that history does not repeat itself (D'Antonio and Meyerson, 2002).

Limitations of Current Strategies and Application of Ecological Understanding

In spite of several decades of practice and research associated with revegetating weed-infested plant communities, revegetation more often results in failure than success (James and Svejcar, 2010). This may be due to the predominant use of traditional agronomic practices that may not be effective on landscapes that vary considerably across time and space. For example, a standard revegetation operation on weed-infested rangeland in the western USA would likely include a herbicide application, seedbed preparation with a tiller, disk, harrow, or fire, and seeding either with a broadcast seeder or drill (Sheley *et al.*, 2006; Mangold *et al.*, 2007a; Davies, 2010). This may be effective in cropping systems, but less so on extensive, natural lands, where the ecological processes directing vegetation change and plant species traits are different. For example, in cropping systems fertilizer and other soil amendments are used, crops are fast growing and highly

productive, there are fairly uniform soil and topographic conditions across a field, and plant community dynamics are driven primarily by manmade disturbances. In contrast to cropping systems, natural systems vary considerably in soil characteristics, elevation, aspect, plant community types, etc., across any given area (Pickett and Cadenasso, 1995), and a variety of disturbances (both manmade and natural) direct vegetation change. Furthermore, plants native to such systems typically are slow growing and not highly productive in comparison to cropping systems (MacArthur, 1962; Tilman and Wedin, 1991). Until such differences between natural systems and cropping systems are recognized, and methods are developed to address variation in biotic and abiotic traits and processes across natural systems, revegetation will probably continue to result in more failures than successes.

In most cases, invasion and subsequent loss of native vegetation is a slow process, taking years to decades to occur (Kowarik, 1995). At the same time, when revegetation is implemented, it usually occurs over a time period of 1 to 2 years. Adequate control of invasive plant propagules prior to introducing propagules of desired species may take multiple years, from 2 to 5, depending on the target species and its abundance in the seed bank. Furthermore, legacy effects of invasive plant dominance on other ecosystem components like soil microbes (Stinson *et al.*, 2006) and nutrient cycling (Mack and D'Antonio, 2003) may take several years to decades to diminish before native species reestablishment and dominance are encouraged.

Not only should pre-revegetation efforts be extended beyond the typical 1 to 2 year timeframes, but the evaluation of revegetation outcomes should as well. Again, revegetation is typically monitored for 1 to 2 years, but experience suggests that shifts in plant community composition, including that imposed through management, take much longer and initial trends may or may not be indicative of longer-term outcomes (Roche *et al.*, 2008; Rinella *et al.*, 2011). Along with follow-up management to boost the performance of seeded species, monitoring and evaluation should extend beyond a few years.

New Approaches for Revegetation that Integrate Our Best Understanding of Ecological Processes

The lack of success in revegetation of weed-infested plant communities and the ever-increasing need for the restoration of lands degraded by weeds suggests new approaches are necessary. Combining our existing knowledge and past experiences with novel approaches that consider the ecological processes associated with site availability, species availability, and species performance will move us towards restoring plant communities that meet management objectives and maintain ecosystem services in the near and distant future.

Site availability

The nature of natural systems is such that we must address a mosaic of variable conditions across a large landscape. It will not suffice to treat the landscape as a uniform field across which large pieces of equipment are driven to implement a standardized disturbance treatment. Instead, a site should be carefully assessed prior to any revegetation efforts to ascertain ecological processes in need of repair, and how site availability, species availability, and species performance may be modified to remove invasive plants and create safe sites for desired plant communities. Then disturbance treatments should parallel the mosaic of biotic and abiotic conditions, capitalizing on site and species availability and species performance where they are intact and amending them with treatments when necessary. For example, Sheley *et al.* (2009) tested 'augmentative restoration' in which they identified and selectively repaired or replaced damaged processes (in the context of site availability, species availability, and species performance) at

three sites in northwestern Montana, USA. Their research sites occurred on ephemeral wetlands that were dominated by invasive plants and had varying levels of disturbance by meadow voles (*Microtus pennsylvanicus* Ord.) (site availability), remnant native plants (species availability), and water availability (species performance). At two of the three sites, using augmentative restoration improved revegetation outcomes by combining treatments that would maximize seedling establishment while minimizing costs.

When designing disturbance to create safe sites, abiotic and biotic soil characteristics must be considered. Soil ecological knowledge (SEK), which acknowledges the interactions among principal components of the soil as well as feedback between aboveground and belowground ecosystem processes, is gaining recognition as an important factor to consider during restoration (Heneghan *et al.*, 2008) (see Morris, Chapter 3, and Eviner and Hawkes, Chapter 7, this volume). Invasive plants can alter physical (e.g. water infiltration capacity), chemical (e.g. nitrogen mineralization, accumulation of allelopathic compounds), and biological (i.e. mycorrhizal fungi associations) soil characteristics. Recognizing feedbacks between invading plant species and soil may be crucial to restoring plant communities to a more desired condition (Heneghan *et al.*, 2008). For example, Vinton and Goergen (2006) documented positive feedback between litter quality and nitrogen mineralization when restoring grasslands invaded by smooth brome, an introduced grass that demands more nitrogen than native grasses. Others have found that soil legacies from plants that previously occupied the soil can contribute to strong priority effects of invasive plants on native plants. Grman and Suding (2010) found that soil legacies created by invasive plants decreased native colonizer biomass by 74%. Such legacies could be due to changes in resource availability (Mack and D'Antonio, 2003), buildup of allelochemicals (Orr *et al.*, 2005), or alterations in soil microbial communities including pathogens and mycorrhizal fungi (Kourtev *et al.*, 2002; Hawkes *et al.*, 2006).

If soil legacies are present, site preparation must include actions to remediate them to favor site availability free from such legacies. When conditions of high soil nitrogen concentrations exist, it may be necessary to implement approaches for lowering nitrogen availability into site preparation, such as soil carbon addition, burning, grazing, topsoil removal, and/or biomass removal (Perry *et al.*, 2010). If allelopathic compounds are present and inhibiting the establishment of seeded species, activated carbon could be incorporated into the soil prior to seeding to sequester such compounds (Kulmatiski and Beard, 2006). Solarization, which is the capturing of radiant heat from the sun to cause biological, chemical, and physical changes to the soil, has been shown to kill weed seeds or pathogens and improve seedling establishment (Moyes *et al.*, 2005). While it may be too expensive to implement such site preparations over the entire landscape, critical areas could be identified and more intensively prepared for seeding than others. Most importantly, assessing variation in biotic and abiotic character-istics such as soil texture and surface structure, nutrient and water cycling, and decomposition rates across the site as part of site preparation will allow the implementation of vital and appropriate treatments.

Unfortunately, many invasive plant-infested lands occur on steep and rocky terrain that is not amenable to designed disturbance via standard equipment. Just as using traditional agronomic practices often fails where sites *are* accessible, developing site preparation and seeding methods for revegetating rough, inaccessible terrain has fallen short. Researchers and practitioners should be encouraged to develop novel techniques for situations where revegetation equipment cannot be used. The use of domestic livestock as a method of designed disturbance warrants further investigation. Soil surface depressions created by hoof prints can roughen the soil surface and cause a decrease in overland water flow (Gutterman, 2003), thus accumulating and retaining moisture for seeds and serving as a

microsite for colonization (Isselin-Nondedeu *et al.*, 2006). Domestic sheep and goats may also impact species performance and limit invasive plant species availability by grazing invasive forbs (Launchbaugh, 2006). At the same time domestic livestock can also aggravate erosion processes and increase soil compaction, so care would need to be taken if they were to be used in site preparation prior to aerial seeding on steep and rocky slopes.

Extensive energy and resources have been devoted to developing and improving seeding technology so that propagules are directly placed into a safe site. Equipment and methods for seeding natural systems in the western USA were being developed as early as the 1940s as ranchers and land managers attempted to increase forage production and rehabilitate overgrazed landscapes. The standard rangeland drill seeder was designed in the early 1950s and remained the industry standard for over 50 years. Some newer rangeland drills, such as the Truax Rough Rider™, combine technologies from the original rangeland drill with those of more modern no-till agricultural drills (Kees, 2006). While much progress has been made, more is necessary. For example, controlling seeding depth, ensuring good seed-to-soil contact, and proper calibration for mixed seed varieties, have all been identified as problem areas (Kees, 2006). James and Svejcar (2010) compared seedling establish-ment across field plots where seeds were planted using a rangeland drill or seeded by hand. In the hand-seeding treatments, a furrow similar in width and depth to the drill was made; seeds were placed, covered with 1 cm soil, and gently compacted. Irrigation and weeding treatments were also applied across seeding treatments. Seeding method was the only factor that limited establishment with seedling density over sevenfold higher in the hand-seed compared to the drill-seeded treatments. The authors concluded that even modest improvements in seeding technology might yield sub-stantial increases in seeding success (James and Svejcar, 2010).

As discussed earlier in this chapter, recommended seeding rates for revegetating invasive plant-infested sites can be two to three times higher than standard recommendations (Sheley *et al.*, 2008). However, increasing seeding rate will only be useful if there are adequate safe sites available, created through designed disturbance, to accommodate all seeds. There has been debate about whether seedling recruitment is naturally limited by the number of seeds and their dispersal capabilities or the number of safe sites (Satterthwaite, 2007). When designing revegetation programs, theoretically, we should have complete control over site availability; however, we have a fairly limited understanding of what constitutes a safe site and the relationship among safe site, invasive plant propagules, and desired plant propagules (in other words, which species will capture site availability) (Parker, 2000). Further research into this question is needed.

Species availability

Compared to advancements for designing disturbance to address site availability, methods for addressing species availability through species selection have experienced greater advances and more novel approaches. For example, the quantity and quality of plant material for revegetation in the USA has increased substantially in the past decade (Shaw and Pellant, 2010) and will likely continue to increase as researchers and practitioners continue to recognize the need for revegetation on invasive plant-infested landscapes. While I will not go into a discussion about origin of plant material and maintenance of genetic integrity, researchers and practitioners are increasingly stressing the importance of seed origin so that habitats of restoration and donor sites can be appropriately matched and genetic integrity can be maintained (Vander Mijnsbrugge *et al.*, 2010). I will also not address the use of native plant material versus introduced plant material, but will instead present some novel approaches that integrate both native and introduced species into the revegetation scheme.

Based on research that suggests low diversity systems are more susceptible to invasion (Tilman, 1997; Levine, 2000; Dukes, 2002; Pokorny *et al.*, 2005), selecting species for revegetation should focus on choosing species that are morphologically and functionally diverse. The rationale behind this is that species-rich communities will be less invasible because they are occupying a larger proportion of available niches and maximizing resource uptake. Therefore, resources are less available for use by potentially invading species (Elton, 1958). Studies that have actually quantified niche differentiation found invasion decreased when species richness and niche occupation increased (Jacobs and Sheley, 1999; Carpinelli *et al.*, 2004). Other research suggests that successfully establishing a species-rich seed mix that contains at least one desired species of similar morphology to the target invasive plant may limit reinvasion (Mangold *et al.*, 2007a).

Environmental conditions such as precipitation timing and amount, temperature, and solar radiation, during the year of seeding, may not match the ecological requirements of each individual species selected for seeding (Wirth and Pyke, 2003). Therefore, diverse seed mixes could improve seedling establishment by increasing the probability that environmental conditions will match the requirements of at least one species. Sheley and Half (2006) compared seeding monocultures of six different native forbs with seeding a mix of all six forbs under two different watering regimes. Establishment of the mixture, as measured by density of seeded species, was similar to the average density for individual species, and the forb mixture was seven times more competitive with the invasive forb spotted knapweed than a single native forb was when seeded alone. Because species differ in traits and ranges of tolerances, seeding mixtures with greater species richness are more likely to contain a species within a functional group that will germinate and emerge under varying and unpredictable environmental conditions (Tilman, 1994).

Even though some research suggests low diversity systems are more susceptible to invasion (see above), other studies suggest native species diversity and invasive plant diversity are positively correlated (Stohlgren et al., 1999; Symstad, 2000; Bruno et al., 2004). While this discrepancy may be due to the scale upon which the diversity-susceptibility to invasion question is examined, positive coupling between native and invasive plant diversity can also be linked to productivity; that is, the highest native plant diversity and invasion should occur on productive sites, or 'hot spots' of native diversity (Stohlgren et al., 1999). Highly productive sites likely meet plant resource requirements and are extremely desirable for plant growth. Therefore, such areas typically support many different species and may be especially vulnerable to invasion, especially if moderate disturbances free resources for invasive plants (Radosevich et al., 2007). If this postulate is true, selecting species for revegetation that are highly productive may go further to prevent reinvasion than focusing on species-rich seed mixes because highly productive species will maximize resource use. Evidence certainly exists to suggest this may be the case. For example, crested wheatgrass, a productive, introduced perennial grass, has been observed to resist invasion by knapweed and cheatgrass (Berube and Myers, 1982; D'Antonio and Vitousek, 1992). James et al. (2008) compared native bunchgrasses, perennial forbs, and annual forbs in their soil nitrogen use and resistance to annual grass invasion. They found that native bunchgrasses, the dominant plant functional group at their study sites, acquired the most nitrogen from all nitrogen pools, accumulated the most biomass, and was the only functional group that inhibited annual grass establishment.

Ideally, revegetation mixes will be diverse *and* productive. Recent and future developments in the field of plant materials will help this to become a reality. Because plant functional traits such as biomass production, canopy size, relative growth rate, and specific leaf area have been shown to be correlated with competitive per-formance of a species (James and Drenovsky, 2007; Isselin-Nondedeu et al., 2006),

quantifying such traits for the broad array of currently available plant materials may be very useful in designing seed mixes that are both diverse and productive. In the future revegetation species will be described by their traits and requirements for establish-ment, such as growth form, preferred soil type, minimum precipitation requirements, and seeding rate, but may include functional traits such as relative growth rate and specific leaf area as well. Such information would be highly useful to practitioners who are striving to design seed mixes that are invasion resistant.

Just as designed disturbance that creates safe sites for desired species requires flexibility across the landscape to reflect variable conditions, there should also be flexibility in choosing species appropriate for various locations on the landscape. All too often a species mix is created based on land management goals, and the same species mix is then seeded across the entire project area. Alternatively, species mixes could vary according to soil resource availability and texture, topography, degree of infestation severity, potential seed dispersal routes (i.e. livestock and wildlife trails, roads, waterways, etc.), and other factors commonly found in an ecological site description (ESD). It might also be useful to identify areas on the landscape where successful establishment of seeded species is critical, and then seed islands of chosen species (Elstein, 2004). Species seeded in islands could be chosen based on their dispersal capabilities, with the intent of choosing species that would have a high likelihood of dispersing to non-seeded areas over time. Because areas seeded to islands would be small relative to the entire landscape, they could be intensely managed for successful establishment of seeded species. In short, flexibility and creativity in selecting species and placing them on the landscape should be encouraged.

If disturbances are designed to potentially favor desired species, then the benefit of high seeding rates of desired species to overwhelm the pool of available propagules (including those of invasive plants) and occupy the majority of safe sites can be realized. As stated earlier in this chapter, seeding rates 5 and 25 times the recommended rate have been shown to increase establishment of perennial grasses in the short term in spotted knapweed-infested grasslands (2 years after seeding) (Sheley et al., 1999). But did high seeding rates result in favorable outcomes in the long-term? Six years after seeding, only the highest seeding rate (12,500 seeds m^{-2}) produced wheatgrass density higher than that of the non-seeded control (Sheley et al., 2005). A designed disturbance (tillage or herbicide application) combined with high seeding rates doubled wheatgrass density compared to seeding in the absence of disturbance, and spotted knapweed biomass tended to be lower where wheatgrass density and biomass was highest (Sheley et al., 2005). Fifteen years after revegetation, the effect of high seeding rates on wheatgrass establishment was less pronounced as wheatgrass density and biomass at lower rates had increased to similar levels; however, the effect of well-established wheatgrass on reducing reinvasion by spotted knapweed was more dramatic than at 2 or 6 years, with an 86% reduction at the highest seeding rate (Rinella et al., 2011). This long-term data set suggests high seeding rates can improve revegetation outcomes by reducing reinvasion.

Revegetation has been dominated by seeding relatively high rates of grasses in comparison to forbs. This is likely due to plant material availability and cost, the effectiveness of grasses to reduce soil erosion (Boyd, 1942; Jelinski and Kulakow, 1996), and the belief that highly productive grasses can control weedy species and improve soil characteristics (Dickson and Busby, 2009). While grasses are important in reducing reinvasion (e.g. James et al., 2008), other research suggests forbs may be very important for reducing reinvasion by weedy forbs (Pokorny et al., 2005; Mangold et al., 2007a). Seeded grasses can impede the establishment of seeded forbs (Kindscher and Fraser, 2000; Parkinson, 2008). Therefore, it may be useful to alter the proportion of desired grasses relative to desired forbs when considering seeding rate.

Dickson and Busby (2009) varied grass and forb seeding rates with the objective of finding the density of grass and forb seeds that would maximize forb abundance and richness during prairie restoration in Kansas, USA. They found that higher seeding densities of grasses resulted in decreased forb cover, biomass, and richness, and higher seeding densities of forbs increased forb richness. Exotic species biomass dramatically declined at all seeding rate combinations. Seeding islands of forb-rich mixes (Elstein, 2004) could eliminate unwanted competition between desired forbs and grasses, would help to mimic the natural patchiness of species usually found across natural landscapes (Dickson and Busby, 2009), and would allow for the follow-up management of reinvading weedy forbs with a broadleaf herbicide across the majority of the site (minus the forb-rich islands) without harming seeded, desired forbs.

It is vital that we continue to investigate and identify the key ecological processes at a site that are contributing to propagule availability and establishment of desired species. Seed and seedling herbivory is one ecological process that has received some attention, and conclusions thus far suggest it may have far more significant impacts on seedling establishment than previously acknowledged (Hulme, 1994; Fenner and Thompson, 2005). For example, Watts (2010) found that underground foraging by pocket gophers (Thomomys spp.) limited recruitment and establishment of native bunchgrasses in California, USA, grasslands. Bunchgrasses that attained a certain size were able to survive the effects of predation, suggesting that short periods (3–5 years) of gopher exclusion could increase the chance of survival for perennial bunchgrasses (Watts, 2010). In another study, grasshopper herbivory negatively impacted native plant diversity in crested wheatgrass stands in North Dakota, USA; when abundant, grasshoppers and other invertebrate herbivores could impede native species establishment, especially that of grasses (Branson and Sword, 2007). Accounting for processes, like herbivory, that present obstacles to seedling establishment, and then adjusting seeding rates accordingly, will require a deeper understanding of how such processes function.

Usually revegetation occurs by seeding a mix of species that meet the long-term management goals for the site. The seed mix is typically drill- or broadcast-seeded in fall, and invasive-plant control measures are implemented shortly after (Sheley *et al.*, 2008). Because they are chosen to meet long-term management objectives, the seeded species are commonly slow-growing, nutrient conservative, long-lived, late-seral species. However, soil characteristics (e.g. nutrient concentrations, organic matter, and microbiota) at the site may not be conducive to their establishment and growth (Firn *et al.*, 2009). For example, the existing propagule pool likely contains invasive plants, which are fast-growing and nutrient exploitative; the recent disturbance designed to control invasive plant performance and increase site availability has likely resulted in an increase in soil resources and ambient light, both of which are favorable for colonizing species (i.e. weeds).

Revegetating with multiple planting phases using bridge species (cover crops) that transition the plant community from one successional stage to the next (early-, mid-, and late-seral) addresses the need to seed species into environmental situations that are more conducive to their growth as described above. For example, initially fast-growing, short-lived species could be seeded to provide immediate and direct competition with invasive plants that may be regenerating from the seed bank (i.e. nurse or cover crops) (Vasquez *et al.*, 2008; Perry *et al.*, 2009). These species may be seeded as an individual phase or as a mix with mid-seral species that are intermediate in their growth rate, nutrient use, and longevity. These initial and mid-phase species could very likely be exotic species (D'Antonio and Meyerson, 2002). The first one to two revegetation phases would function to return conditions at the site, such as soil nutrient concentrations, organic matter and microbial communities, to a condition more appropriate (e.g. low nutrient

concentrations, high organic matter, and well-developed and complex microbial communities) for the establishment and survival of native, late-seral species. Finally, late-seral species would be seeded that meet the long-term management objectives for the site. From a more practical perspective, when revegetating invasive forb-infested plant communities, it may be necessary to first establish grasses, while hindering the performance of the invasive forb with a follow-up broadleaf herbicide application, then inter-seed with desired forbs and shrubs during a second seeding phase. Inter-seeding may allow the established grasses to facilitate the establishment of desired forbs and shrubs and reduce interspecific competition between seedlings of different plant functional groups (Gunnell *et al.*, 2010).

A similar approach has been proposed and tested for restoring plant communities dominated by annual grasses like cheatgrass and medusahead (*Taeniatherum caput-medusae* (L.) Nevski). In 'assisted succession' a site is moved from annual plant to perennial plant-domination by first seeding a competitive, aggressive perennial grass, followed by creation of site availability and reinsertion of species native to the pre-invasion plant community (Cox and Anderson, 2004). An aggressive grass, such as crested wheatgrass, has a higher likelihood of establishing in a primarily monotypic stand of invasive annual grass than do native species. Over time the reestablishment of perennial vegetation versus annual vegetation helps to restore ecological processes like nutrient cycling and fire regimes to a condition more favorable for perennial native species. Further development and testing of this unique idea is needed. One potential obstacle to its implementation is adequate control of the competitive, aggressive grass so that native species can ultimately be reestablished (Fansler and Mangold, 2011). If seeding methods are going to include multiple seeding phases, care must be taken to ensure the bridge species of choice do not become problematic themselves (D'Antonio and Meyerson, 2002).

Because the seedling stage is arguably the most vulnerable stage in a plant's life history (Otsus and Zobel, 2002; Clark and Wilson, 2003), developing seeding methods that assist a plant through this stage is critical. One method may be to transplant juvenile plants instead of planting seeds. Transplanted seedlings have a higher rate of survival than seedlings germinating from seed in the field (Page and Bork, 2005; Middleton et al., 2010). Transplanting seedlings across large landscapes would be very labor and time intensive. Therefore, similar to the island seeding concept (Elstein, 2004), transplanting could be applied by identifying areas on the landscape where successful establishment of seeded species is critical and transplant small patches of desired species. Integrating transplants along with seeding would ensure that dispersal and establishment of certain species is guaranteed, thus improving the overall richness, diversity, and quality of outcomes (Middleton et al., 2010).

Species performance

As stated earlier, during revegetation weedy species must be reduced or eliminated so that desired species can establish and grow with minimal competition from undesired species. When considering revegetation strategies, it is necessary to carefully examine invasive and other nuisance weedy species in the context of disturbance, succession, and ultimately restoration. D'Antonio and Meyerson (2002) state that when setting priorities and goals for revegetation, it is essential to understand the potential transience of invasive and weedy species at a site and the role they might play in altering processes that influence the course of succession. They also suggest that there may be some situations where a nuisance weedy species is short-lived and successional to native species, thus not requiring control in the post-disturbance environment; in contrast, long-lived invasive plants or tenacious invasive annuals that are both good colonizers and persistent community members should be a top management priority. In a like manner, populations of invasive plants may vary in their degree of 'invasiveness' based on the biology (e.g. seed production and vegetative spread, recruitment, and growth rates) of the species and whether habitat requirements are being fully met for populations to reach their maximum rates of growth and spread. Managers should prioritize managing the performance of more invasive populations that will increase in size and density at a faster rate than less invasive populations, which may serve as a significant source of propagules (Lehnhoff et al., 2008; Maxwell et al., 2009).

Controlling invasive plant performance will take time, and time should be readily factored into revegetation plans. The length of time since introduction of the invasive plant and the biology and life history of the species (i.e. prolific seed producer, annual versus perennial species, and reproductive strategy) can play a big role in the effectiveness and longevity of control measures, because they influence propagule availability and reinvasion potential (Reinhardt Adams and Galatowitsch, 2008). This, in turn, can determine the extent to which seeded desirable species must compete with invasive species during their establishment phase. As seed production and seed longevity of an invasive species increases, the more time it will take to adequately control its performance. Because plant community composition is somewhat predictable based on seed bank composition (van der Valk and Pederson, 1989), sampling of the seed bank prior to designed disturbance will provide insight into whether invasive species are likely to remain dominant for some time (D'Antonio and Meyerson, 2002), and if other weedy species are present that may increase following any type of soil disturbance (Baskin and Baskin, 1998). If the proportion of invasive species propagules to desired species propagules is very high, then multiple years of control may be necessary prior to reintroducing desired species (Fansler and Mangold, 2011). While this may add time and expense to the revegetation program, the probability for success may increase.

Following revegetation the threat of reinvasion is substantial due to the high likelihood that a large number of invasive species propagules still exist on the site. The likelihood of seedling mortality is also high. Monitoring to assess desired species establishment compared to reinvading species will allow the timely intervention and follow-up management to control species performance. If grasses have been seeded first in a multi-phase seeding process as described above, applying control methods (e.g. herbicide application or grazing with sheep) to hinder the performance of reinvading weedy forbs will be necessary. If desired species seedlings appear to be underperforming, intervention to improve their performance may be necessary. Intervention may include supplemental irrigation, precision fertilization, or over-seeding (broadcast) to increase the number of desired species propagules for subsequent germination, emergence, and establishment when conditions are more amenable. While such activities will require more time and money, the benefits may be substantial. Furthermore, if critical areas for seedling establishment are identified early in the planning and revegetation process, then strategies can be concentrated there rather than being implemented over the entire area. Most ecological processes like nutrient and water cycling, decomposition rates, propagule dispersal patterns, and plant–plant interference, will not heal within a few years, therefore management should be focused on directly influencing their functionality. Evaluating plant community composition (e.g. cover and frequency) and seed bank composition, sampling the soil for organic matter and nutrient concentrations, and noting evidence of higher trophic level interactions (e.g. herbivory) over time are a few measures that may be taken to assess whether the plant community is transitioning in a desirable direction.

Monitoring and evaluating the success or failure of our revegetation efforts must be a long-term process and measured in the context of plant community successional trajectories. It also takes dedicated individuals who are interested not only in implementing revegetation, but evaluating its effectiveness across short-, mid-, and long-term timeframes. Very few long-term studies exist, but those that do suggest initial trends may not be indicative of long-term outcomes (Ferrell et al., 1998; Roche et al., 2008; Rinella et al., 2011). Rinella et al. (2011) re-sampled four projects that integrated various forms of invasive plant control and seeding in invasive forb-infested grasslands in western Montana, USA. They measured seeded species and invasive forb density and biomass 15 years after seeding. Some seeded populations remained very small for 6 or more years but then became highly productive and greatly suppressed the invader (Fig. 10.3). Other populations maintained high densities for 3 or more years but then became exceedingly rare or extinct. The results suggest that seeding sometimes provides appreciable long-term benefits, but at other times fails, and that short-term data can both over- and under-estimate the lasting benefits of revegetation. Additional long-term studies are needed to assess the likelihood of favorable seeding outcomes, to identify good seeded species traits, and refine ecologically based strategies for controlling site availability, species availability, and species performance during revegetation.

Conclusion

Severely degraded plant communities may require the modification of site availability, species availability, and species performance to improve the success of revegetation and return plant communities to a desired state. Revegetation is a challenging prospect that often results in less than optimum outcomes. In a relatively brief manner, this chapter has attempted to describe standard practices, identify limitations to those practices, and propose new approaches for improving revegetation in the future. The ideas and approaches here are certainly not an exhaustive list, but rather some ideas for which there is supporting data. It is my hope that these

(a)

(b)

(c)

Fig. 10.3. Spotted knapweed (*Centaurea stoebe* L.)-infested grasslands where no seeding (a) or seeding with bluebunch wheatgrass (*Pseudoroegneria spicata* (Pursh) A.Love) (b) or intermediate wheatgrass (*Thinopyrum intermedium* (Host) Barkworth & D.R. Dewey) (c) had occurred 15 years earlier. Notice persistence of spotted knapweed in the non-seeded plots and lack of spotted knapweed in the seeded plots. Photos courtesy of Jane Mangold, July 2010.

ideas prompt further thought, discussion, and research into their application as well as the development of additional approaches.

References

Baskin, C.C. and Baskin, J.M. (1998) *Seeds: Ecology, Biogeography, and Evolution of Dormancy and Germination.* Academic Press, New York.

Berger, J.J. (1993) Ecological restoration and nonindigenous plant species: a review. *Restoration Ecology* 1, 74–82.

Berube, D.E. and Myers, J.H. (1982) Suppression of knapweed invasion by crested wheatgrass in the dry interior of British Columbia. *Journal of Range Management* 35, 459–461.

Borman, M.M., Krueger, W.C. and Johnson, D.E. (1991) Effects of established perennial grasses on yields of associated annual weeds. *Journal of Range Management* 44, 318–322.

Boyd, I.L. (1942) Evaluation of species of prairie grasses as interplanting ground covers on eroded soils. *Transactions of the Kansas Academy of Sciences* 45, 55–58.

Branson, D.H. and Sword, G.A. (2007) Grasshopper herbivory affects native plant diversity and abundance in a grassland dominated by the exotic grass *Agropyron cristatum*. *Restoration Ecology* 17, 89–96.

Brooks, M.L. (2008) Plant invasions and fire regimes. In: Zouhar, K., Smith, J.K., Sutherland, S. and Brooks, M.L. (eds) *Wildland Fire in Ecosystems: Fire and nonnative invasive plants.* US Department of Agriculture, Forest Service, Rocky Mountain Research Station, Ogden, Utah, pp. 33–60.

Bruno, J.F., Kennedy, C.W., Rand, T.A. and Grant, M.B. (2004) Landscape-scale patterns of biological invasions in shorelines plant communities. *Oikos* 107, 531–540.

Carpinelli, M.F., Sheley, R.L. and Maxwell, B.D. (2004) Revegetating weed-infested rangeland with niche-differentiated desirable species. *Journal of Range Management* 57, 97–103.

Clark, D.L. and Wilson, M.V. (2003) Post-dispersal seed fates of four prairie species. *American Journal of Botany* 90, 730–735.

Cox, R.D. and Anderson, V.J. (2004) Increasing native diversity of cheatgrass-dominated rangeland through assisted succession. *Journal of Range Management* 57, 203–210.

D'Antonio, C. and Meyerson, L.A. (2002) Exotic plant species as problems and solutions in ecological restoration: a synthesis. *Restoration Ecology* 10, 703–713.

D'Antonio, C.M. and Vitousek, P.M. (1992) Biological invasions by exotic grasses, the grass/fire cycle, and global change. *Annual Review of Ecology and Systematics* 23, 63–87.

Davies, K.W. (2010) Revegetating medushead-invaded sagebrush steppe. *Rangeland Ecology and Management* 63, 564–571.

Davis, M.A., Grime, J.P. and Thompson, K. (2000) Fluctuating resources in plant communities: a general theory of invasibility. *Journal of Ecology* 88, 528–534.

Davison, J.C. and Smith, E.G. (2007) Imazapic provides 2-year control of weedy annuals in a seeded Great Basin fuelbreak. *Native Plants* 8, 91–95.

Dickson, T.L. and Busby, W.H. (2009) Forb species establishment increases with decreased grass seeding density and with increased forb seeding density in a northeast Kansas, USA, experimental prairie restoration. *Restoration Ecology* 17, 597–605.

Di Iomaso, J.M. (2000) Invasive weeds in rangelands: species, impacts, and management. *Weed Science* 48, 255–265.

Dixon, R.M. (1988) Land imprinting for vegetative restoration. *Restoration and Management Notes* 6, 24–25.

Dukes, J.S. (2002) Species composition and diversity affect grassland susceptibility and response to invasion. *Ecological Applications* 12, 602–617.

Ehrenfeld, J.G. (2003) Effects of exotic plant invasions on soil nutrient cycling processes. *Ecosystems* 6, 503–523.

Elstein, D. (2004) A friendly solution to restoration. *Agricultural Research Magazine* 52, 20–21.

Elton, C.S. (1958) *The Ecology of Invasions by Animals and Plants.* University of Chicago Press, Chicago, Illinois.

Emery, S.M. and Gross, K.L. (2005) Effects of timing of prescribed fire on the demography of an invasive plant, spotted knapweed (*Centaurea maculosa*). *Journal of Applied Ecology* 42, 60–69.

Fansler, V.A. and Mangold, J.M. (2011) Restoring native plants to crested wheatgrass stands. *Restoration Ecology* 19, 16–23.

Fenner, M. and Thompson, K. (2005) *The Ecology of Seeds*, 2nd edn. Cambridge University Press, Cambridge, UK.

Ferrell, M.A., Whitson, T.D., Koch, D.W. and Gade, A.E. (1998) Leafy spurge (*Euphorbia esula*) control with several grass species. *Weed Technology* 12, 374–380.

Firn, J., House, A.P.N. and Buckley, Y.M. (2009)

Alternative states models provide an effective framework for invasive species control and restoration of native communities. *Journal of Applied Ecology* 47, 96–105.

Grman, E. and Suding, K.N. (2010) Within-year soil legacies contribute to strong priority effects of exotics on native California grassland communities. *Restoration Ecology* 18, 664–670.

Gunnell, K.L., Monaco, T.A., Call, C.A. and Ransom, C.V. (2010) Seedling interference and niche differentiation between crested wheatgrass and contrasting native Great Basin species. *Rangeland Ecology and Management* 63, 443–449.

Gutterman, Y. (2003) The influences of animal diggings and runoff water on the vegetation in the Negev Desert of Israel. *Israeli Journal of Plant Science* 51, 161–171.

Harper, J.L., Williams, J.T. and Sagar, G.R. (1965) The behavior of seeds in soil I. The heterogeneity of soil surfaces and its role in determining the establishment of plants from seed. *Journal of Ecology* 53, 273–286.

Hastings, M.S. and DiTomaso, J.M. (1996) Fire controls yellowstar thistle in California grasslands. *Restoration and Management Notes* 14, 124–128.

Hawkes, C.V., Belnap, J., D'Antonio, C. and Firestone, M.K. (2006) Arbuscular mycorrhizal assemblages in native plant roots change in the presence of invasive exotic grasses. *Plant and Soil* 281, 369–380.

Heneghan, L., Miller, S.P., Baer, S., Callaham, M.A., Jr, Montgomery, J., Pavao-Zuckerman, M., Rhoades, C. and Richardson, S. (2008) Integrating soil ecological knowledge into restoration management. *Restoration Ecology* 16, 608–617.

Hobbs, R.J. and Huenneke, L.F. (1992) Disturbance, diversity, and invasion: implications for conservation. *Conservation Biology* 6, 324–337.

Huenneke, L.F., Hamburg, S.P., Koide, R., Mooney, H.A. and Vitousek, P.M. (1990) Effects of soil resources on plant invasion and community structure in Californian serpentine grassland. *Ecology* 71, 478–491.

Hulme, P.E. (1994) Seedling herbivory in grassland: relative impact of vertebrate and invertebrate herbivores. *Journal of Ecology* 82, 873–880.

Isselin-Nondedeu, F., Rey, F. and Bedecarrats, A. (2006) Contributions of vegetation cover and cattle hoof prints towards seed runoff control on ski pistes. *Ecological Engineering* 27, 193–201.

Jacobs, J.S. and Sheley, R.L. (1999) Competition and niche partitioning among *Pseudoroegneria spicata*, *Hedysarum boreale*, and *Centaurea maculosa*. *Great Basin Naturalist* 59, 175–181.

James, J.J. and Drenovsky, R.E. (2007) A basis for relative growth rate differences between native and invasive forb seedlings. *Rangeland Ecology and Management* 60, 395–400.

James, J.J. and Svejcar, T. (2010) Limitations to postfire seedling establishment: the role of seeding technology, water availability, and invasive plant abundance. *Rangeland Ecology and Management* 63, 491–495.

James, J.J., Davies, K.W., Sheley, R.L. and Aanderud, Z.T. (2008) Linking nitrogen partitioning and species abundance to invasion resistance in the Great Basin. *Oecologia* 156, 637–648.

Jelinski, D.E. and Kulakow, P.A. (1996) The Conservation Reserve Program: opportunities for research in landscape-scale restoration. *Restoration and Management Notes* 14, 137–139.

Jessop, B.D. and Anderson, V.J. (2007) Cheatgrass invasion in salt desert shrublands: benefits of postfire reclamation. *Rangeland Ecology and Management* 60, 235–243.

Kees, G. (2006) Rangeland drills: can seed placement be improved? *Tech Tip 0622–2345–MTDC*. US Department of Agriculture Forest Service, Missoula Technology and Development Center, Missoula, Montana. Available at: www.fs.fed.us/t-d/pubs/pdfpubs/pdf06222345/pdf06222345dpi72.pdf (accessed 20 January 2011).

Kindscher, K. and Fraser, A. (2000) Planting forbs first provides greater species diversity in tallgrass prairie restorations (Kansas). *Ecological Restoration* 18, 115–116.

Kourtev, P.S., Ehrenfeld, J.G. and Haggblom, M. (2002) Exotic plant species alter the microbial community structure and function in the soil. *Ecology* 83, 3152–3166.

Kowarik, I. (1995) Time lags in biological invasions with regard to the success and failure of alien species. In: Pysek, P., Prach, K., Rejmanek, M. and Wade, M. (eds) *Plant Invasions: General Aspects and Special Problems*. SPB Academic, Amsterdam, the Netherlands, pp. 15–38.

Krueger-Mangold, J.M., Sheley, R.L. and Svejcar, T.J. (2006) Toward ecologically-based invasive plant management on rangeland. *Weed Science* 54, 597–605.

Kulmatiski, A. and Beard, K.H. (2006) Activated carbon as a restoration tool: potential for control of invasive plants in abandoned agricultural fields. *Restoration Ecology* 14, 251–257.

Kyser, G.B. and DiTomaso, J.M. (2002) Instability in a grassland community after the control of yellow starthistle (*Centaurea solstitialis*) with prescribed burning. *Weed Science* 50, 648–657.

Laufenberg, S.M., Sheley, R.L., Jacobs, J.S. and Borkowski, J. (2005) Herbicide effects on density and biomass of Russian knapweed (*Acroptilon repens*) and associated plant species. *Weed Technology* 19, 62–72.

Launchbaugh, K. (ed.) (2006) *Targeted Grazing: A Natural Approach to Vegetation Management and Landscape Enhancement.* Cottrell Printing, Centennial, Colorado.

Lehnhoff, E.A., Rew, L.J., Maxwell, B.D. and Taper, M.L. (2008) Quantifying invasiveness of plants: a test case with yellow toadflax (*Linaria vulgaris*). *Invasive Plant Science and Management* 1, 319–325.

Levine, J.M. (2000) Species diversity and biological invasions: relating local process to community pattern. *Science* 288, 852–854.

Lozon, J.D. and MacIsaac, H.J. (1997) Biological invasions: are they dependent on disturbance? *Environmental Review* 5, 131–144.

Lym, R.G. and Tober, D.A. (1997) Competitive grasses for leafy spurge (*Euphorbia esula*) reduction. *Weed Technology* 11, 787–792.

MacArthur, R.H. (1962) Generalized theorems of natural selection. *Proceedings of the National Academy of Sciences* 48, 1893–1897.

Mack, M.C. and D'Antonio, C.M. (2003) Exotic grasses alter controls over soil nitrogen dynamics in a Hawaiian woodland. *Ecological Applications* 13, 154–166.

Mangold, J.M., Poulsen, C.L. and Carpinelli, M.F. (2007a) Revegetating Russian knapweed (*Acroptilon repens*) infestations using morphologically diverse species and seedbed preparation. *Rangeland Ecology and Management* 60, 378–385.

Mangold, J.M., James, J.J. and Sheley, R.L. (2007b) Presence of soil surface depressions increases water uptake by native grass seeds. *Ecological Restoration* 25, 278–279.

Maxwell, B.D., Lehnhoff, E. and Rew, L.J. (2009) The rationale for monitoring invasive plant populations as a crucial step for management. *Invasive Plant Science and Management* 2, 1–9.

Middleton, E.L., Bever, J.D. and Schultz, P.A. (2010) The effect of restoration methods on the quality of the restoration and resistance to invasion by exotics. *Restoration Ecology* 18, 181–187.

Monaco, T.A., Osmond, T.M. and Dewey, S.A. (2005) Medusahead control with fall- and spring-applied herbicides on northern Utah foothills. *Weed Technology* 19, 653–658.

Monsen, S.B. and Stevens, R. (2004) Seedbed preparation and seeding practices. In: Monsen, S.B., Stevens, R. and Shaw, N.L. (eds) *Restoring Western Ranges and Wildlands.* Gen.

Tech. Rep. RMRS-GTR-136-vol-1. USDA Forest Service, Rocky Mountain Research Station, Fort Collins, Colorado, pp. 121–154.

Moyes, A.B., Witter, M.S. and Gamon, J.A. (2005) Restoration of native perennials in a California annual grassland after prescribed spring burning and solarization. *Restoration Ecology* 13, 659–666.

O'Driscoll, P. (2007) Seeds contain hope of healing the Great Basin. *USA Today*

Orr, S.P., Rudgers, J.A. and Clay, K. (2005) Invasive plants can inhibit native tree seedlings: testing potential allelopathic mechanisms. *Plant Ecology* 181, 153–165.

Otsus, M. and Zobel, M. (2002) Small-scale turnover in a calcareous grassland, its pattern and components. *Journal of Vegetation Science* 13, 199–206.

Page, H.N. and Bork, E.W. (2005) Effect of planting season, bunchgrass species, and neighbor control on the success of transplants for grassland restoration. *Restoration Ecology* 13, 651–658.

Parker, I.M. (2000) Safe site and seed limitation in *Cytisus scoparius* (Scotch broom), invasibility, disturbance, and the role of cryptogams in a glacial outwash prairie. *Biological Invasions* 3, 323–332.

Parkinson, H.A. (2008) Impacts of native grasses and cheatgrass on Great Basin forb development. MSc Thesis, Montana State University, Bozeman, Montana.

Perry, L.G., Cronin, S.A. and Paschke, M.W. (2009) native cover crops suppress exotic annuals and favor native perennials in a greenhouse competition experiment. *Plant Ecology* 204, 247–259.

Perry, L.G., Blumenthal, D.M., Monaco, T.A., Paschke, M.W. and Redente, E.F. (2010) Immobilizing nitrogen to control plant invasion. *Oecologia* 163, 13–24.

Pickett, S.T.A. and Cadenasso, M.L. (1995) Landscape ecology: spacial heterogeneity in ecological systems. *Science* 269, 331–334.

Pickett, S.T.A., Collins, S.L. and Armestor, J.J. (1987) Models, mechanisms and pathways of succession. *Botanical Review* 53, 335–371.

Pokorny, M.L., Sheley, R.L., Zabinski, C.A., Engel, R.E., Svejcar, T.J. and Borkowski, J.J. (2005) Plant functional group diversity as a mechanism for invasion-resistance. *Restoration Ecology* 13, 448–459.

Pokorny, M.L., Mangold, J.M., Hafer, J. and Denny, M.K. (2010) Managing spotted knapweed (*Centaurea stoebe*) in infested rangeland after wildfire. *Invasive Plant Science and Management* 3, 182–189.

Radosevich, S.R., Holt, J.S. and Ghersa, C.M. (2007) *Ecology of Weeds and Invasive Plants*, 3rd edn. John Wiley & Sons, Hoboken, New Jersey.

Reinhardt Adams, C. and Galatowitsch, S.M. (2008) The transition from invasive species control to native species promotion and its dependence on seed density thresholds. *Applied Vegetation Science* 11, 131–138.

Rinella, M.J., Mangold, J.M., Espeland, E.K., Sheley, R.L. and Jacobs, J.S. (2011) Long-term population dynamics of seeded plants in invaded grasslands. Ecological Applications. Accepted.

Roche, C.T., Sheley, R.L. and Korfhage, R.C. (2008) Native species replace introduced grass cultivars seeded after wildfire. *Ecological Restoration* 26, 321–330.

Rodgers, V.L., Finzi, A.C., Werden, L.K. and Wolfe, B.E. (2008) The invasive species *Alliaria petiolata* (garlic mustard) increases soil nutrient availability in northern hardwood-conifer forests. *Oecologia* 157, 459–471.

Satterthwaite, W.H. (2007) The importance of dispersal in determining seed versus safe site limitation of plant populations. *Plant Ecology* 193, 113–130.

Shaw, N. and Pellant, M. (2010) Great Basin Native Plant Selection and Increase Project. *FY 2009 Progress Report*. USDA Forest Service, Rocky Mountain Research Station, Fort Collins, Colorado.

Shaw, N., Lambert, S., Debolt, A. and Pellant, M. (2005) Increasing native forb seed supplies for the Great Basin. In: Dumroese, R.K., Riley, L.E. and Landis, T.D. (eds) *National Proceedings: Forest and Conservation Nursery Associations-2004*. USDA Forest Service Proceedings RMRS-P-35, pp. 94–102.

Sheley, R.L. and Half, M.L. (2006) Enhancing native forb establishment and persistence using a rich seed mixture. *Restoration Ecology* 14, 627–635.

Sheley, R.L., Svejcar, T.J. and Maxwell, B.D. (1996) A theoretical framework for developing successional weed management strategies on rangeland. *Weed Technology* 10, 766–773.

Sheley, R.L., Jacobs, J.S. and Velagala, R.P. (1999) Enhancing intermediate wheatgrass establishment in spotted knapweed infested rangeland. *Journal of Range Management* 52, 68–74.

Sheley, R.L., Duncan, C.A., Halstvedt, M.B. and Jacobs, J.S. (2000) Spotted knapweed and grass response to herbicide treatments. *Journal of Range Management* 53, 176–182.

Sheley, R.L., Jacobs, J.S. and Lucas, D.E. (2001) Revegetating spotted knapweed infested rangeland in a single entry. *Journal of Range Management* 54, 144–151.

Sheley, R.L., Jacobs, J.S. and Svejcar, T.J. (2005) Integrating disturbance and colonization during rehabilitation of invasive weed-dominated grasslands. *Weed Science* 53, 307–314.

Sheley, R.L., Mangold, J.M. and Anderson, J.L. (2006) Potential for successional theory to guide restoration of invasive-plant-dominated rangeland. *Ecological Monographs* 76, 365–379.

Sheley, R.L., Carpinelli, M.F. and Morghan, K.J.R. (2007) Effects of imazapic on target and nontarget vegetation during revegetation. *Weed Technology* 21, 1071–1081.

Sheley, R.L., Mangold, J.M., Goodwin, K.M. and Marks, J. (2008) *Revegetation guidelines for the Great Basin: considering invasive weeds*. US Department of Agriculture, Agricultural Research Service, Washington, DC.

Sheley, R.L., James, J.J. and Bard, E.C. (2009) Augmentative restoration: repairing damaged ecological processes during restoration of heterogeneous environments. *Invasive Plant Science and Management* 2, 10–21.

Stinson, K.A., Campbell, S.A., Powell, J.R., Wolfe, B.E., Callaway, R.M., Thelen, G.C., Hallett, S.G., Prati, D. and Klironomos, J.N. (2006) Invasive plant suppresses the growth of native tree seedlings by disrupting belowground mutualisms. *PLoS Biology* 4, e140.

Stohlgren, T.J., Binkley, D., Chong, G.W., Kalkhan, M.A., Schell, L.D., Bull, K.A., Otsuki, Y., Newman, G., Bashkin, M. and Son, Y. (1999) Exotic plant species invade hot spots of native plant diversity. *Ecological Monographs*, 69, 25–46.

Story, J.M., Callan, N.W., Corn, J.G. and White, L.J. (2006) Decline of spotted knapweed density at two sites in western Montana with large populations of the introduced root weevil, *Cyphocleonus achates* (Fahraeus). *Biological Control* 38, 227–232.

Symstad, A.J. (2000) A test of the effects of functional group richness and composition on grassland invasibility. *Ecology* 81, 99–109.

Thompson, T.W., Roundy, B.A., Mcarthur, E.D., Jessop, B.D., Waldron, B. and Davis, J.N. (2006) Fire rehabilitation using native and introduced species: a landscape trial. *Rangeland Ecology and Management* 59, 237–248.

Tilman, D. (1994) Competition and biodiversity in spatially structured habitats. *Ecology* 75, 2–16.

Tilman, D. (1997) Community invasibility, recruitment limitation, and grassland biodiversity. *Ecology* 78, 81–92.

Tilman, D. and Wedin, D. (1991) Dynamics of nitrogen competition between successional grasses. *Ecology* 72, 1038–1049.

United States Department of the Interior, United States Fish and Wildlife Service (USDI USFWS) (2001) *National Wildlife Refuge System: biological integrity, diversity, and environmental health.* 601 FW 3. Washington, DC.

United States Department of Interior (USDI) and United State Department of Agriculture (USDA) (2002) *Report to Congress* – interagency program to supply and manage native plant materials for restoration and rehabilitation on Federal lands. Washington, DC.

Vander Mijnsbrugge, K., Bischoff, A. and Smith, B. (2010) A question of origin: where and how to collect seed for ecological restoration. *Basic and Applied Ecology* 11, 300–311.

van der Valk, A.G. and Pederson, R.L. (1989) Seed banks and the management and restoration of natural vegetation. In: Leck, M.A., Parker, V.T. and Simpson, R.L. (eds) *Ecology of Soil Seed Banks.* Academic Press, New York.

Vasquez, E., Sheley, R.L. and Svejcar, T.J. (2008) Creating invasion resistant soils via nitrogen management. *Invasive Plant Science and Management* 1, 304–314.

Vinton, M.A. and Goergen, E.M. (2006) Plant-soil feedbacks contribute to the persistence of *Bromus inermis* in tallgrass prairie. *Ecosystems* 9, 967–976.

Watts, S.M. (2010) Pocket gophers and the invasion and restoration of native bunchgrass communities. *Restoration Ecology* 18, 34–40.

Wilson, M.V. and Ingersoll, C.A. (2004) Why pest plant control and native plant establishment failed: a restoration autopsy. *Natural Areas Journal* 24, 23–31.

Wirth, T.A. and Pyke, D.A. (2003) Restoring forbs for sage grouse habitat: fire, microsites, and establishment methods. *Restoration Ecology* 11, 370–377.

Index